An Introduction to Genetics for Language Scientists

During the last few decades we have discovered an enormous amount about our genomes, their evolution and, importantly for linguists and language scientists, the genetic foundations of language and speech.

Accessible and readable, this introduction is designed specifically for students and researchers working in language and linguistics. It carefully focuses on the most relevant concepts, methods and findings in the genetics of language and speech, and covers a wide range of topics such as heritability, the molecular mechanisms through which genes influence our language, and the evolutionary forces affecting them.

Filling a large gap in the literature, this essential guide explores relevant examples including hearing loss, stuttering, dyslexia, brain growth and development, as well as the normal range of variation. It also contains a helpful glossary of terms and a wide range of references so the reader can pursue topics of interest in more depth.

DAN DEDIU is Senior Investigator in the Language and Genetics Department at the Max Planck Institute for Psycholinguistics in Nijmegen, The Netherlands.

An Introduction to Genetics for Language Scientists

Current concepts, methods and findings

Dan Dediu

CAMBRIDGE
UNIVERSITY PRESS

CAMBRIDGE
UNIVERSITY PRESS

University Printing House, Cambridge CB2 8BS, United Kingdom

Cambridge University Press is part of the University of Cambridge.

It furthers the University's mission by disseminating knowledge in the pursuit of education, learning and research at the highest international levels of excellence.

www.cambridge.org
Information on this title: www.cambridge.org/9781107001299

First published 2015

A catalogue record for this publication is available from the British Library

ISBN 978-1-107-00129-9 Hardback

Dedicated to my mother, Maria Dediu, who managed to foster curiosity, interest in science and optimism even in the darkest of times.

Contents

Illustrations

Acknowledgements

This book is the result of a long process that began more than five years ago, with Helen Barton visiting the University of Edinburgh and convincing the young and naïve freshly minted doctor that I was to transform his PhD thesis (a monster of some 460 pages) into a readable book. The idea was that there were no introductions to genetics written for scientists interested in language and speech and the available general introductions were either too long and dense or did not really appeal to the interests of these readers.

Soon after this meeting I moved to Nijmegen, the Netherlands, joining the world-renowned Max Planck Institute for Psycholinguistics with the specific task to begin working on the genetics of language and speech, trying to build upon the enormous amount of expertise scattered across the various departments of the institute, paving the way as it were for the upcoming new department of Language and Genetics. This task was extremely interesting as it effectively meant bridging the gaps between specialties, and I worked with neuroscientists, "classical" psycholinguists, quantitative historical linguists and typologists (for lack of a better name), specialists in child language acquisition and even the odd field linguist. This taught me invaluable lessons about what is interesting, relevant and accessible to all these scientists, and made me discover topics that I did not even know existed and are potentially so important for the genetics of language and speech (such as the emerging village sign languages and the genetics of hearing deficits; thank you Connie). Likewise, teaching several introductory courses in various summer schools and giving invited lectures at conferences and workshops put me in direct contact with the maddeningly complicated and devilishly smart audience this book is designed for, and forced me to sharpen my explanations, pick the right examples and hedge my theoretical positions.

But it was no easy job. Trying to convey the importance of *differences* (as opposed to the always assumed ideal participant that one approximates by having dozens of imperfect replicas in an experimental condition) was a main stumbling block, as was dispelling the fear that what I am "really" interested in was to classify people based on their genes (no, geneticists are some of the most vehement opponents of classifications as they experience first-hand, day-by-day the gradualness and complexity of human variation), or, equally worse, to find genetic differences that are – therefore one would say – unchangeable, reinforcing the differences between those who have and those who don't (of course, if modern genetics and evolutionary theory teaches us anything it is that the environment is *essential* to everything genetic to the point of blurring this apparently clear-cut distinction).

Of course, as any cursory look at my CV shows, I am no geneticist by training, but I have always been interested in biology (literally, I remember reading about evolution in my 4th or 5th grade thanks to my mother, a biology teacher) and I have been working on questions to do with evolution, genetics and language for the past 10 years or so. The good news is that you don't need to have had biology classes to be able to understand it, but the downside is that, as a self-taught geneticist, you will always have gaps in your knowledge. Thus, I am particularly indebted to my new colleagues – the "real geneticists", as I sometimes call them – in the Language and Genetics department here, at the Max Planck Institute for Psycholinguistics (whose work oft-times fills me with awe), especially to Prof. Simon Fisher, Dr. Sonja Vernes, Dr. Clyde Francks, Dr. Sarah Graham, Dr. Tulio Guadalupe, Dr. Nicolas Brucato, Paolo Devanna, Alessandro Gialluisi, and Amaia Carrión Castillo.

The Max Planck Institute for Psycholinguistics in Nijmegen is an amazing place for inter-disciplinary work where we somehow manage to talk across discipline boundaries and effectively collaborate with each other, a feat that is much more difficult to accomplish than one would naïvely think. It has also financially supported me through most of the time spent writing this book and its library allowed me to access all the journals, papers and books I needed to; many thanks to the librarians and especially to Karin Kastens. My current project here, the so-called "G[ɜ]bils" (Genetic biases in language and speech, funded by the Netherlands Organisation for Scientific Research [NWO] through the VIDI project 276-70-022, together with Dr. Scott Moisik and Rick Janssen) is a living example of how taking into account genetics in an inter-disciplinary framework can shed light on patterns of diversity in phonetics and phonology.

I have learned a lot from discussions with many colleagues across the years, but I want to mention in particular Prof. Stephen C. Levinson, Prof. Simon Fisher and Prof. D. Robert Ladd (University of Edinburgh, UK), who very nicely illustrate how one can successfully bridge multiple disciplines at once and yet still manage to be at the forefront of one's own scientific fields.

Several colleagues took their time to check parts of the manuscript, make many valuable suggestions and point me in the right direction, and I would like to thank Prof. Simon Fisher, Dr. Sonja Vernes, Dr. Clyde Francks, Dr. Sarah Graham, Dr. Tulio Guadalupe, Dr. Nicolas Brucato, Alessandro Gialluisi,

Amaia Carrión Castillo, Dr. Scott Moisik and Dr. Connie de Vos. I also want to thank one anonymous reviewer who had very important suggestions for improving the book, the most notable being a better coverage of quantitative genetics. Of course, any remaining errors and infelicities are mine!

This would have not been possible without the support of my wife, Dr. Alexandra Dima, who in fact got me to learn statistics in the first place and with whom I would often discuss not only data analysis strategies, but the deeper meaning and assumptions behind various statistical techniques and what *not* to read in a reported result (despite what is claimed in the abstract and conclusions), a very important skill to have. At times she encouraged (and even forced) me to continue working on this project and her suggestions and feedback made this a much better book than it would have otherwise been. Thank you so much!

A special "Dank jullie wel!" to Bernadette and Herman Arts without whom our adaptation to the Netherlands would have been so much more difficult (and we would have indubitably discovered the local cherries from Bemmel much later).

Finally, this book would have been much more difficult to write without all the *Open-Source* software I used throughout the years and I wish to thank all those who have contributed, one way or the other, to specific software packages and to the philosophy that makes this possible. My operating system has always been a flavour of `GNU/Linux`, including `Debian` (`http://www.debian.org`), `Ubuntu` (`http://www.ubuntu.com`), `ArchLinux` (`http://www.ubuntu.com`), `OpenSuse` (`http://www.opensuse.org`) and `CentOS` (`http://www.centos.org`). The book was written in LaTeX using initially `Kile` (`http://kile.sourceforge.net`) but mostly `TexMaker` (`http://www.xm1math.net/texmaker`) and `TexStudio` (`http://texstudio.sourceforge.net`) as editors/environments. All data analyses for this book used `R` (R Development Core Team, 2010), an open-source programming environment focused on statistics available free of charge at `http://www.r-project.org`. The graphs were generated using a wide variety of tools, including `R`, `LaTeXDraw` (`http://latexdraw.sourceforge.net`), `Inkscape` (`http://inkscape.org`), the LaTeX packages `pstricks`, `pst-pdgr` (pedigrees), `mhchem` (chemical formulas), `xy` (blocks and arrows diagrams), and `amssymb`, `MnSymbol`, `wasysym` and `tipa` (for various symbols including IPA), and even with `LibreOffice's` (`http://www.libreoffice.org`) own `Draw` component.

1 Introduction

For those of us interested in the scientific study of language and speech (whom I will call here, for want of a better term, *language scientists*), keeping abreast with the relevant knowledge and scientific disciplines is becoming more and more difficult. Besides linguistics "proper" (and its classic sub-disciplines phonetics, phonology, morpho-syntax, semantics and pragmatics) now we must know some psychology, we have to be conversant in the cognitive neurosciences, understand something about language disorders and, lately, be on friendly terms with genetics. Familiarity with this latter discipline is increasingly necessary for meaningful discussions about *language origins and evolution*, its *acquisition* by children and the design of individually tailored effective *second-language learning* curricula, the structure of our *capacity for language*, and to address language and speech *impediments*, to mention just a few.

Unfortunately, fundamental notions of genetics are not yet part of the standard training in the language sciences, and this means that the interested language scientist must either ignore it at his/her own peril, acquire it piecemeal from heterogeneous popularization sources (with their associated uneven quality, reliability and relevance) or plunge head-on into the dense, confusing and exponentially growing primary literature. Another possibility would be to read one or more of the existing excellent introductions to genetics, genomics, biochemistry, population genetics, evolutionary theory, etc., but these are in general too broad, they address a very different audience and cover much too much material, most of it uninteresting and not directly relevant for the language scientist.

This book aims to fill this gap by offering an introduction to selected aspects of modern genetics and genomics, tailored for scientists involved in the study of language and speech. It tries to provide a condensed selection of relevant topics, briefly introducing the needed concepts, methods and results, and using – as far as possible – examples directly related to language, speech and hearing, while constantly pointing the interested reader towards important papers and recent developments and trends in these areas. I hope that after finishing this relatively small book you will have a deeper appreciation of what genetics is, how it can be used in your work, and how to interpret findings that have a genetic component. Moreover, you should be comfortable addressing the primary literature and navigating the new developments and findings that will keep arriving at an ever increasing rate.

But most important for me is that you should be able to actively participate in inter-disciplinary research involving genetics: to properly study this enormously complex phenomenon – human language – our varied expertise is essential, as essential as that brought in by geneticists, neuroscientists, statisticians and psychometricians, among others, as equal partners in a dynamic and creative dialogue. To make this work, each participant needs to grasp to an acceptable degree what the others are thinking and doing and, more generally, to try to see the world through their eyes.

Thus, this book aims to invite us, the language scientists, to see the world through genetic lenses as it were, making us capable not only of judging the relevance of genetic findings and methods, but of actively participating in the adaptation of existing methods – and the invention of new ones – appropriate to the questions we are interested in answering.

Writing this book has been particularly difficult for a number of reasons. Genetics is an extremely broad, complex and very rapidly evolving field, where quite a sizeable proportion of publications from 10 years ago are literally old and their assumptions, methods and findings were amended or even invalidated by newer publications. A field where new directions and research questions continually pop up, where more often than not there is a real race between multiple teams to publish similar results in high-ranking journals such as *Science* and *Nature*, where hair-raising ethical issues are combined with tremendous pressures emanating from the health industry and political agendas. Where the lone genius is more and more a rarity being replaced by labs of tens of people and networks of tens of such labs producing papers with tens or even hundreds of authors. Where the funds required for a single project go beyond the wildest dreams of most social scientists, not to mention those working in the humanities...

On the other hand, as the title tries to convey, there probably is no unified "Science of Language" to speak of but a plethora of fields of research springing from different historical roots, using quite different methods and having different goals and standards of explanation, which I assume that you, the reader, are painfully familiar with. Thus, besides the several schools within theoretical linguistics proper, there are typologists, historical linguists, sociolinguists, morphologists, syntacticians, semanticists, dialectologists, psycholinguists, phoneticians, phonologists, cognitive scientists, neuroscientists, speech pathologists, engineers working on speech comprehension and synthesis, neurologists and psychiatrists dealing with speech and language, and philosophers of language, to mention just a few and glossing over the differences hidden behind such convenient labels.

I wrote this book trying to keep in mind the varied needs and interests of all of them when it comes to genetics. Some will want to know about the genetics relevant to language evolution, others to understand how the genes and the environmental factors manage to build competent language users by the billions, while yet others try to find parallels between the patterns of linguistic and genetic diversities. Some will certainly want a conceptual, generic, bird's-eye view of genetics while others will feel disillusioned if actual methodology and mathematics are lacking. And some would prefer cutting-edge research and results while others understandably will want to first build a solid basis on those results that have stood the test of time (of which there are plenty in genetics) from which to confidently start exploring. I hope I have managed to address all these issues and I hope that for each reader there is something useful here, hopefully more than a lone paragraph buried among pages of useless prose, tables, equations and figures.

The book tries to be as modular as possible, but still the best approach is to read it sequentially given that concepts, methods and findings are introduced as needed.

First (Chapter 2), we begin by addressing the various approaches to the nature-nurture question focusing on heritability and the amazing complexity behind seemingly simple concepts such as "innate" and "acquired". This topic, of what is due to "nature" and what to "nurture", is an important one in the language sciences, but unfortunately the manner in which it is sometimes approached feels rooted in the past, disconnected and impervious to recent advances in genetics, developmental and evolutionary biology. This chapter tries to offer an updated view of the concepts, findings and methods, and to ensure a proper understanding. We then (Chapter 3) encounter the actual reality of how genetic information is stored, transmitted and expressed, discussing such processes as replication, transcription and translation, and the structure of genes. Chapter 4 focuses on patterns of inheritance, exemplifying them with some examples relevant to language and speech such as a dominant pathology affecting speech, recessive hearing loss that resulted in the emergence of a new sign language and the sex-linked transmission of colour-perception deficits. We will discover how genes are actually found in Chapter 5 where we encounter association and linkage studies and see some examples of genes discovered using these methods, while the next chapter (6) gives some actual examples of how genes work. This chapter is very important not only because it describes real-world cases of genes affecting phenotypes relevant for speech and language but also because it dispels any simplistic notions about how genes do their jobs. A very short Chapter 7 discusses the promises of whole exome and genome sequencing, but for now this potential has not been used for language and speech. Also here I put together quantitative and molecular genetics and try to illustrate what to expect about the genetic architecture of speech and language. I dedicate a special chapter (8) to population and evolutionary genetics, discussing the forces that shape the genetic structure of human populations and their relevance for understanding human history and the patterns and processes shaping linguistic diversity. This should provide the foundations necessary for discussing in their context the issues surrounding language origins and evolution, on one hand, and the biological background to the patterning of linguistic diversity, on the other. This leads naturally (Chapter 9) to discussing recent advances in the understanding of the fascinating interactions between culture and the biological bases that make it possible. I review several cases of gene–culture co-evolution with a special significance for language and speech such as the spontaneous emergence of new sign languages in communities with a high incidence of hereditary hearing loss and the proposal that our genetic background might bias the process of language change, thus influencing linguistic diversity. Finally, as a guide for the interested reader, the conclusions (Chapter 10) contain a list of further readings and other resources (such as online databases and tools) that can be consulted for further information.

Throughout the book there are references to both reviews and introductory texts, on one hand, and to primary research (fundamental findings, description of methodology or cutting-edge reports) on the other, allowing the interested reader to continue on their own and deepen their expertise. The Appendix provides the actual `R` code implementing some points discussed in the book, while abundant footnotes clarify and give technical detail and actual snippets of code, as needed. Finally, a Glossary provides short definitions of the most important new terms and abbreviations.

Box 1: Boxes with technical details

Technical discussions and mathematical details are included in boxes such as this one and can be safely skipped. However, they are still recommended for a fuller understanding of the topic under scrutiny.

But before we start this journey of discovery, I must try to answer a fundamental question that I heard several times being explicitly formulated, and many more times lurking implicitly behind comments, suggestions, questions and discussions over a pint of beer: why should I, as a student or scientist interested in language and speech, care about genetics and evolutionary theory?

1.1 Why is genetics relevant for me?

Indeed, why? Why invest precious time and effort in reading this 300+ pages book?

This is a frequent thought (if not a frequently asked question) when genetics is introduced to language scientists. The same might be said about statistics, evolution or electrical engineering! But while a weak argument can be made for the last one (it's used in some approximations of the acoustics, not to mention building and maintaining experimental equipment), it is much easier to argue for the others. Indeed, statistics is not only useful as a tool for testing the difference between two conditions in an experimental design, but offers a surprising and at times inspiring manner of viewing the world. Likewise, evolutionary thought puts things in a much wider perspective and gives meaning to many otherwise incomprehensible phenomena, the emergence of language being just one of them.

Genetics is relevant on many levels to those studying language and speech. A useful distinction can be made here between three such levels of complexity and associated time scales (Smith and Binder, 2013): the *individual*, the *population* and the whole *species*. The individual level and the associated **ontogenetic** processes concern the build-up of the machinery necessary for learning and using language. The population level and the **glossogenetic** timescale (Hurford, 1990) refer to the supra-individual processes acting over longer periods of time and involved in language change and the patterning of linguistic diversity. Lastly, the species level involves **phylogenetic** processes shaping the emergence and evolution of language on even longer time scales.

Necessarily, these distinctions are artificial and all these levels, processes and time scales are continuously interacting, defining each other in the process. Thus, it makes no sense to discuss isolated individuals (where would they acquire language from?) or groups abstracted away from the people composing them (who is doing the talking?) without the evolutionary context that produced both the capacity for language and the actual languages we speak (which is most probably not a coincidence). Nevertheless, these three levels offer a useful first approximation and should be kept in mind as we think about the genetic foundations of language.

To reiterate, an understanding of genetics is relevant to all of them. First, it should be obvious that the development and maintenance of a language user are rooted in genetic mechanisms that, in intimate and continuous interaction with environmental factors including the general cultural and the linguistic, ensure the development of the organs and systems necessary for perceiving, producing, processing and learning language. As we will see, this is not an encapsulated phase of "development" where the genome is "read" (as a recipe for making bread would be) and then archived and forgotten until the next generation needs it in the womb; quite the opposite, our genome is a dynamic and active thing continuously being expressed and involved in complex regulatory cycles, reacting to changes within and without our bodies on the level of the millisecond, allowing us to adapt to our continuously changing environment and to learn.

Likewise, nobody would deny that changes in our genomes were necessary to make our species develop (invent?) and use language, but the exact nature of these changes, the reasons they happened, and when they did are very contentious issues. Understanding the structure of our genome and how it is expressed helps constrain the range of plausible accounts for the emergence and evolution of language and current breathtaking advances quickly transform armchair speculation into testable hypotheses. Famously, the Société de Linguistique in Paris banned discussion about the origins of language in 1866, arguing, convincingly, that the evidence simply is not there to test the proposals, but we are quickly reaching an age where after some 150 years, thanks mostly to genetics, this ban can be safely lifted.

Probably the hardest to justify is why genetics would be relevant at the population, glossogenetic level: what would a typologist, historical linguist or field linguist working in the Amazon gain from understanding genetics? One answer is that, for language as for biology or any other entity shaped by historical processes, the past is the key for understanding the present and more and more understanding the past of language means understanding the past of its speakers, where a major role is played by genetics (Cavalli-Sforza et al., 1994; Jobling et al., 2013). We will only touch on this fascinating topic here as there are very good introductory works available, with Jobling et al. (2013) being highly recommended, but the key insight is that this type of correlation (or lack thereof) between languages and genes is purely accidental, being caused by a shared causal factor, namely historical processes affecting populations. Thus, there's nothing in the genes of the populations speaking Chinese languages (such as Mandarin, Cantonese, or Wu) that makes them speak such languages; any correlations there might be between their genes and their languages are purely an accident of history.

But there might also be causal links between a population's genetic make-up and the language(s) it speaks (Dediu, 2011a, 2013) in the sense that a genetic background (dis)favours the presence of certain structural (or typological) linguistic features, such as the use of variations in voice pitch to convey not only intonation but also distinctions between words or grammatical information (what is called **linguistic tone**; see for example Yip, 2002). For example, there could be something in the distribution of genetic diversity within South-East Asia that makes the presence of tone languages (such as the Chinese languages but also Vietnamese and Thai) much more probable than say in Europe (Dediu and Ladd, 2007). Such *genetic biases* are very weak at the individual level but get amplified through language use and transmission, such that they influence the trajectory of language change and, ultimately, the distribution of linguistic diversity (Ladd et al., 2008).

Therefore, understanding how our genome is structured and how it works is indeed relevant for most kinds of language scientists, and opens up new perspectives on the nature of language, its evolution and change.

2 Nature, nurture, and heritability

In this chapter we approach, at a fairly abstract level, the fundamental questions concerning the relationships between the phenotype (the observable properties of individuals), the genotype and the environment. We discuss the paramount importance of variation in studying these relationships and we define, estimate and discuss the meanings and misinterpretations of heritability. Far from being a simple concept, heritability will turn out to have some non-intuitive properties that make the interpretation of heritability estimates quite a tricky exercise. Likewise, we will discover that, in fact, all the related concepts and distinctions, such as innate and acquired, or nature and nurture, are fuzzy and far from their apparent clarity in everyday discourse. We will end with a very brief survey of heritability studies in speech and language. This chapter also introduces several fundamental concepts of statistics that are necessary for a proper understanding of many topics covered in this book.

2.1 Phenotype, genotype and environment

It is unquestionable that both "nature" and "nurture" are required for the development of a linguistic human being. Lacking "nature" will limit language development no matter how much "nurture" there might be, as many a pet owner can easily confirm. This is seemingly supported by studies of chimps (such as Nim Chimpsky and Washoe) reared in conditions similar to those experienced by human babies and infants, but which nevertheless fail to go beyond a rather limited level of language usage. On the other hand, having "nature" but lacking "nurture" is equally devastating, as shown by the cases of children who, for various reasons, have not been exposed to language during the so-called **critical period** for language acquisition (a well-known case being Genie) and who fail to develop full-blown language despite considerable efforts.

Thus, if we denote, in a highly abstract manner, the "nature" as G (from genetics) and the "nurture" as E (for environment), then we can attempt to write down a symbolic equation describing how these two factors relate and interact in producing the phenomenon of interest, P (for **phenotype**). For us, P will usually mean some aspect of language and speech or some other relevant cognitive process, but it can mean virtually any feature an organism possesses. Thus, it can refer to what we may loosely conceptualize as individual features, such as a person's height (as measured with a meter from the top of the head to the feet while standing), to the eye colour subjectively placed in categories such as "blue", "green", "brown" or "black", to molecular aspects such as the speed with which a certain enzyme breaks down a given molecule in the body, or to relational phenomena such as pair-bonding or the use of language and speech. Of course, these levels are far from clear-cut and fixed, but they prove useful in understanding complex phenomena such as those of interest here. Moreover, a major enterprise in modern science is to be able to understand how lower-level phenomena interact in order to produce higher-level ones which, in this context, means that we would like to provide a "full story" ranging from molecules to social networks and language change, without necessarily implying a strong a priori reductionism.

The simplest form of such an equation would be (i) $P = G$ or (ii) $P = E$, which would represent the cases where a phenotype is purely the product of the genes or of the environment. What would qualify here, though? Maybe (i) could describe those aspects of an individual which are "purely" biological while (ii) would be applicable to those which are shaped by the environment alone. Like having a heart, which all humans do, or dyeing your hair, which clearly depends on someone's culture. In what sense is it meaningful to think that (i) or (ii) would hold? What is the basis for the intuition that having a heart might be an instance of (i) while dyeing one's hair is an instance of (ii)? It seems that we think G is behind humans having a heart because all humans do and there are other things which do not and this seems quite stable across environments, cultures, historical periods, etc. Analogously, we think that E is behind hair dyeing because some do and others don't and this critically depends on one's local culture, available technology, etc. Thus, all these judgements essentially rest on *the patterns of variation* in P: while hearts seem to follow biology and disregard the environment, hair dyeing seems to do the opposite.

However, taking a closer look at those phenotypes which might seem fully determined by "nature", on one hand, and at those fully determined by "nurture", it soon becomes clear that these cases are pure abstractions, lacking any reality or meaningfulness. There are people without a heart (not only in the metaphorical sense!), but they are simply aborted at an early stage of development. Moreover, there are other milder conditions compatible with intra-uterine development, birth and even life to maturity, characterized by various defects of the heart[1] and, while for some there is a clear genetic component, for others there is an as clear environmental causation, including maternal pre-gestational diabetes, infection with rubella or exposure to the drug Thalidomide during gestation, among others (Jenkins et al., 2007; van der Bom et al., 2010). On the other hand, hair dyeing basically needs some hair to start with, hair development, its patterning on the body, the right biochemical properties, etc., which all involve a lot of genetics (Shimomura and Christiano, 2010; Törnqvist et al., 2010). Moreover, one's personality, gender and other factors, all having some genetic component, have a role to play in the actual behaviour under consideration. Thus, we can safely conclude that the equations (i) and (ii) cannot hold and we need to consider more complex ones involving both G and E.

But before delving into the complexities of heritability it is instructive to briefly overview the fascinating debates surrounding seemingly simple and familiar concepts such as "innate", "acquired", "inborn", "learned", "nature" and "nurture".

2.2 Innateness, a slippery and complex concept

Many fundamental arguments in the language sciences (and cognitive sciences in general) seem to revolve around the twin notions of "innate" and "acquired" (or the related "nature" and "nurture"). To some leading linguists, Noam Chomsky included, various properties of language such as the patterning of linguistic diversity with the apparent existence of **universals** (Newmeyer, 2005; Comrie, 1989; Croft, 2003), the process of language acquisition seemingly capitalizing on information not present in the primary data (the so-called **poverty of the stimulus** argument; Chomsky, 1980), and the computational data structures and machinery postulated for processing language (Chomsky, 1965, 1980) seem to point to "innateness". However, what exactly is "innate" and the exact nature of this "innateness" are extremely vague (see, for example, Pullum and Scholz, 2002; Mameli and Bateson, 2011; Mameli, 2008; Cowie, 1999; Kiikeri and Kokkonen, 2007; or Bateson and Mameli, 2007) and in need of clarification.

In fact, when a proper analysis of these apparently obvious concepts is done in the light of modern data and theory from the biological and cognitive sciences, one is left with a collection of not necessarily consistent proposals and properties. Mameli and Bateson (2006) (see also Mameli and Bateson, 2011; Mameli, 2008; Bateson and Mameli, 2007) conduct such an analysis of "innateness" and identify not less than 26 manners in which this concept has been (explicitly or implicitly) defined and used in the scientific literature

[1] Encompassed under "congenital heart disease" (Hoffman and Kaplan, 2002).

(Table 1 on page 177 in Mameli and Bateson, 2006, gives a summary). For example, one might define an "innate" character as being not acquired, but a quick look at development reveals that basically all characters are acquired given that the just-fertilized egg is quite different from the adult. Likewise, the "obvious" presence at birth fails to capture things like sexual characteristics and parental behaviour, while its refinement requiring the reliable appearance at a well-defined life stage fails to account for learning/cultural influences.

Genetic determinism, influence and encoding are increasingly refined versions of a familiar argument but fail to explain much, given that genes do not, in general, deterministically dictate development but arguably all traits are influenced by genes, while genetic information/encoding is a very difficult notion. In this same vein falls Chomsky's poverty of the stimulus argument which, in a nutshell, proposes that some traits do not extract information from the environment (in the particular case of language acquisition, supposedly some evidence is simply not in the data the child sees). However, while intuitively appealing (if this information is not provided by the environment it must come from somewhere else, namely the genes), as always, nature is much more complex and subtle. For example, scars and calluses do not seem to extract information from the environment for their formation but are not really "innate", while the complexity of the interactions between genes and environmental factors during development seems to rule out such simple dichotomies (we will encounter later in this book several examples of such fascinating interactions). An example used by Mameli and Bateson (2006) is the classic work by Gottlieb (e.g., 1991, 1997) on mallard ducks showing that the exposure of ducklings to their own calls while still inside the egg facilitates the later recognition of the (otherwise quite different) adult species-specific calls.

One final type of definition that we will discuss here is the widespread idea that something is "innate" if it is species-specific or species-typical. However, for almost all interesting characteristics there are some members of the species that lack it (or have "deviant" versions of it), raising the question of how to define the norm and the exceptions. This problem of defining the norm is extremely difficult to solve, as shown by the complex issues faced by medicine and psychiatry, for example. Another issue is that for many approaches individuals suffering from genetic pathologies – such as the developmental speech dyspraxia (**DVD**) experienced by members of the KE family affected by a mutation in the ***FOXP2*** gene (see Sections 4.2 and 6.7 for details) – must be considered atypical, implying that these pathologies are not "innate" on this account. Finally, characteristics that are learned by all normal individuals, with reading, writing and using a mobile phone rapidly becoming valid examples, must be considered as "innate".

The bottom line of this extremely instructive exercise is that no single definition of "innate" (or "acquired", "nature, "nurture") seems to capture all aspects and complexities of the natural world. One solution would be to declare this exercise invalid and pick one definition as being the correct one (or arguing that there is no need to actually stick with a definition at all) while dismissing all counter-examples as freaks of nature irrelevant to the enterprise at hand (here, understanding language). However, I feel that such an approach is disingenuous and non-scientific and should be avoided at all costs. One possible way out would be to apply a milder version of this, which is to explicitly pick one such definition (as opposed to implicitly using it) and discuss its pluses and minuses for the task at hand, warning the reader that the data collection, methods, conclusions and inferences depend on it.

A better alternative (in my view) was proposed by Mameli and Bateson, namely to consider some of these definitions (selected based on their coverage and explanatory power) as capturing some aspects of a complex reality, and score each trait of interest on each of these *i-properties* (innateness properties), resulting in a sort of innateness score. Some characteristics would score on all i-properties (e.g., grooming behaviour in rats or babbling – vocal and/or gestural – in infants) while others on none (e.g., the belief that the Earth is round or using the word /dɒg/ to designate the domestic dog *Canis lupus familiaris*).[2] Mameli and Bateson propose to transform the theoretical discussion about "innateness" into an empirical enterprise where we would score many characteristics on these i-properties and see if there are bundles of i-properties that tend to correlate and what these patterns of correlations would look like. At one extreme it could be the case that all i-properties tend to score the same across characteristics (the so-called *cluster hypothesis*), suggesting that there is something to "innateness" behind these disparate definitions, while the other would be that there are no meaningful clusters of definitions (the *clutter hypothesis*), suggesting that there's nothing to "innateness" as a general concept and we should be specific about which i-property(ies) we intend in each particular case.

It is currently not settled which of these models fits reality better (Mameli and Bateson, 2011), but probably something more similar to the clutter hypothesis than the cluster hypothesis holds, whereby certain i-properties might inter-correlate better than others. This has important consequences for our theories and discussions, as showing that a trait (say language) fits some i-properties does not necessarily warrant automatic inferences about others. For example, even assuming there are language universals – a highly contested topic and better conceptualized as statistical tendencies of various strengths (e.g., Evans and Levinson, 2009; Levinson and Evans, 2010; Dunn et al., 2011; Dediu et al., 2013; Dediu and Levinson, 2012) – and that the

[2] Such an analysis should be conducted for various properties of language and speech but it goes beyond the scope of this book.

poverty of the stimulus holds – again, hotly contested and probably wrong (e.g., Pullum and Scholz, 2002) – this would not automatically entail conclusions about there being one gene (or a few) specific for language (Chater et al., 2009; Christiansen and Chater, 2008) or its recent origins and evolution (Dediu and Levinson, 2013).

To conclude, it is important to be aware that not only is there no "nature versus nurture" but that the reality is much more complex, subtle and fascinating than any motto can capture, there being many interactions on many levels between what could be conceptualized as genes and environments (we will discuss these further in this book), but that the even apparently simple concepts involved ("nature", "nurture", "innate", "acquired", "gene", "environment") are in fact vague and, more often than not, misleading.

2.3 Some basic notions of statistics

The concept of **heritability** is of special importance in framing the discussions about the genetic bases of language and speech and these sections aim at clarifying it. However, for a proper understanding of this rather technical concept, we need to briefly introduce fundamental notions of statistics such as sample and population, mean, variance, co-variance and correlation, and statistical testing. These notions will prove useful not only in this section but also later in the book when we consider more complex problems such as how to know, given some data, if a gene is "really" associated with a trait of interest or if the apparent association could be simply due to chance. This introduction should allow readers without prior statistical training to understand the concepts and computations discussed here, but for good general introductory books the interested reader could consult for example Coolican (2004), Field (2009), Crawley (2005) or, for a Bayesian approach, Berry (1996).

Let us take a simple example: height – measuring this phenotype is easy and very reliable, and it seems intuitively to have a large genetic basis in that the children of tall people tend to be tall and those of short parents tend to be short. There seems also to be some sort of "blending inheritance" at play, as the offspring tend to fall in between their parents' heights. In fact, through the work of Francis Galton, Karl Pearson and especially Ronald Fisher at the turn of the twentieth century, height has played a crucial role in the development of modern genetics and evolutionary biology as well as statistical theory (Lettre, 2009; Halliburton, 2004; Lynch and Walsh, 1998). There is extensive variation in height at many levels: within the same person across time (development), between genders (males tend to be taller than females), between individuals at the same developmental stage and gender within a population (inter-individual variation) and between populations (people of East Asian origin tend to be shorter than those of European origin).

So, let's see how we can use some real-world data on height to introduce some useful statistical notions. First, we need a *sample* of individuals (here, just a set of people) and let's say there are N of them (where N can be 1, 10 or 10,000). If we measure their height (using a measuring instrument such as a meter), we will end up with N numbers, each representing the height of one individual. Let us denote these numbers as $h_1, h_2, \ldots h_N$, where h_i is the height of the ith individual; in this case the actual order of the individuals is not important and the indices 1, 2,... N can simply be seen as a shorthand for the individuals' name. To make things as concrete as possible, I will use a dataset collected in 1993 on about 25,000 children in Hong Kong (Leung et al., 1996) and publicly available on Statistics Online Computational Resource (SOCR)'s website (`http://wiki.stat.ucla.edu/socr/index.php/SOCR_Data_Dinov_020108_HeightsWeights`, downloaded Dec 2010). Such a set of numbers represents the distribution of height in our group and can be depicted graphically using a **histogram** (Figure 2.1). The x (or horizontal) axis represents the range of heights in our sample, from

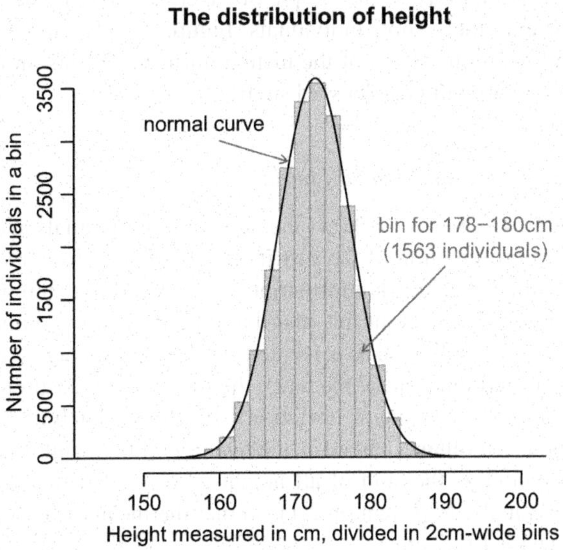

Figure 2.1 The histogram representing the distribution of height in 25,000 children from Hong Kong (Leung et al., 1996). The x (horizontal) axis represents the range of children's heights split in 2 cm-wide bins, while the y (vertical) axis represents the number of children with height falling within each bin. Superimposed is the trace (black curve) of an idealized normal distribution with the same mean (central tendency) and standard deviation (spread of values).

153 cm to 191 cm. I divided this continuous range of 38 = 191 − 153 cm into 2-cm-wide bins (e.g., all heights between 160 and 162 cm belong to the same bin, while those between 162 and 164 belong to the next bin; please note that the bin width of 2 cm is not a given but can be relatively freely varied and we could have chosen instead 1 cm or 10 cm bins, for example). For each such bin, the figure shows a light grey rectangle with darker grey border whose vertical dimension ("tallness") represents the number of individuals in our sample whose height falls within the bin. Thus, for the bin 160–162 cm there are 192 individuals with height between 160 and 162 cm, while for the next bin, 162–164 cm, there are 527 individuals with height between 162 and 164 cm. The histogram is a very useful graphical representation as it gives an "instant" and "holistic" view of the distribution: in our case it shows that most people have a height in the "bulge" of the distribution around 172 cm, with fewer and fewer having heights which are farther away from this central tendency in either direction. The degree to which the heights are spread away from (or clumped around) this central tendency represents the dispersion of the distribution. All of this, of course, fits our intuitions about height, with the vast majority of the people of "average" height and very few extremely tall or extremely short individuals. Figure 2.1 also shows a black curve closely following the shape of the histogram: it represents an idealized normal distribution and will be discussed shortly.

2.3.1 Mean, variance and standard deviation

While the histogram in Figure 2.1 shows the number of individuals in each bin, giving a faithful representation of the distribution, it is hard to work directly with it as there's simply too much information. Luckily, we can summarize the most important information about this distribution with just two numbers (or parameters): where the bulk's peak is located (the central tendency or mean) and how spread around this central tendency the individual measurements are (given by the variance or standard deviation). In a more analytic vein, this sample of heights has a **mean**, defined as the *average* of the heights contained therein, and computed as the sum of all heights, h_i, divided by the number of measured individuals, N (Equation 2.1), so that the mean of our sample of heights is $mean(h) = 172.7$ cm:

$$mean(h) = \frac{\sum_{i=1}^{N} h_i}{N} \tag{2.1}$$

However, it is clear from the histogram that this central tendency does not tell the full story about our distribution: we need to somehow quantify also the spread of the values around the mean. A commonly used such measure is represented by the **variance** of the sample defined as the *average squared deviation from the mean* (Equation 2.2):

$$var(h) = \frac{\sum_{i=1}^{N} (h_i - mean(h))^2}{N - 1} \tag{2.2}$$

which in our case is $var(h) = 23.3\ \text{cm}^2$ (the use of $N - 1$ instead of N in the divisor results in an unbiased estimate, but this is a technical point). Please note that, in opposition to the mean, the variance is not measured in the same units as the sample, cm versus cm^2, making it a bit awkward to interpret. Therefore, in most cases we will work with the square root of the variance, named the **standard deviation** (Equation 2.3):

$$sd(h) = \sqrt{var(h)} = \sqrt{\frac{\sum_{i=1}^{N} (h_i - mean(h))^2}{N - 1}} \tag{2.3}$$

and, for our sample, $sd(h) = 4.8$ cm. Less spread distributions have smaller standard deviations. Thus, our sample of heights can be summarized by its mean $mean(h) = 172.7$ cm and standard deviation $sd(h) = 4.8$ cm. But is this information sufficient?

2.3.2 *The normal distribution*

The distribution of height in our sample has some special properties: first, it is approximately symmetrical, in that there are equal numbers of people taller and shorter than the mean by a given amount. To make things clearer, let's look at the histogram in Figure 2.1: the 2 cm bin containing the mean $mean(h) = 172.7$ cm ranges between 172 and 174 cm and contains the maximum number of persons in any such bin, namely 3542. Moving one bin away from the mean on both sides, we have the bins 170–172 cm and 174–176 cm containing 3365 and 3230 persons, respectively, while the next more distant bins 168–170 cm and 176–178 contain 2736 and 2374 persons, respectively. However, this depends on the actual bin width we have chosen, and a much better approach is to forget about the histogram and look directly at the number of people with height within a given distance from $mean(h)$ on both sides: thus, there are 1793 people at least 1 cm shorter than the mean and 1764 at most 1 cm taller than the mean, 3514 people at least 2 cm shorter than the mean and 3535 at most 2 cm taller than the mean, 7653 and 7614 for a deviation of 5 cm or less and so on.

There are many types of symmetric distributions, but there is one with a special status in statistics, namely the **normal distribution**, which has, among others, the important property that it is completely described by its mean and standard deviation. It is a bell-shaped distribution and for our mean of 172.7 cm and standard deviation of 4.8 cm it is represented in Figure 2.1 by the black curve (the shape of this curve is reflected in the well-known popular name of the normal distribution, "the bell curve"). One other important property is that approximately 68% of the data (persons, in our case) falls within one standard deviation from the mean, approximately 95% of the data falls within two standard deviation, and most of it (approximately 99.7%) is within three standard deviations. Given that our data looks pretty normally distributed (our histogram closely follows the normal curve), let's check if these properties hold for us: one standard deviation from the mean represents all heights between $mean(h) - sd(h) = 167.9$ and $mean(h) + sd(h) = 177.5$ cm and there are $14,915$ people with these heights, representing 68.3% of the $21,842$ in the database. Within two standard deviations ($mean(h) - 2sd(h) = 163.0$ to $mean(h) + 2sd(h) = 182.3$ cm) there are $20,856$ people (95.5%) and within three standard deviations $mean(h) - 3sd(h) = 158.2$ to $mean(h) + 3sd(h) = 187.2$ cm) there are $21,799$ people (99.8%).

The normal distribution is extremely important in the biological and social sciences (among others) as it is very commonly encountered in practice: usually, the phenomena of interest turn out to be approximately normally distributed. One factor explaining this is that most interesting phenomena are the result of many factors working together, and of the so-called Central Limit Theorem, which guarantees (under certain general conditions) that the sum of many identically distributed random variables tends to a normal distribution.

To make this clear using genetic concepts, let us imagine that a person's height is purely genetically determined and that many identical and independent genes are involved. This last condition is not far from the truth: we already know of hundreds of genes involved in height (Allen et al., 2010) – with many more remaining to be discovered (Yang et al., 2010) – each contributing very little. Each such gene can come in two versions (or **alleles**), let's call them *A* and *a*, and each gene has the same effect: if it has form *A* it adds a very small amount to the person's baseline height (let's say 1 mm) and if it is in form *a* it adds nothing; furthermore, consider that each person has the same baseline height (take it to be 1 m). For the vast majority of their genes – or **loci** – humans have two copies of each of them: one inherited from the mother and one from the father. Thus, one individual can have two copies of the *a* allele, another two copies of the *A* allele, while yet another one copy of an *a* and one of the *A* (please note that which one comes from the mother and which one from the father does not usually matter). This arrangement whereby each individual has two copies of each gene is called diploidy (the individual itself is called a **diploid**), and more details will be discussed in Chapter 3.

Box 2: Properties of the normal distribution

The exact shape of the normal distribution is given by the following formula:

$$f(x) = \frac{1}{2\pi\sigma^2} e^{-\frac{(x-\mu)^2}{2\sigma^2}} \tag{2.4}$$

where $\pi = 3.1415\ldots$ is the circle circumference to diameter ratio, $e = 2.7182\ldots$ is Euler's number, μ and σ are the distribution's mean and standard deviation and x is a specific value. $f(.)$ is called the *probability density function* (or *pdf*) and for any two values, a and b, the area under the curve represented by plotting $f(.)$ between a and b is equal to the probability of observing a value x between them ($a \leq x \leq b$). For example, let's take $\mu = mean(h) = 172.7$ cm, $\sigma = sd(h) = 4.8$ cm and $a = mean(h) - sd(h) = 167.9$ cm and $b = mean(h) + sd(h) = 177.5$. This is shown in Figure 2.2 where the shaded area represents the probability of someone having a height between 167.9 and 177.5 cm, $P(167.9 \leq x \leq 177.5) = 0.68$. More precisely, $P(a \leq x \leq b) = \int_a^b f(x)\,dx$.

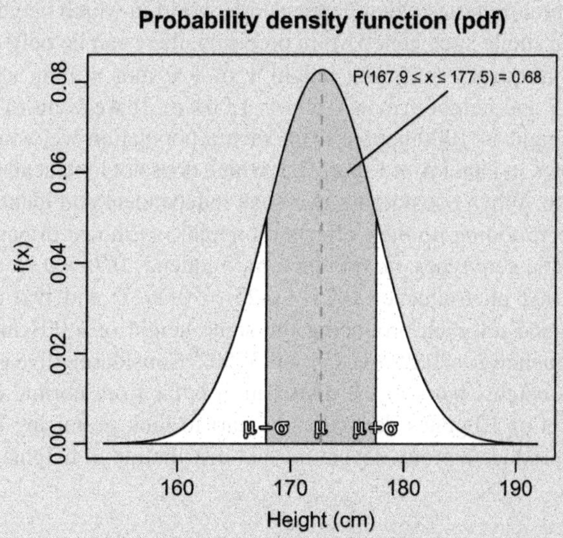

Figure 2.2 The probability density function (pdf) of a normal distribution with the same parameters (mean and standard deviation) as our height sample. The x axis shows the possible heights, while the y axis represents probability. The shaded area is the probability of a value x being at most one standard deviation from the mean (between $\mu - \sigma$ and $\mu + \sigma$).

However, to make things as simple as possible, we will assume here that we are referring to a fictional species of **haploid** humans which have only one allele for every gene; therefore there are only two possible types of people when speaking of this particular locus: those with an *a* and those with an *A*. Further, let us assume that there are multiple such loci within an individual, each locus being capable of accommodating only one *a* or *A*; thus, a haploid person with five such loci of the form *A* (thus with **genotype** *AAAAA*) will have a height 1 m + 5 × 1 mm = 1.005 m, while a person with genotype *AaaAA* with be 1.003 m tall. As we simply sum over their effects, the order of the loci does not matter, such that all individuals with genotypes *aaAAA*, *aAaAA*, *aAAaA*, …(there are 10 genotypes with 2 *a*s and 3 *A*s[3]) will have the same height, namely 1.003 m; thus this specific height can be realized in 10 ways. Finally, the two alleles *a* and *A* might not necessarily have the same frequency in the population; let's say *a* has a frequency of 0.75 (75% of the people have it for each gene independently) and the frequency of the other allele, *A* is 1 − 0.75 = 0.25 (25%). Given that the five loci we are considering are independent, the probability of a person having genotype *aaaaa* is $0.75 \times 0.75 \times 0.75 \times 0.75 \times 0.75 = 0.24$, while genotype *AaaAA* has probability $0.25 \times 0.75 \times 0.75 \times 0.25 \times 0.25 = 0.0088$.

With these preliminaries, what happens in a world in which height is under the control of a single such gene? More precisely, there can be only two genotypes, *a* with frequency 0.75 and height 1 m + 0 mm = 1 m, and *A* with frequency 0.25 and height 1 m + 1 mm = 1.001 m. If we were to randomly measure the height of 1000 people from such a population we would get the histogram shown in Panel **A** of Figure 2.3, which does not look at all like a normal distribution. When considering two such independent and identical genes acting in concert adding up their effects on height (as shown in panel **B**), we get three possible genotypes: *aa* (height 1 m, frequency $0.75 \times 0.75 = 0.5625$), *AA* (height 1.002 m, frequency $0.25 \times 0.25 = 0.0625$) and two equivalent genotypes *Aa* and *aA* each producing the same height of 1.001 m and with combined frequency $2 \times (0.25 \times 0.75) = 0.3750$.[4] Considering five genes adds more possible heights but still the distribution is far from normal (panel **C**), while the effect of 10 genes does already start to look promising (panel **D**), and 50 genes produce a pretty decent normal distribution of heights (panels **E** and **F**).

[3] In general, the number of genotypes having k a alleles for $n > k$ genes (thus having $n - k$ A alleles) is given by the number of k-combinations of a set of n elements, $\binom{n}{k} = \frac{n!}{k! \times (n-k)!}$, where the factorial of an integer m is defined as $m! = 1 \times 2 \times \ldots \times (m-1) \times m$. Thus, in our case $n = 5$ genes and $k = 2$ a alleles, resulting in $\binom{5}{2} = \frac{5!}{2! \times 3!} = \frac{1 \times 2 \times 3 \times 4 \times 5}{1 \times 2 \times 1 \times 2 \times 3} = \frac{120}{12} = 10$ possible genotypes.

[4] The frequencies of these three possible genotypes add up to 1.0 as expected.

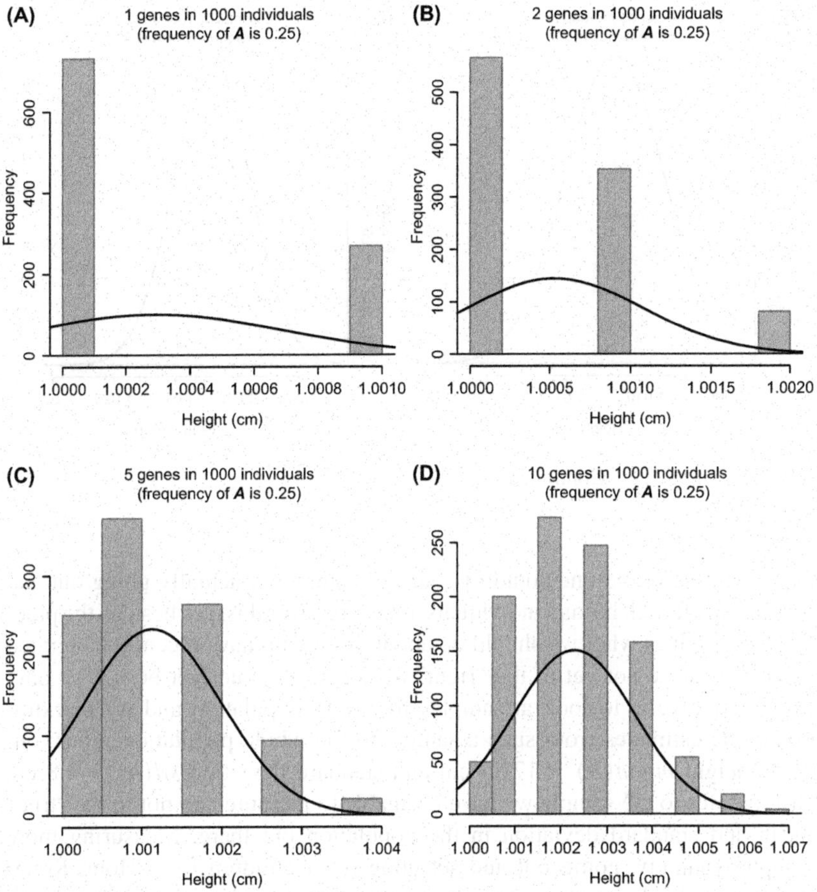

Figure 2.3 An illustration of the Central Limit Theorem using a set of identical and independent genes influencing height. These genes have two alleles, *a* with a population frequency of 75% and no effect on height and another allele *A* with frequency 0.25% which adds 1 mm to the person's height. Panel **A** show the case of a single such gene sampled in 1000 individuals, while panels **B** to **E** show that when more such genes add their effects, the distribution of height in the population resembles more and more a normal distribution (shown by the black lines). Panel **F** shows the effect of sampling more individuals.

2.3.3 *Statistical populations and samples*

Of course, our interest is not in the particular *sample* we happen to have collected (i.e., the 25,000 children being measured) but in the distribution of height in all the children from Hong Kong (the statistical

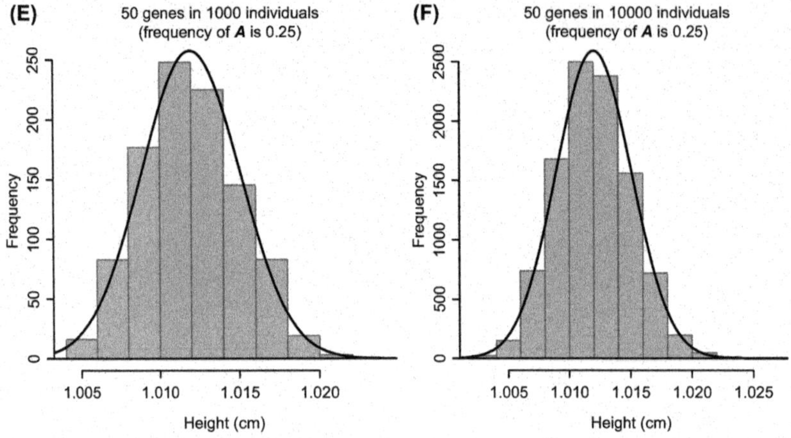

Figure 2.3 (Continued)

population[5]). While one could of course dream about actually going out and measuring all such human individuals, there are several issues besides the practicalities such as what we should do about those who have recently emigrated or died or those not yet of the "right" age or, indeed, not yet born. The only real alternative is to look at samples from this population and try to extrapolate the estimates from such a sample to the whole population. Thus, our mean height $mean(h) = 172.7$ cm and standard deviation $sd(h) = 4.8$ cm strictly refer to the sample we have been able to measure and not to the "true" mean and standard deviation in the population out there. Measuring more samples (namely, going out and selecting other children from Hong Kong) would very likely result in other means and standard deviations, all estimating the true mean and standard deviation but with an error. In theory we could measure lots and lots of such samples, resulting in a set of such sample means on one hand, and another set of sample standard deviations, and it is these sets that become closer and closer to the true mean and standard deviation.

To make the distinction between samples and populations clear, the population parameters are sometimes denoted using Greek letters: μ for the mean and σ for the standard deviation. Of course, we cannot know for sure the population parameters (such as μ and σ) as we seldom can, even in principle, go out and measure all members of the population, and all we can do is guess them

[5] Please note that the same term – population – is used both in the statistical sense (detailed below) and in the genetic sense (discussed later), but these two should not be confused.

based on the estimates we can actually obtain (such as $mean(h)$ and $sd(h)$), and put limits on our confidence in their true value. For example, what is the probability, having observed this sample of 25,000 children, that the real mean μ is higher than say 1.5 m? Most of statistics is in fact concerned with this type of qualified guess, and understanding how this works is essential for most of genetics as we must, most of the time, draw inferences from limited samples and quantify our degree of trust in these inferences. Add to this the uncertainty due to errors in sampling, measurement, etc. and it is clear we need principled ways of quantifying our uncertainty about the population of interest.

2.3.4 *Covariance and correlation*

Rarely are we interested in a single parameter, such as height. The most interesting questions in science concern the relationships between several parameters and the simplest set of such questions refers to how the variation in one parameter is linked to the variation in the other. Luckily, our dataset of 25,000 children in Hong Kong (Leung et al., 1996) also contains the weights of the children, denoted w and distributed as shown in Figure 2.4: the mean weight in this sample is $mean(w) = 57.7$ kg and the standard deviation $sd(w) = 5.3$ kg.

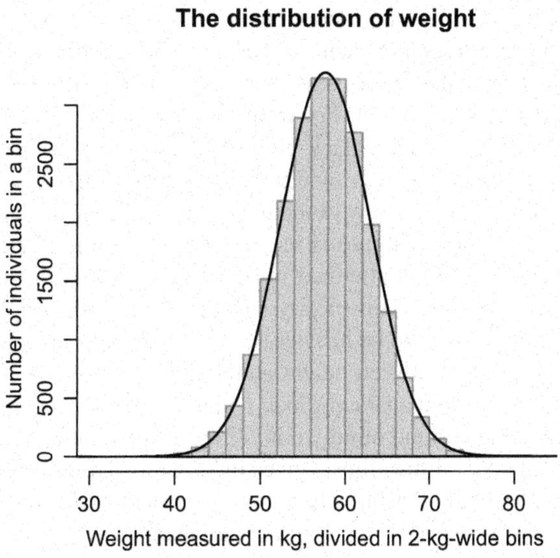

Figure 2.4 The histogram representing the distribution of weight in the same sample of 25,000 children from Hong Kong for which height was represented in Figure 2.1.

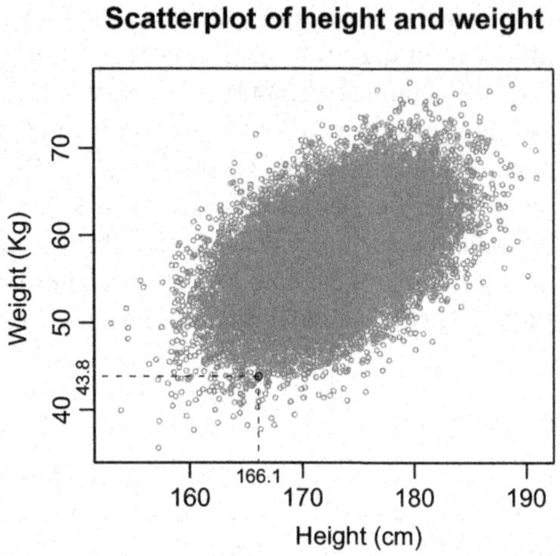

Figure 2.5 Scatterplot of height (horizontal axis) and weight (vertical axis) in 25,000 children from Hong Kong. Highlighted in black is one random person.

But is there any relationship between the height and the weight of these children? More precisely, do taller children also tend to be heavier than shorter ones (as intuition might suggest) and if so, how do we graphically represent such a relationship? Equally importantly, how do we quantify it? Figure 2.5 shows a **scatterplot** representing each child as a small grey circle, where each dot's value on the horizontal (the x) axis represents its height and its value on the vertical (the y) axis its weight. For example, one random child (number 6181) is represented as a larger black circle and dotted black lines explicitly show his/her height (166.6 cm) and weight (43.8 kg). Because there are $25,000$ such small circles, very few of them can be seen individually but what is really important about this figure is its overall pattern: the circles are not randomly scattered around but clumped in an almost perfect ellipsis with its long axis tilted at about 45°. This clearly suggests that taller people (bigger height goes to the right) tend to be heavier (larger weights go up). A perfectly deterministic relationship between height and weight would have resulted in all the grey circles falling on a straight line, while no relationship would have been shown by a circular cloud of grey dots. If there is a relationship, the orientation of the long axis of the ellipsis gives its direction: if it is tilted like $\diagup$ it represents a *positive* relationship where both parameters increase in unison (our case), while $\diagdown$ represents a *negative* relationship where one parameter increases where the other one decreases.

Quantitatively, such a relationship between two variables, h and w, is captured by the **covariance** between them, denoted $cov(h, w)$ and defined as the average of the products of the deviations from the means.

$$cov(h, w) = mean[(h - mean(h)) \times (w - mean(w))] = \frac{\sum_{i=1}^{N} [(h_i - mean(h))(w_i - mean(w))]}{N} \quad (2.5)$$

In our case, $mean(h) = 172.7$ cm, and the first two heights are $h_1 = 167.1$ cm and $h_2 = 181.7$ cm, resulting in the deviations $h_1 - mean(h) = -5.6$ cm and $h_2 - mean(h) = 9.0$ cm, respectively, while for weights we have $mean(w) = 5.3$ kg, $w_1 = 51.3$ kg and $w_2 = 61.9$ kg, resulting in the deviations $w_1 - mean(w) = -6.4$ kg and $w_2 - mean(w) = 4.3$ kg, respectively. This translates into the first two products: $(h_1 - mean(h)) \times (w_1 - mean(w)) = -5.6 \times -6.4 = 35.84$ and $(h_2 - mean(h)) \times (w_2 - mean(w)) = 9.0 \times 4.3 = 38.70$. Doing the computations for all children and taking the average gives the covariation $cov(h, w) = 12.8$, which is positive, as expected from the scatterplot.

However, it is difficult to judge the strength of this association, as the covariance has no natural scale and depends on the actual values of the variables. To address this shortcoming, we normalize the covariance so that it ranges between −1 and +1 by dividing it by the product of the standard deviations of the two variables. This is the **Pearson correlation** between the two variables:

$$cor(h, w) = \frac{cov(h, w)}{sd(h) \times sd(w)} \quad (2.6)$$

For our data, $sd(h) = 4.8$ and $sd(w) = 5.3$, so that the correlation between height and weight in these children is $cor(h, w) = \frac{12.8}{4.8 \times 5.3} = 0.50$, which is quite large given that the maximum possible is 1.0.

When two variables are inversely related, the increase in one being linked to a decrease in the other, their covariance and correlation will be negative, while when the two variables increase at the same time, their covariation and correlation will be positive. A covariation and correlation of 0.0 could mean that the variables are *independent*.[6] A rule of thumb could be that correlations between −0.1 and 0.1 are very weak (or non-existent), between 0.1 and 0.3 (or between −0.1 and −0.3) are weak, between 0.3 and 0.5 (or between −0.3 and −0.5) are moderate, and between 0.5 and 1.0 (or between −0.5 and −1.0) are strong.

Before we continue, let us note a property of covariance which will prove essential to the study of heritability, discussed in detail in the next sections. How do we compute the variance of a sum of variables? This is important

[6] Another possibility is that the variables are related (i.e., not really independent) but the relationship between them is not linear. This is a more general issue with the Pearson correlation as it only measures the strength of the linear relationship between variables and there are various ways to deal with this by using, for example, non-parametric correlation coefficients such as Spearman's ρ. See Anscombe (1973) for some classic examples and Field (2013) or any other introductory textbook for more discussion.

as we will need to be able to think about not only the variance of the genetic factors themselves and that of the environmental factors separately (as both the genes and the environment affect language and speech) but also the variance of their acting together (as usually both genes and environment are required for development). More exactly, what is the variance of $h + w$, the sum of the height and weight for our children?[7] It is easy to compute this from our data ($var(h + w) = 76.9$), but there is a more general result, namely:

$$var(h + w) = var(h) + var(w) + 2cov(h, w) \tag{2.7}$$

which, as expected, produces the same result: $var(h + w) = 23.3 + 28.0 + 2 \times 12.8 = 76.9$. The take-home message is that the variance of a sum equals the sum of the variances if and only if the variables are not correlated.

2.3.5 *Regression*

A related notion is that of **regression** and given that many methods use some form of regression, on one hand, and that many other statistical techniques are based on it, on the other, we will briefly investigate it here. Let's go back to the relationship between height and weight but this time we will use a different dataset that records the height and weight but also the age of 1035 Major League baseball players (this dataset is available for download from `http://wiki.stat.ucla.edu/socr/index.php/SOCR_Data_MLB_HeightsWeights`). The relationship between height and weight is represented by the scatterplot in Figure 2.6 and it can be seen that there is a clear positive relationship between the two, confirmed by the Pearson's $r = 0.53$.

However, the figure also features a black oblique line that seems to summarize the shape of the cloud of points pretty well: this is the graphical representation of the linear regression of weight on height. The fundamental idea is that we would like to find the best linear relationship between weight and height of the form $w = \alpha + \beta \times h$, where w and h are the weight and height respectively and α and β some coefficients we estimate from the data; α is called the *intercept* and β the *slope* of the regression line. Of course, just by looking at the scatterplot it is clear there is no way we can reduce all this cloud of points to a single line, but we can try our best, namely to find the line that minimizes the total prediction error.

Given a point (let's name it i) on the scatterplot, its *residual* ϵ_i is defined as the difference between the real weight of the point, w_i, and its weight as given by the regression line, namely the value $\alpha + \beta \times h_i$: thus $\epsilon_i = w_i - (\alpha + \beta \times h_i)$.

[7] Adding heights and weights might strike one as akin to the proverbial mixture of apples and pears but we are using it only as an illustration of a general principle here. When discussing heritability we will consider genetic and environmental factors acting together to produce a trait of interest, and their sum will represent a certain type of relationship.

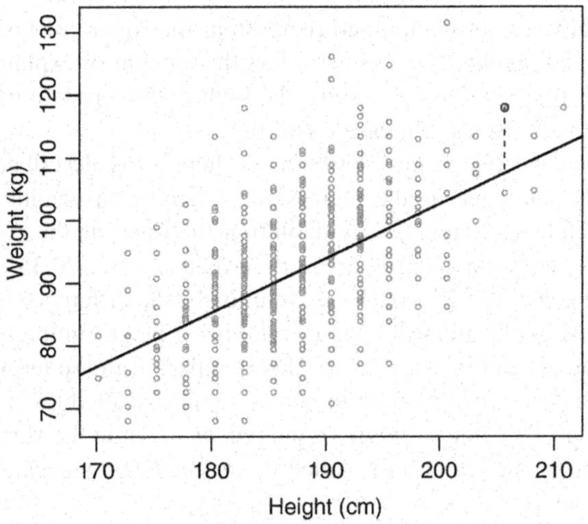

Figure 2.6 Scatterplot of height (horizontal axis) and weight (vertical axis) in 1035 Major League baseball players. The black line represents the regression of the weight on height and the dark grey dashed line from the highlighted individual to the regression line represents the error for this particular individual.

For example, let's take the 63rd player in the dataset (the highlighted dot in Figure 2.6), who has a height $h_{63} = 205.74$ cm and weight $w_{63} = 117.93$ kg (a pretty big guy!); for argument's sake let's look at a boring line with $\alpha = 0$ and $\beta = 1$ (the diagonal line passing through the origin), and for this we would have the residual $\epsilon_{63} = 117.93 - (0 + 1 \times 205.74) = -87.81$, an underestimate of almost 88 kg! This residual can be visualized as the vertical distance between the point and the line. Doing this computation for all datapoints and adding up the squared residuals $\sum_{i=1}^{1035} \epsilon_i^2$, we obtain the total error for this particular line described by $\alpha = 0$ and $\beta = 1$. In fact, the line that minimizes this sum of squared residuals is given by $\alpha = -70.35$ and $\beta = 0.87$ and is actually shown in Figure 2.6; for this line, the residual of the 63rd player is only about 9.28 kg (shown by the vertical dashed line). The linear regression also allows us to *predict* the weight of new, unseen individuals for whom we have measured height; for example, the best prediction for somebody who's 182.5 cm tall is that he would weigh 87.43 kg.

Intuitively, some datasets might be better described by a line than others, but there is a way of quantifying how well linear regression explains the data using the measure denoted R^2 and varying between 0.0 (really bad fit, no explanatory power) and 1.0 (the data fits the regression line perfectly without any errors); in our case, $R^2 = 0.28$, a low to moderately good fit. As you can imagine, there is a link between correlation and regression: the square of Pearson's r between weight and height, $r = 0.53$, is in fact the amount of explained variation: $r^2 = 0.53 \times 0.53 = 0.28 = R^2$. Thus, the higher the correlation between two variables the better linear regression will fit.[8]

Please note that the regression of weight on height (as described above) is not necessarily the same as the regression of height on weight (but the correlation of weight and height is by definition the same as the correlation of height and weight): the first regression is described as $w = -70.35 + 0.87 \times h$ while the second is $h = 157.25 + 0.33 \times w$. Intuitively, the difference is that we try to find the best explanation of weight in terms of height in the first case, and the other way around in the second. This is reflected in the terminology: the variable that is explained (weight in our example) is called the *dependent variable* (shorthand *DV*), the *response* or the *outcome*, while the variable that does the explanation (height here) is called the *independent variable* (or *IV*), the *predictor* or the *covariate*.

Finally, it is important to realize that regression is not causation! Finding that weight can be relatively well predicted by height does not necessarily mean that we have uncovered a causal relationship between height and weight. This is purely a statistical relationship that may (or may not) say something about causal mechanisms linking the two. This should be obvious in our case where it is not height per se that determines somebody's weight but they are both related through more fundamental developmental constraints and laws of physics. In general, inferring causality from correlational designs is a complicated, controversial and active research topic (see for example Pearl, 2000; Spirtes et al., 2000; or Hernán and Robins, manuscript).

Things get a little more interesting when we have more than two variables and we want to predict one (say, weight) using the others (here height and age); now we have what is called linear multiple regression, which describes how one DV is predicted by multiple IVs. The pair-wise correlations between weight (w), height (h) and age (a) are: $r_{wh} = 0.53$ (as we've seen above), $r_{wa} = 0.16$ and $r_{ha} = -0.07$, and the scatterplots are given in Figure 2.7.

So, we might have the feeling that weight is explained mostly by height but also by age: how are we going to quantify this relationship? A good starting point is to realize that we need to go beyond the pair-wise relationships and look directly at the distribution of weight given height and age. Luckily we can draw this as a tri-dimensional scatterplot where the vertical (z) axis is the weight and the two horizontal axes (x and y) age and height respectively; Figure 2.8 shows this (as a 2D projection; for now please ignore the grey plane and the different colours of the spheres). Each little sphere represents

[8] The link between correlation and regression is deeper, in the sense that, with our notations here, the slope $\beta = r\frac{sd(w)}{sd(h)}$.

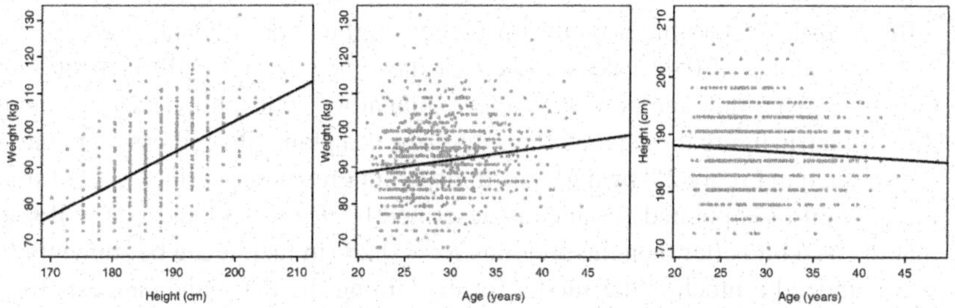

Figure 2.7 Scatterplots of weight × height (left), weight × age (middle) and height × age (right) with their respective regression lines.

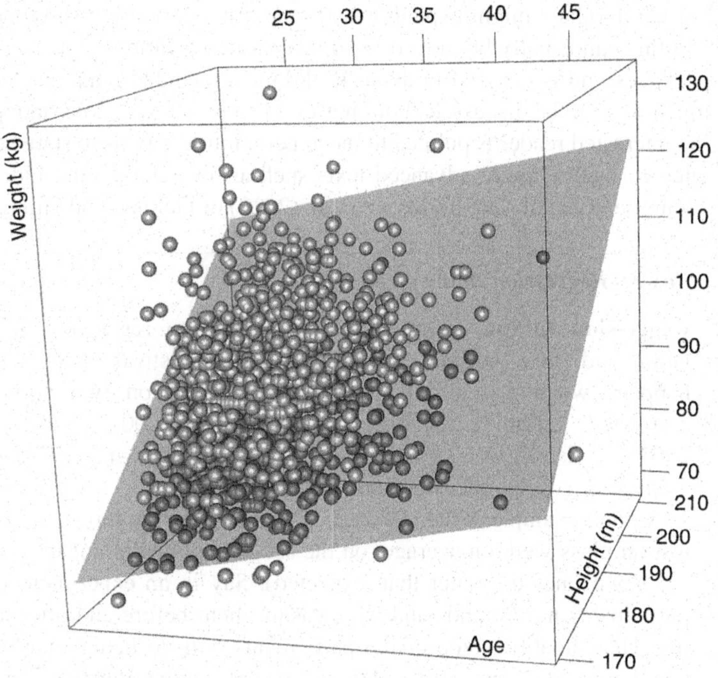

Figure 2.8 Scatterplot of weight (vertical axis), height (y axis going away from the observer) and age (x axis from left to right) in the 1035 Major League baseball players. Each sphere represents one player. The oblique semi-transparent grey plane is the regression plane of weight on height and age and darker spheres are below this plane, while lighter ones are above it.

one individual and the whole cloud of spheres seems to suggest that some sort of relationship holds between weight, age and height.

We can regress weight simultaneously on age and height by finding the parameters α, β and γ that make the predicted weight $w = \alpha + \beta \times h + \gamma \times a$

differ as little as possible from the observed weights. For our data, $\alpha = -87.39$, $\beta = 0.89$ and $\gamma = 0.44$ (please note that these new α and β differ from the old ones, given that age now makes an explanatory contribution as well), and the geometric figure these parameters describe is a plane (the regression plane), represented in Figure 2.8 by the semi-transparent grey plane surface. As before, the amount of explained variance is described by $R^2 = 0.32$ and you will see that this is a bit better than the previous $R^2 = 0.28$ (in fact, it can be shown that this is indeed a much better model for explaining the data) and suggests that weight is positively related to height and age.

When having multiple IVs it becomes important to know which ones are really making a contribution to the linear model (some might simply add nothing while others might explain exactly the same variation that other IVs do), and this can usually be judged from the output of various statistical software packages and is covered in detail in the recommended readings. There is so much more to be discussed about regression and its extensions and quirks, but the interested reader is pointed to the general introductions to statistics already cited as well as more advanced texts such as Tabachnick and Fidell (2001), Montgomery et al. (2013), Myers et al. (2012) or Gelman and Hill (2006).

Box 3: Regression to the mean

Regression to the mean is closely related to regression and correlation and, in a nutshell, this is a purely statistical phenomenon that happens when a selected sample is measured on two non-perfectly correlated variables (Morton and Torgerson, 2003; Barnett et al., 2005; see also `http://www.socialresearchmethods.net/kb/regrmean.php`, accessed June 2014).

A good example is the observation that usually the top performers on a test fare less well, on average, on the second application of the test (while the worst ones do better than expected). Say in an experiment we teach participants new sounds and we measure them before and after the training. Unavoidably, some of the participants will do better than others on the "before" test, and let's say the test scores can vary between 0 and 100 and the whole group's mean score is 50, but the best 10 and the worst 10 participants had an average of 80 and 20, respectively. Next we train the participants and measure them again; this time we see a group improvement from the old 50 to 60 points, but amazingly, when looking at the previous 10 best scorers now their mean has improved only slightly to 83 (so only 3 points better while for the whole group this improvement is 10), while the 10 worst ones did actually better than the group to an amazing 40 (thus some 20 points improvement)! Moreover, the new top 10 are not necessarily the same that were in the top 10 before (and likewise for the bottom ones).

This is a familiar pattern and many explanations can be proposed: the best 10 are already close to the ceiling and there's only so much better they can get while there is a lot of improvement possible for the bottom ones, etc. But the truth is that even if the intervention (here, teaching) doesn't do anything at all, this sort of change is to be expected: the best will regress towards the mean, just as the worst will improve towards it. And this is so simply because in any measurement involving randomness, the best are best also because they were lucky the first time (likewise, the worst were probably hindered by bad luck as well). But the second time luck will most probably change, dragging some of the top performers down, and some of the bottom ones upwards towards the mean.

The regression to the mean is pervasive and concerns not only repeated measurements (as in the example above or sports performance in successive seasons or games) but any two imperfectly correlated measures. In the baseball players example (Figure 2.9), the whole sample's average height and weight are 187.19 cm and 91.49 kg, while for the tallest 5% these are 197.79 cm and 100.24 kg (greater than the whole average but not as big as expected, the 5% cut-off point for weight being 108.86 kg) and for the shortest 5% 176.561 cm and 82.47 kg (compared to the 5% cut-off of 77.11 kg). Regression to the mean is an important phenomenon to have in mind when designing experiments or trying to interpret their results, as it is very widespread and can have very subtle effects.

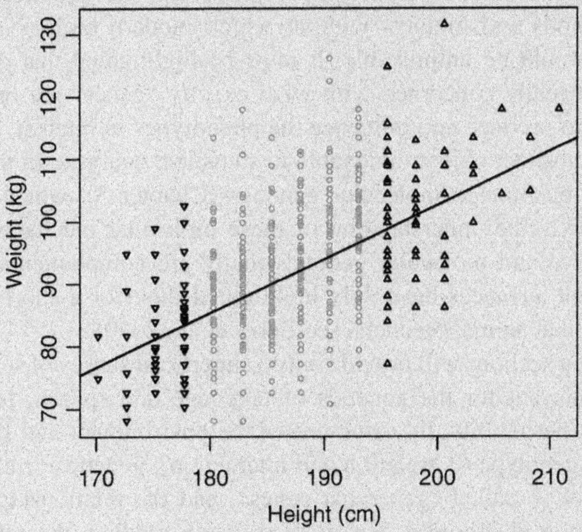

Figure 2.9 The top (Δ) and bottom 5% (∇) baseball players by height are clearly not that extreme by weight as well.

2.4 Basic notions of quantitative genetics

Quantitative genetics is concerned with the understanding of continuously varying features (or so-called **quantitative traits**) which show much more than two possible character states (e.g., disease present versus absent or red versus white flowers). In contrast, classic **Mendelian genetics** looks at discrete characters that are transmitted in a simple (or *Mendelian*) manner across generations, good examples being the speech and language disorder presented by certain members of the KE family (see Section 4.2.1), many types of congenital deafness (Section 4.2.2), and all sorts of spectacular mutations in the fruit fly *Drosophila* such as *Antennapedia*, which converts antennas into legs (see FlyBase `http://flybase.org/` for many more). At the beginning of the last century, it was felt that the major phenotypic effects that seemed to be produced by mutations in the fruit fly discovered by the early geneticists (such as T.H. Morgan) and transmitted in an all-or-nothing Mendelian manner were incompatible with Darwin's theory of evolution, which seemed to require continuous variation and small, gradual changes. However, when R.A. Fisher showed that continuously varying traits can be the result of many loci each making a small contribution to the trait and inherited according to Mendel's simple laws (see Section 3.5), it not only became clear that (Mendelian) genetics and Darwin's evolution were compatible, but also resulted in the birth of quantitative genetics.

In the meantime, quantitative genetics has developed into a very successful but highly mathematical and complex field that has produced essential concepts, methods and findings without which modern biology and cognitive sciences would be unthinkable. It must be highlighted that quantitative genetics is not really concerned with what exactly "genes" are or how they actually work to produce and influence the phenotypes of interest, but instead treats them as abstract objects amenable to statistical analysis. In this respect, it stands in opposition to molecular genetics (Chapter 3), which has as its main goal exactly the understanding of these molecular mechanisms. However, quantitative and molecular genetics really are complementary to each other and recent advances have only highlighted this (for a short and funny opinion targeted at animal breeders, see Berry et al., 2010).

The following sections will provide only a superficial survey of some aspects of this field relevant for the genetics of language and speech, focusing on the concept of heritability, the structure of the environment and its relationships with the genotype (correlation and interaction), genetic correlations and the existence of so-called "generalist genes", and the relationships between rare and common phenotypes. For more in-depth reading, the reference for quantitative genetics is probably Lynch and Walsh (1998), but this might prove a hard read for newcomers, for whom Plomin et al. (2008) would probably be a much gentler introduction, with an accent on behaviour and cognition.

Whenever appropriate, I will use height as a particularly well-suited concept, given that it is a quantitative trait *par excellence*, its measurement is very cheap and reliable, it is one of those traits that clearly has a strong genetic basis and where, due to these features, we have extensive genetic studies using hundreds of thousands of participants to uncover a large number of genes each participating in its own manner in this apparently simple outcome.

2.4.1 *Partitioning the phenotype and broad-sense heritability, H^2*

With the basic statistical concepts introduced above, we now turn to understanding how a phenotype of interest (such as vocabulary size, ability to learn a second language or height) can be described in terms of the contribution of both genetic and environmental factors. In our symbolic notation, we can write that $P = G + E$, meaning that *both* nature and nurture participate in expressing the phenotype. However, as discussed above, in order to study this relationship, we need to look at the patterns of variation in P, G and E and ask how much of the variance observed in P is accounted for by the variance in G and by the variance in E, respectively.[9] More precisely, the phenotype (P) can be partitioned into the genetic (G) and environmental (E) effects, $P = G + E$, but what we are really interested in is the relationship between variation in the phenotype, $var(P)$, the genome, $var(G)$ and the environment, $var(E)$:

$$var(P) = var(G + E) = var(G) + var(E) + 2cov(G, E) \tag{2.8}$$

In general, the genotype G and the environment E do not act independently on the phenotype P, but interact constantly during the development and maintenance of the phenotype; moreover, the more we understand about the molecular bases of life and its evolution the less clear the distinction between genotype and environment becomes (more on these topics in Section 2.2 and Chapter 6), but certain simplifying assumptions are necessary in order to understand the concept of heritability.

The proportion of variation in the phenotype, $var(P)$, accounted for by the variation in the genotype, $var(G)$, represents the (broad sense) **heritability** of this phenotype,

$$H^2 = \frac{var(G)}{var(P)} \tag{2.9}$$

[9] This is a very brief introduction and discussion of basic quantitative genetics concepts and methods and for more information the interested reader should consult the exhaustive Lynch and Walsh (1998), Plomin et al.'s (2008) introduction to behaviour genetics, or papers such as Visscher et al. (2008) and Stromswold (2001).

Intuitively, a phenotype for which all the differences between individuals are fully explained by the genetic differences between them should have a very high heritability (in fact, the maximum possible one). This is confirmed by the formula above, as $H^2 = var(G)/var(P) = var(G)/var(G) = 1$ because the variability in phenotype, $var(P)$, is in fact identical to the variability in genotype, $var(G)$. Conversely, a phenotype where genetics "does not play any role", meaning that the differences between individuals, $var(P)$ are fully accounted by the differences in environment these individuals have experienced, $var(E)$, should have a very low heritability (in fact, the minimum possible one):[10] $H^2 = var(G)/var(P) = 0/var(P) = 0$ because $var(G) = var(P) - var(E) = var(P) - var(P) = 0$. Thus, the heritability H^2 can vary between a minimum of 0 and a maximum of 1. All these issues will be discussed in more detail later (Section 2.4.5) after we have a better understanding of the way heritability is defined and measured.

2.4.2 *Partitioning the genotype and narrow-sense heritability, h^2*

Broad-sense heritability, H^2, treats the genotype as an undifferentiated whole, G, which cannot be further from the truth. In fact, as we will explore in detail in the following sections, G has its own structure. As a first approximation, we can think of our genetic information as being composed of individual, atomic **loci** (singular *locus*) which can hold one of a number of variants or **alleles**. To simplify, let us consider three biallelic loci, where one locus L_1 can have alleles *A* and *a*, while another locus L_2 has alleles *B* and *b* and L_3 with alleles *C* and *c*. Given that we are **diploid** organisms, the vast majority of our genetic loci come in two copies, not necessarily holding identical genes. Thus, one individual can have genotype *AA* at locus L_1, *Bb* at L_2 and *cc* at L_3 (which can be written for short *AABbcc*), while another can be *aAbbCC* and yet another *aabbcc*, for example. Let us assume that these loci are involved in determining height.

We say that there are *additive genetic effects* when each allele *A*, *a*, *B*, *b*, *C* and *c* contributes to the resulting phenotype independently of the others. For example, an *A* adds 1 mm to the baseline height, an *a* adds nothing (0 mm), *B* subtracts 1 mm, *b* adds 2 mm, *C* adds 1 mm and *c* subtracts 2 mm: with these, an individual with genotype *AABbcc* will be 1 mm shorter than the baseline $(1 + 1 - 1 + 2 - 2 - 2 = -1)$, and one with *aAbbCC* will be 7 mm taller $(0 + 1 + 2 + 2 + 1 + 1 = 7)$.

[10] But, as discussed later in Section 2.4.5, this does not mean at all that genes are not important in this case but simply that variation in genetics has no impact on phenotypic variation. For example, some genes are so important that they virtually show no variation in humans (presumably because individuals with other forms of these genes fail to be born at all), and, therefore, they would not be included in heritability estimates despite (or because of) their paramount importance.

Besides these simple additive effects, sometimes alleles at the same locus interact with each other, producing *dominance genetic effects*. In this case, the individual effects of the two alleles at the same locus do not add up. Let us consider again the locus L_1 with possible alleles A and a but now A is **dominant** over a (or equivalently, a is **recessive** with respect to A), which means that the effects of the genotypes *AA* and *Aa* are identical and different from *aa*. In some cases, the dominance/recessiveness is incomplete in that the genotypes *AA* and *Aa* are not completely identical (but also not the result of a simple additive effect of *A* and *a*).

Finally, when alleles at different loci interact in non-additive ways, we have **epistasis** or interaction genetic variance (Cordell, 2002). Thus, the effect of the alleles at one locus depends on those at a second locus, as happens for example in **gene regulation**, where one gene controls the effects of its target genes, or when the products of two genes interact in biochemical pathways.

With these, we can partition the genotype G into its additive, A, dominance, D, and epistatic, I, effects,

$$G = A + D + I \tag{2.10}$$

It is important to note that in sexual organisms (such as us), only the additive effects A are passed faithfully from parent to offspring, as the additive effect of an allele is by definition independent of the rest of the genotype. By contrast, the dominance/recessive effects D depend explicitly on the two alleles at the same locus and given that the offspring inherits one allele from the father and one from the mother its two alleles (and thus the dominance effect) might differ from those in its parents. Likewise, the epistatic effects I depend on even larger genetic contexts that are affected by sexual reproduction. Therefore, the only genetic effect that is faithfully inherited across generations is the additive genetic effect, A, resulting in the definition of the **narrow-sense heritability** h^2 as the proportion of phenotypic variation that is due to additive genetic variation:

$$h^2 = \frac{var(A)}{var(P)} \tag{2.11}$$

2.4.3 *Partitioning the environment*

To make matters even more interesting, the environment E is not a monolithic entity but can be likewise partitioned into a *shared* (or common) environment, denoted C, and a *non-shared* (or idiosyncratic) environment, denoted here e (or sometimes, confusingly, E), symbolically written as $E = C + e$.

The shared environment C represents those aspects that influence all members of the group of interest (e.g. the family members or both twins) such as the house and neighbourhood they live in, the kind of food they eat, the health and educational systems they experience, and type of intellectual stimulation they might receive. This shared environment has a very important role in making people more similar than expected: for example, bad food habits increase their similarity in the risk of obesity and diabetes and a good education increases their similarity in later job opportunities and socio-economic status (SES). On the other hand, different shared environments between groups increases their dissimilarity: again, the differences in obesity and diabetes rates between countries are in large part down to different food (and exercise) habits, and differences in the educational systems have long-lasting effects on social inequalities. Thus, we must consider the profound effects of the shared environment when pondering genetics: the fact that a group of people suffer from obesity might have a genetic component but shared food habits might inflate this similarity.

By contrast, the non-shared environment e represents those aspects that are specific to an individual, such as illnesses, the particular friends one has, the idiosyncratic experiences one had in school, or the odd book one has read as a teenager that changed one's life. These experiences decrease the similarity between people, even among persons with the same genotype such as monozygotic (or identical) twins: one twin might fall ill or experience psychological trauma the other twin did not, and these events have life-long consequences resulting in differences in health and SES between the twins. Thus, e hides the genetic similarity between people.

2.4.4 *Estimating heritability*

The heritability of a phenotype of interest P is impossible to measure directly and instead must be estimated using various methods (Lynch and Walsh, 1998; Falconer and Mackay, 1996). Probably the best-known is represented by **twin studies** whereby pairs of **monozygotic** (or identical) twins (MZ), who share the same genotype, are compared to **dizygotic** (or fraternal) twins (DZ), who share on average only half their genomes, just as regular siblings do. Assuming that there are no fundamental differences in the environment that the MZ and DZ twins experience (but one cannot safely assume this with regard to non-twin siblings), then any differences between MZ and DZ twins must be down to the differences in their genes. Formally, given a set of N MZ twin pairs where each pair $i \in 1..N$ is composed of two twins denoted MZ_i^1 and MZ_i^2, and M DZ twin pairs $j \in 1..M$ composed of two twins denoted DZ_j^1 and DZ_j^2 (which twin is denoted ^{1} and which ^{2} is irrelevant), we can measure the phenotype of interest P in each MZ twin resulting in $2N$ measures $P_{\mathrm{MZ}_i^1}$ and $P_{\mathrm{MZ}_i^2}$, and $2M$ measures $P_{\mathrm{DZ}_j^1}$ and $P_{\mathrm{DZ}_j^2}$. Then we compute the correlation between MZ twins,

$$r_{\mathrm{MZ}} = cor(P_{\mathrm{MZ}_i^1}, P_{\mathrm{MZ}_i^2}) \tag{2.12}$$

and the correlation between DZ twins,

$$r_{\mathrm{DZ}} = cor(P_{\mathrm{DZ}_j^1}, P_{\mathrm{DZ}_j^2}) \tag{2.13}$$

For example, Figure 2.10 shows simulated data for $N = 100$ MZ and $M = 120$ DZ twin pairs, here $r_{\mathrm{MZ}} = 0.80$ and $r_{\mathrm{DZ}} = 0.60$.

Usually, MZ twins should have more similar phenotypic measures than the DZ twins as their genotypes G are more similar than for the DZ twins (assuming the environments E are similar for the two types of twins), resulting in an estimate of P's heritability

$$h^2 = 2 \times (r_{\mathrm{MZ}} - r_{\mathrm{DZ}}) \tag{2.14}$$

in our case, $h^2 = 2 \times (0.80 - 0.60) = 2 \times 0.20 = 0.40$.

Twin studies in particular (probably due to their popularity and apparent simplicity) have been heavily criticized for a number of reasons (see Charney, 2012, for a recent comprehensive account). For example, the assumption that MZ twins have identical genomes has been called into question as **somatic mutations** occurring after the separation of the two developing embryos can occur separately in each twin (Li et al., 2013) and influence various phenotypes, including brain development. Furthermore, while DZ twins share on average 50% of their genomes, there is wide variation between pairs in the exact amount and identity of shared genetic variants. More importantly, the assumption of similarity in the environments experienced by MZ and DZ twins might be too simplistic, as even the pre-natal environments of MZ and DZ twins might differ strongly with MZ twins both sharing resources and potentially competing for them.[11] The post-natal environments might also differ with the obviously greater similarity between MZ twins prompting either more similar, or alternatively artificially dissimilar, treatment. **Epigenetic** differences between MZ twins, ultimately driven by environmental differences, could also play an important role (Fraga et al., 2005) in explaining greater-than-expected differences between MZ twins, also in language-related tasks (Stromswold, 2006). Finally, twins might not be representative of the population at large, but it is unclear what influence this might have given that heritability estimates obtained using other designs tend to agree.

Another classic design is represented by **adoption studies**, which compare adoptees with their foster and biological families, respectively. The comparison with the foster relatives controls for the environmental influences that these

[11] Even the term "MZ twins" hides a lot of complexity as there are several types of twins that might or might not share the same *chorion* and/or *amniotic sac*.

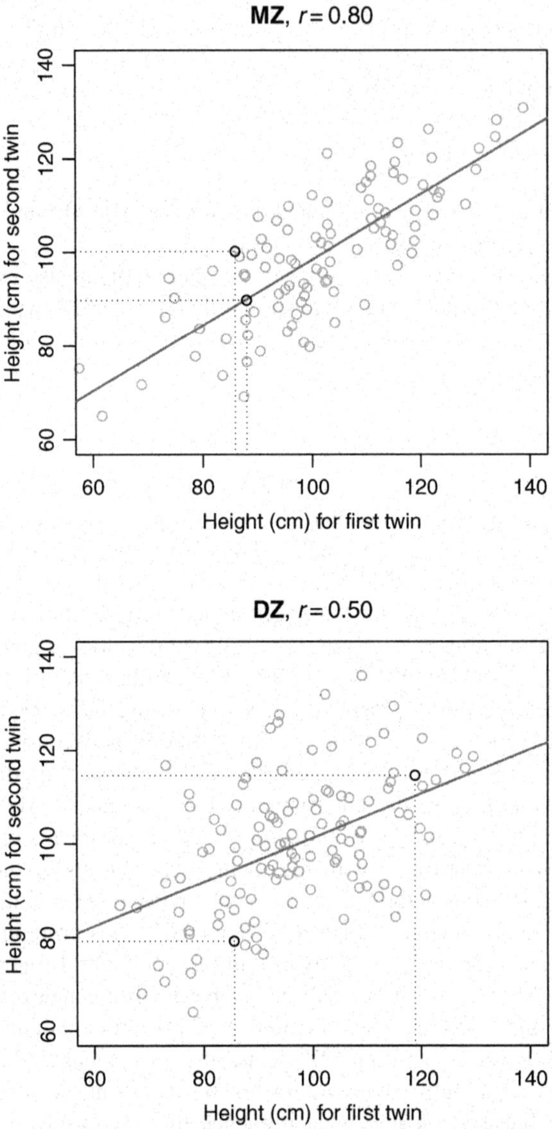

Figure 2.10 The correlation between co-twin heights for a set of $N = 100$ MZ and $M = 120$ DZ twin pairs. The top figure represents the scatterplot with regression line for the MZ twins with two random pairs highlighted, while the bottom figure is the equivalent for the DZ twins. This is based on simulated data.

adopted children share with them, while the comparison with the biological family accounts for the shared genetic influences.

Currently much more powerful methods based on Structural Equation Modelling (SEM) that can accommodate more complex designs and sources of variance are used to estimate the heritability of traits of interest (Lynch and Walsh, 1998; Posthuma, 2009). Recently methods such as **Genome-wide Complex Trait Analysis** (**GCTA**; Yang et al., 2011) that can estimate heritability from very large sets of unrelated individuals with genome-wide genotype data (such as collected for **Genome-Wide Association Studies** or **GWAS**; see Section 5.3) have been used to estimate the heritability of various traits such as intelligence (Plomin et al., 2013).

However, it is important to highlight that, despite their limitations, heritability estimates still play important roles in understanding the genetics of complex traits. For example (Section 7.2) they provide the benchmark against which the amount of variation explained by the actually identified loci is evaluated. Moreover, the fact that usually the heritability estimates tend to agree across studies, populations and different methods reinforces the idea that heritability reflects a real phenomenon. Finally, these family designs have utilities beyond estimating the heritability and help in studying the genetic correlation between traits, looking at the effects of the environment and various types of gene–environment interactions (see Sections 2.4.7 and 2.4.8 for details).

2.4.5 *Heritability: what it does and does not mean*

It is extremely easy to misinterpret and misrepresent heritability estimates and, unfortunately, these can have profound and real consequences on how groups of people are educated (or not), how they are regarded and discriminated by the larger majority and, consequently, their access to various services. To clarify what heritability as defined above and used in heritability studies actually means, we will discuss, in no particular order, some of its apparently counter-intuitive properties (see also Visscher et al., 2008, Plomin et al., 2001, and Lynch and Walsh, 1998, among others).

Given that heritability is a ratio of variances, $h^2 = var(A)/var(P)$ or $H^2 = var(G)/var(P)$, it cannot be estimated when there is no variation in the phenotype, $var(P) = 0$. However, there are some uniform traits, such as the incapacity to pass through concrete walls or to teleport to Mars and back, that intuitively have nothing to do with genetic inheritance, being simply consequences of the fundamental properties of matter, energy and information. Other uniform traits, such as having one heart, intuitively have a lot to do with genetics. The main difference between these traits for which heritability cannot be estimated is that there is (or isn't) variation at higher levels (for example, there are other species with no heart at all or with multiple hearts), but heritability fails to capture it. Likewise, genes that are so important as to be fixed across the population (there is only one allele present) will not contribute to $var(G)$, making no difference to the heritability of the trait which would not even exist without them.

Interestingly, the heritability of the same trait can vary with the environment it is measured in. For example, muscle strength varies among individuals and is moderately heritable (Cardinale et al., 2011; Silventoinen et al., 2008). However, if everybody started exercising the exact same amount and in exactly the same fashion or nobody exercised at all (two ways of making the environment constant), then the remaining differences in strength would be entirely due to differences in genotype, resulting in a higher heritability. Conversely, if a random half of the population exercised intensely and the other not at all, then the differences in strength would be mostly due to this environmental intervention, resulting in a lower heritability estimate.

Moreover, the heritability of a trait is not necessarily constant across age either (Bergen et al., 2007). For example, the heritability of intelligence increases dramatically with age apparently in a linear manner (Davis et al., 2009; Haworth et al., 2009), raising interesting theoretical questions about the mechanism (probably involving correlations between genotype and environment whereby individuals actively choose and shape their environment in ways partially influenced by their genotypes; Haworth et al., 2009) but also with practical consequences for educational policies.

Finally, heritability estimates cannot be immediately generalized to other groups and populations and do not necessarily say anything about the *inter*-group differences either. This point is extremely important and often overlooked: for example, the heritability of height is very large (consistently greater than 0.8; Lettre, 2009) and recent association studies have uncovered hundreds of genetic loci involved in this phenotype (Allen et al., 2010; Yang et al., 2010), and there are well-known differences in height between populations. However, the jump to the conclusion that these differences in height between populations are genetic in nature ("innate") is not warranted. The significant increase in height during roughly the last century and a half in Europe and the USA (the so-called secular trend; Cole, 2003) seems to suggest that environmental factors, especially improvements in diet, hygiene and health services, play a major role in these inter-population differences. Moreover, while this increase in height in western Europe and especially in the USA has ceased (Danubio and Sanna, 2008; Larnkaer et al., 2006; Komlos and Lauderdale, 2007), it is currently taking place in other parts of the world such as China (Zong et al., 2010; Chen and Ji, 2013), reflecting changes in the relevant environmental factors and suggesting that the current differences in height might even disappear.

Thus, finding that a phenotype, such as height or intelligence, has a moderate or high heritability does not warrant judgements related to its lack of malleability and the uselessness of any interventions. As shown by examples such as height, environmental changes can have a massive impact and, in fact, jumping from heritability estimates to policies that assume the unchangeability of the trait is wrong and betrays a view of development, genes and environment that has been relegated to the rubbish dump of history by recent advances in genetics, evolutionary biology and the cognitive sciences (see Section 2.2). Nevertheless, despite its conceptual and methodological problems, heritability studies are important in that a phenotype which systematically produces moderate or high heritability estimates warrants more expensive and targeted studies looking for the actual genetic architecture underlying it.

2.4.6 *The heritability of language and speech*

As opposed to height, traits related to language and speech are very difficult to define and measuring them reliably and validly[12] represents an enormous challenge, especially when we need to measure relatively large samples. Nevertheless, there is a wealth of studies reported in the literature that estimate the heritability of various aspects of speech and language, both normal and pathological, usually making use of twin designs.

Despite its age, the classic review by Stromswold (2001) still provides a good overview and introduction to the heritability of language and speech and should be consulted by anyone interested. For example, the heritability of **Specific Language Impairment** (or **SLI**) is large (Bishop, 2003; Barry et al., 2007) but seems to depend on the diagnostic criteria and the specific psychometric tests used (Bishop and Hayiou-Thomas, 2008; Hayiou-Thomas, 2008). Likewise, *dyslexia* (or *Reading Disorder*) and various related tasks have high heritabilities (e.g., Hensler et al., 2010), and the liability to stuttering is also large ($h^2 = 0.70$; Felsenfeld, 2002). Not only pathologies show high heritability, but also aspects of normal variation such as the size of the expressive vocabulary (Stromswold, 2001) and the acquisition of a second language in school (Dale et al., 2010), but it should be noted that more work is needed in designing valid and reliable tasks to be used in studying the normal variation. Various parameters of the vocal tract such as its cross-sectional area (Patel et al., 2008), the volume of various soft structures (Schwab et al., 2006) and the dimensions of the hard palate (Townsend et al., 1990) also seem to be heritable,

[12] For a thorough discussion of **reliability**, **validity** and other extremely important concepts pertaining to measurement and variation, one should consult any good introduction to **psychometrics** such as Nunnally and Bernstein, 1994, or Furr and Bacharach, 2007, and for more in-depth issues Borsboom, 2009 is a good guide.

but the sample sizes are generally too small and the studies are not replicated for any strong conclusions to be drawn.

2.4.7 *Relationships between genotype and environment*

As discussed above in Section 2.4.1, the partitioning of the phenotype into its genetic and environmental components, $P = G + E$, allows us to derive the partitioning of the variances, $var(P) = var(G + E) = var(G) + var(E) + 2cov(G, E)$. If the environment and genotype are independent, then $cov(G, E) = 0$ and can be safely ignored from the discussion, leaving the clean and simple additive decomposition $P = G + E$. However, in many cases the two are not independent, and their relationships are sometimes complex and fascinating (Plomin et al., 1977).

First, there are *genotype–environment interactions* (denoted $G \times E$), whereby different genotypes might react differently to the environment. An example will help clarify: phenylketonuria (or PKU; OMIM[13] 261600) is a genetic disease that can result in cognitive impairment unless diet is controlled (by removing food that contains phenylalanine, for example protein); thus, to simplify, there are two types of genotype (normal, G_n and abnormal, G_p) and two types of environment (normal, featuring food with phenylalanine, E_n, and a modified environment without phenylalanine-containing food, E_m). There are four possibilities: a person with normal genotype inhabiting a normal environment ($G_n \times E_n$) and resulting in a normal phenotype, a normal genotype inhabiting a modified environment ($G_n \times E_m$) resulting in a normal phenotype, an abnormal genotype inhabiting a normal environment ($G_p \times E_n$) resulting in the disease PKU, and an abnormal genotype inhabiting a modified environment ($G_p \times E_m$) resulting in a normal phenotype. There are many examples of such interactions but probably the most important take-home message is that a "genetic" phenotype (such as PKU here, which results from mutations in the *PAH* gene) can be addressed through appropriate environmental manipulation. Thus it is completely unwarranted to assume that if something is "genetic" (such as PKU) then nothing can be done about it (short of changing somebody's genome and replacing the faulty gene, a hope actively pursued by various types of *gene therapy*); changing the environment in the right manner might go a long way. This is extremely important when it comes to social attitudes and policy decisions having to do with cognition, where it is sometimes wrongly assumed that given that, for example, intelligence has a genetic component (usually derived from heritability studies; Section 2.4.5) and some people have a poorer genetic endowment of it, then nothing can be done to help them.

[13] Online Mendelian Inheritance in Man (OMIM) is an extensive online catalogue of genetic disorders and genes, available at `http://www.omim.org`; each entry is identified through a unique number.

Second, there are *genotype–environment correlations* (sometimes denoted $r_{G,E}$), which refer to the greater-than-expected co-occurrence of certain genotypes and environments. For example, we might find that tall people tend to be over-represented in basketball teams while short people are under-represented, and that linguistically gifted individuals are linguists more often than expected. There are three main types of such correlations (Plomin et al., 1977), so-called passive, active and evocative.

Passive genotype–environment correlations occur because people tend to inherit not only their genes from their parents but also the environment, an environment that, in some respects at least, reflects the parents' own genotypes. For example, parents with high verbal abilities will not only construct a familial environment where linguistic resources are available but they will also provide more opportunities for their children to experience language and, if verbal abilities are partially heritable, their children will inherit both genes and environments related to high verbal abilities, generating a positive correlation between the two. It is important to realize that the children themselves did nothing to generate these correlations, but are passive benefactors (or not, depending on the phenotype of interest) of this simultaneous inheritance.

In contrast, **active genotype–environment correlations** are due to the individuals actively shaping their environment to reflect their genotype. To continue our example, people with high verbal abilities will look for all types of opportunities and experiences involving language such as reading, writing and learning new languages. Likewise, tall people will lean towards sports where their tallness is rewarding, such as basketball, which short people will tend to find non-rewarding and consequently avoid.

Finally, people with different genotypes might make other people react differently, producing **evocative genotype–environment correlations**. Those with high verbal abilities might, for various reasons, make their language teachers treat them differently by giving them more feedback and in general investing more time and attention, reinforcing their initial better abilities.

It is important to note that in many cases all three types of genotype–environment correlations tend to occur: a genotype favouring high verbal abilities will tend to find itself in a nurturing environment constructed by the parents, in school will probably impress the language teachers, creating even more opportunities, and it will also seek out and exploit any opportunity to experience language. Thus, I am personally not surprised to find that not a few of my linguist friends (me included) come from families of people that enjoyed language in one way or another, were treated differently when it came to language and actively continued building a world of words around themselves. However, this spiral can have less desirable consequences as well, as apparently might be the case for substance abuse, violence and some psychiatric disorders (e.g., McGue, 1997; Jaffee and Price, 2008; Knafo and Jaffee, 2013; but more research is needed before drawing strong conclusions).

2.4.8 *Sharing genes between phenotypes: genetic correlations and "generalist genes"*

So far we have considered only single phenotypes and we discussed their heritability and the relationships that may exist between their genes and the environment. However, we are often interested in several phenotypes simultaneously and the relationships between them, either between domains (such as intelligence and language) or within domains (e.g., vocabulary size and phonological working memory). Let's consider two such phenotypes, P_1 and P_2; we can compute the correlation between their values in the sample and obtain the phenotypic correlation r_{12}. However, this correlation does not tell us anything about the relationships between the genetic infrastructures of the two phenotypes, but it can be conceptualized as having two components: one due to shared genetic infrastructure (the *genetic correlation*) and one due to environmental factors.

Thus, the genetic correlation r_{G12} (Plomin et al., 2008) represents the sharing of genetic factors between the two phenotypes and, importantly, is independent of the heritability of the two phenotypes. For example, two weakly heritable traits can still have a very large genetic correlation close to 1.0, meaning that even if variance in genetic factors does not explain much of the variance in the two phenotypes independently, these genetic factors tend to be shared among the traits. Conversely, very heritable traits do not necessarily show a high genetic correlation if each one is influenced by specific genetic factors. For example (Plomin and Kovas, 2005; Kovas and Plomin, 2007), large genetic correlations have been found between various measures of reading and language (between 0.67 and 1.0), reading and mathematics (between 0.47 and 0.98) and language and mathematics (between 0.59 and 0.98), and the genetic correlations between various aspects of the same large domain are also large. These findings of big genetic correlations seem to suggest that most genetic factors are shared within and between cognitive domains, a phenomenon named the *Generalist Genes Hypothesis* (Plomin and Kovas, 2005; Haworth and Plomin, 2010). However, these genetic correlations are not 1.0, suggesting that some specialist genes exist as well, genetic factors whose variation accounts for variation in individual phenotypes and are not shared with other phenotypes.

The existence of such generalist genetic factors has important consequences for the search for the actual genes involved in cognition as it suggests that most will have effects across the board, making genes discovered in one domain interesting candidates for the others. Likewise, it implies that the evolution of various aspects of cognition cannot be treated independently from the others as they share a sizeable genetic background. However, it is unclear what exactly the interpretation of these generalist and specialist genetic factors is in terms of actual molecular genomic elements (see Chapter 3), but hopefully advances in molecular genetics will provide interesting answers soon (Haworth and Plomin, 2010; Plomin and Simpson, 2013).

Now it is time to turn to the actual molecular bases of inheritance and the fascinating and intricate ways in which our genetic material influences the emergence of complex phenotypes such as speech and language.

3 The molecular bases of genetics

After encountering the complexities of genetics at an abstract level, it is time to get our hands dirty and discover how the genetic information is actually stored, transmitted and expressed. We will encounter the structure of a typical animal cell, forming the appropriate context for understanding the DNA. We will see that far from being an abstract, mathematical message composed of the four letters A, T, C and G, the DNA molecule is a living, complex entity that not only stores genetic information across generations, but allows it to interact with the internal and external environment on extremely short time-scales. This chapter covers the processes of replication (ensuring the faithful transmission of genetic information), transcription and translation (expressing the genetic message) and we will survey the structure of the chromosomes (discovering recombination, a very important process for generating diversity) and that of genes.

We, as biological organisms, are composed of trillions of specialized cells, organized and coordinated through mechanisms of bewildering complexity and beauty, and any proper understanding of what we are must, at some point, face this level of organization. Far from a brutal reductionism or molecularism, we must understand how molecules interact and organize in increasingly complex structures and how these "levels" of organization continuously "talk" to each other and with the "environment" – including culture – because all these "higher" structures – language, art, culture in general – build upon, and form an organic whole with the molecules which compose us. It is emphatically *not* about explaining, say, the Great Vowel Shift in terms of quantum interactions between carbon, oxygen and nitrogen atoms, but about articulating the appropriate historical, phonological and phonetic explanations in their proper living human context, which imposes specific limitations and creates specific affordances. Ultimately, this will allow a proper understanding of the processes that made culture possible and that continue to shape and remodel it to the present day.

We will introduce here this maddeningly dynamic aspect of genetics as it has already forced massive reappraisals of what we thought we knew and what we were complacently taking as dogma, from the nature of "genetic information" and "innateness" to how evolution works, and how culture and language are shaped by, and shape in turn, our biology. We will try to convey the complexity of the amazing molecular machines and information processors at work, with their intrinsically fuzzy boundaries between "inner" and "outer", between "genetic" and "environmental", having been cobbled together and then perfected to perform functions they were not designed for (and succeeding against all odds).

3.1 We are composed of cells

Without much doubt, a major (and not yet fully understood) event in the history of life on Earth was the emergence of *cells* (Smith and Szathmáry, 1995; Woese, 1998; Glansdorff et al., 2008; Koonin, 2009; Penny and Poole, 1999), which paved the way to the complexification of life to levels unimaginable before. Basically, the cell manages to create and maintain an internal environment separate from – but in continuous and intimate exchange with – the world outside it. This internal environment allows the controlled concentration of molecules required for the organized and predictable interactions that we call **metabolism**. The essential components of a cell comprise the **membrane** (a structure specialized in delimiting the inner from the outer while allowing controlled exchanges between the two), the metabolic machinery (producing energy and cellular components) and the genetic material.

Cells are the fundamental unit of what most of us usually think of as life on Earth, ranging from organisms consisting of a single one (*unicellular* organisms) to those composed of billions of specialized cells (*multicellular* organisms), such as us. However, not all life on Earth is built upon them: there are **viruses** (basically packets of genetic material encased in a protective shell) and **prions** (complex molecules lacking proper genetic material), which need living cells in order to "live" and reproduce.[1] Nevertheless, they are not some irrelevant oddities, as viruses are probably the most widespread form of life on the planet, outnumbering cellular life by many orders of magnitude, play absolutely essential roles in various ecosystems, and have probably been involved even in the emergence of life on Earth (Bergh et al., 1989; Williamson et al., 2008; Villarreal and Witzany, 2010).

[1] Opinions differ about the exact definition of "life" and the status of viruses, naked DNA and prions, with some taking the position that only genetic material enclosed within membranes of lipid bilayers qualifies as living things, while I take a more "liberal" position here.

Cells come in two major kinds: **prokaryotic** and **eukaryotic** cells, differing in their internal organization and complexity, with the fundamental differences being that the latter have a proper nucleus in which the genetic material (the DNA) resides (Snustad and Simmons, 2010). Prokaryotes are smaller, simpler, usually lacking internal sub-structures (or *organelles*) and their genetic material resides directly within the *cytoplasm* (the internal cellular environment). **Bacteria** and **Archaea** are prokaryotes, with *Bacteria* being probably the most widespread and diverse form of cellular life on Earth, and some *Archaea* being famous for their *extremophily* (the capacity to live in extreme environments such as the very hot or chemically aggressive).

Eukaryotes, on the other hand, are larger and show a complex internal structure with many specialized organelles delimited by membranes from the rest of the internal environment. Examples of such organelles are the **mitochondria** (specialized in the production of energy from food), **chloroplasts** (specialized in **photosynthesis**, the use of light to synthesize food), the **endoplasmic reticulum** (specialized for protein synthesis), **lysosomes** (breaking down waste) and the **nucleus** (housing the nuclear genetic material and separating it from the cell environment). Another important structure of the eukaryotic cells is the **cytoskeleton**, a complex of tubes and filaments involved in shaping the cell and helping it move around. Animals, plants and fungi are well-known examples of eukaryotes, with "**protists**" being a catch-all term encompassing many different groups of organisms, mostly unicellular.

Most of the eukaryotic cell's genetic material is housed within the nucleus, where it is split among several linear **chromosomes**, but the mitochondria and the chloroplasts have their own (tiny by comparison) genetic material in the form of a single, circular chromosome. For each cell, there are several mitochondria, ranging from a few to hundreds or thousands (Figure 3.1).

The evolutionary relationships between *Bacteria*, *Archaea* and *Eukaryotes* are complex and controversial but it is probably true that the concept of Prokaryotes is not evolutionarily meaningful in the sense that *Bacteria* and *Archaea* do not share a common ancestor to the exclusion of the *Eukaryotes*. More probable is that, in fact, *Archaea* are more closely related to the *Eukaryotes* than either is to *Bacteria*. Figure 3.2 depicts the currently proposed division of life into three **domains of life** and is based on phylogenetic analyses of very stable and universal pieces of genetic information (Woese et al., 1990; Woese, 2000; Puigbò et al., 2009). The root of this *Universal **Tree of Life*** (or TOL) is known as the **Last Universal Common Ancestor** (or LUCA) and is believed to represent the common ancestor of all cellular life, but its actual nature is heavily debated (Woese, 1998; Penny and Poole, 1999; Glansdorff et al., 2008). However, this represents a highly simplified picture of a very complex and still unclear history which, unexpectedly, seems to show

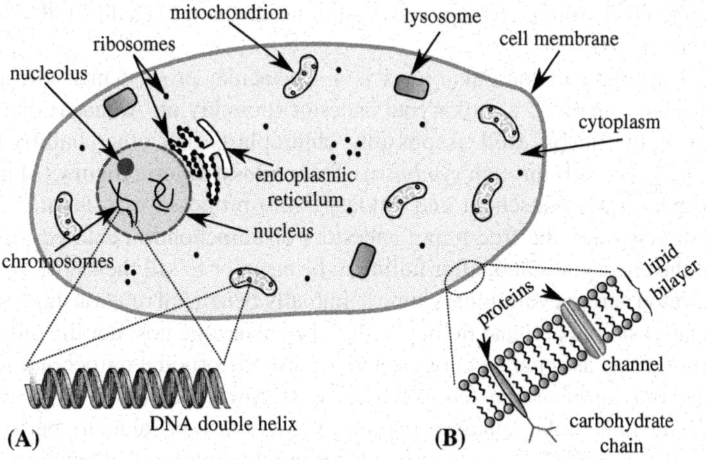

Figure 3.1 The structure of a generalized eukaryotic (animal) cell. The cell is bounded by a membrane (insert **B**) composed of two layers of lipids with various types of embedded proteins allowing matter, energy and information exchanges between the cell and its environment. In the cytoplasm there are various structures such as the nucleus containing most of the genetic material (DNA double helix, insert **A**) structured into several discrete linear chromosomes, mitochondria (with their own tiny genetic material structured as a circular molecule), lysosomes, the endoplasmic reticulum and ribosomes.

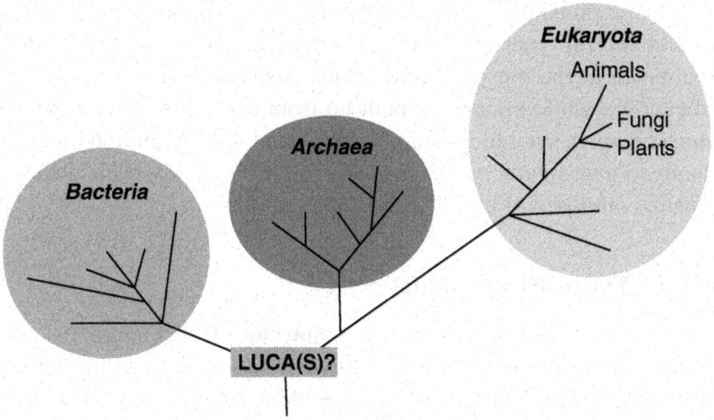

Figure 3.2 The Tree of Life showing the three domains of life and the heavily debated Last Universal Common Ancestor (or Ancestors). Adapted from Woese (2000).

intriguing parallels to language evolution and change (Dediu et al., 2013; Gray et al., 2010).

The origin of the eukaryotes is a spectacular process in itself, as it most probably involved a prokaryote ancestor (possibly an Archaea) that acquired mitochondria (as well as possibly chloroplasts and other hallmarks of the eukaryotic cell) through *symbiosis*. This **endosymbiotic theory** (Margulis and Sagan, 1997; Kutschera and Niklas, 2005) proposes that about 2.2–1.5 billion years ago the free-living ancestors of mitochondria entered the ancestor of the eukaryotes and, after failing to be eaten or to kill their host, managed to develop across evolutionary time a mutually beneficial relationship. Today, this relationship is obligate in that neither the eukaryotic host nor the mitochondria can survive as separate entities, and most of the original mitochondrial genetic material has been transferred into the host's own nucleus.

The unicellular eukaryotes' single cell must be able to perform many functions, such as feeding, moving around, finding food and avoiding dangers and making more copies of itself. However, multicellular organisms have a complex structure, being composed of many cells (ranging from several to billions), differentiated into many **cell types** and organized in **tissues**, **organs** and **systems** of organs. The cells in such a complex body are specialized and there can be hundreds of such cell types (about 200 in humans), such as various kinds of sensory cells, neurons, skin cells or muscular fibres, which excel at performing one or a few functions (such as information processing or neutralizing harmful chemicals) and relying on the whole organism to do the rest (find food, avoid predators, produce offspring, etc.). Tissues are composed of many cells of several types united in performing a certain function at the organism level (such as the muscle tissue, which is specialized in generating mechanical energy), while organs are built up from several tissues (such as the brain, which comprises not only neurons but also blood vessels and other structures). Finally, organs interact synergistically to form systems, such as the circulatory system, composed of blood vessels and the heart.

3.2 The molecules of life

Cells are composed of many types of **molecules** which interact in complex and specific ways. Molecules are composed of **atoms** interacting through **chemical bonds** and, for life, the most important atoms are *carbon*, C, *hydrogen*, H, *oxygen*, O and *nitrogen*, N, with many others present in smaller quantities. These atoms are arranged in specific ways to produce various types of molecules, the most relevant for us being the **lipids**, the **carbohydrates**, the **proteins** and the **nucleic acids**.

Lipids are an essential component of the cellular membranes and energy storage, and carbohydrates are an energy source fundamental to metabolism.

Proteins are a very large and extremely important class of molecules as they do most of the work around the cell, shape it, help it communicate with the outside and enact the genetic information. Their fundamental structural unit is the **amino acid**; there are 20 standard amino acids which the cell uses to build up proteins the same way that Lego blocks can be used to build very complex structures. And, like Lego blocks, amino acids differ in their properties, such as shape, size and electric charge but, unlike Lego blocks, the resulting structures built out of them are linear and can twist and fold in complex ways. These structures – essentially, the proteins – are very large by chemical standards, so large in fact that we can think of them as little machines doing their jobs around the cell. They can, for example, encase an iron (Fe) ion and transport oxygen (O_2) and carbon dioxide (CO_2) around the body (*haemoglobin*), they give shape to the cell (*actin*, *tubulin*) and to the body (*collagen* in cartilage and skin, *keratin* in hair and nails), they move the organism around (*myosin* and *actin* in muscles), etc. They drive and accelerate the very complex chemical reactions necessary for life, which would otherwise not happen fast enough (**enzymes**) and they do so by holding together in their active sites the two (or more) substrates (precursor molecules) in such a way that these substrates are forced to react and result in the product molecule orders of magnitude faster than otherwise. A useful visual image is somebody bringing together in their hands two pieces of iron and binding them together using intense heat, except that enzymes can do the binding without the heat. Yet another essential function proteins play is *informational*: they transport and transform information (in the form of various signalling molecules such as **neurotransmitters** and *hormones*, and/or through changes in their own shape and enzymatic properties) allowing the cell (and the whole body) to be an efficient, functional informational processor. This information processing is essential for the organism's interactions with the environment (sensing information such as light or chemical gradients, and sending out information such as calls and pheromones to other organisms) as well as for coordinating the complex processes within the organism itself (such as reading and interpreting the genetic information).

Thus, we can visualize the cell as an extremely complicated system teeming with purposeful little machines doing all sorts of coordinated things such as building themselves up, producing energy for their functioning and trading and transforming information.

3.3 The genetic material

The genetic information is instantiated in the **nucleic acids**, very large biological molecules composed of **nucleotides** strung together. A nucleotide is composed of a *base*, a *sugar* (these two are known together as a *nucleoside*), and a *phosphate group* which links together two consecutive nucleosides.

There are two types of sugars relevant for the structure of the genetic material: *2′-deoxyribose* and *ribose*, resulting in two types of nucleic acids: **DNA** (*DeoxyriboNucleic Acid*) built out of nucleotides using 2′-deoxyribose, and **RNA** (*RiboNucleic Acid*), using ribose. DNA and RNA share many properties, but they are not identical, making them appropriate for slightly different functions: DNA is a very stable repository of information while RNA is also capable of doing things (it can behave as an enzyme) and is an essential ingredient of several very important cellular components.

There are two types of bases differing in their structure, *pyrimidines* and *purines*. Important are the purines *adenine* (**A**) and *guanine* (**G**), and the pyrimidines *thymine* (**T**), *uracil* (**U**) and *cytosine* (**C**). DNA can contain **A, G, T** and **C**, while RNA replaces **T** by **U**, having thus **A, G, U** and **C**. Thus, a possible DNA molecule could be:

5′ – **TTAGCTAACCGGAAT** – 3′

where the 5′– and –3′ are conventional notations for the two ends of the molecule: these ends are *not* equivalent and the directionality $5' \rightarrow 3'$ is important in many biological processes. It must already be clear that some sort of information could be stored in the pattern of DNA "letters" (nucleotides): for example, **ATA** and **ATC** could mean different things, and we will see shortly that this is, indeed, the case.

3.4 DNA: storing and transmitting information

The bases have a very important property emerging from their chemical structure: they can form pairs through *hydrogen bonds* (a type of relatively weak **chemical bond**). Adenine (**A**) pairs up with thymine (**T**) in DNA and uracil (**U**) in RNA through two hydrogen bonds, **A=T** and **A=U**, respectively, and cytosine (**C**) pairs up with guanine (**G**) in both DNA and RNA through three hydrogen bonds, **C≡G**. Thus, the *single-strand* DNA molecule given above will pair up with a *complementary* single-strand molecule forming a *double-stranded* DNA:

$$\overleftarrow{\hspace{3em}}$$

3′ – **AATCGATTGGCCTTA** – 5′
5′ – **TTAGCTAACCGGAAT** – 3′

$$\overrightarrow{\hspace{3em}}$$

The two strands are oriented in opposing directions (as shown by the arrows, the top one reading in the $5' \rightarrow 3'$ direction from right to left and the bottom one from left to right), and each letter is paired (vertically aligned) with its complement (from left to right, **A** with **T**, **A** with **T**, **T** with **A**, **C** with **G**, and so on).

This linear representation, $\begin{smallmatrix}3' - \text{AATCGATTGGCCTTA} - 5' \\ 5' - \text{TTAGCTAACCGGAAT} - 3'\end{smallmatrix}$, is a strong simplification of the biochemical reality, but useful in understanding how genetic information is transmitted across generations and expressed within the organism. In reality, these two complementary strands are coiled around each other in the familiar double helix, which is in turn folded and coiled at multiple levels so that it can be packed within the tiny cell's nucleus (see Section 3.7 and Figure 3.9 for details). When a cell undergoes *mitotic division* (or **mitosis**), it produces two identical daughter cells, each with its own copy of the genetic material inherited from the mother. This process of copying the genetic information across generations is one of the most important aspects of life and is made possible by the complementarity of the bases of the nucleic acids. Its discovery therefore was essential for understanding how genetic information is transmitted across generations and in suggesting how its meaning was encoded and used.

During this process of DNA **replication**, the helix of the double-stranded molecule is locally unwound, the hydrogen bonds between the complementary bases on the two strands are broken and each strand acts as a template for the assembly of a new complementary strand. As can be seen in Figure 3.3, the replication happens at the *replication fork*, where the two strands of the old double-stranded DNA molecule are separated. These single strands of DNA are then used to bind complementary bases, leading to the creation of two new DNA strands bound to the old ones through new hydrogen bonds. The *elongation* of these new strands happens only in the $5' \rightarrow 3'$ direction, which is the direction in which the replication fork proceeds for the *leading* strand, resulting in a continuous growth of the new strand through the addition of new complementary bases. However, for the complementary, *lagging* strand, the elongation proceeds in bursts: small fragments of the new strand grow in the opposite direction to the replication fork (the lagging strand's own $5' \rightarrow 3'$ direction) and are connected together when they meet (these are called *Okazaki fragments*). This complex process is orchestrated by many enzymes (only one of them, the **DNA polymerase** which binds new complementary bases to the growing strand, being shown here), each with its specific role(s). These enzymes differ to various degrees between eukaryotes and prokaryotes, but the main result is that an old double-stranded DNA molecule results in two daughter double-stranded DNA molecules, each composed of an old strand and a new, complementary, strand; this is known as *semiconservative* replication.

Of course, any copying mechanism has an error rate, no matter how small, so that there is a non-null probability that the copy differs from the original. In the case of DNA replication, this means that sometimes (only about 1 in 10 million) a nucleotide added to the growing new strand is not the complement of the corresponding old nucleotide (e.g., for an old **A** a **C** might be inserted

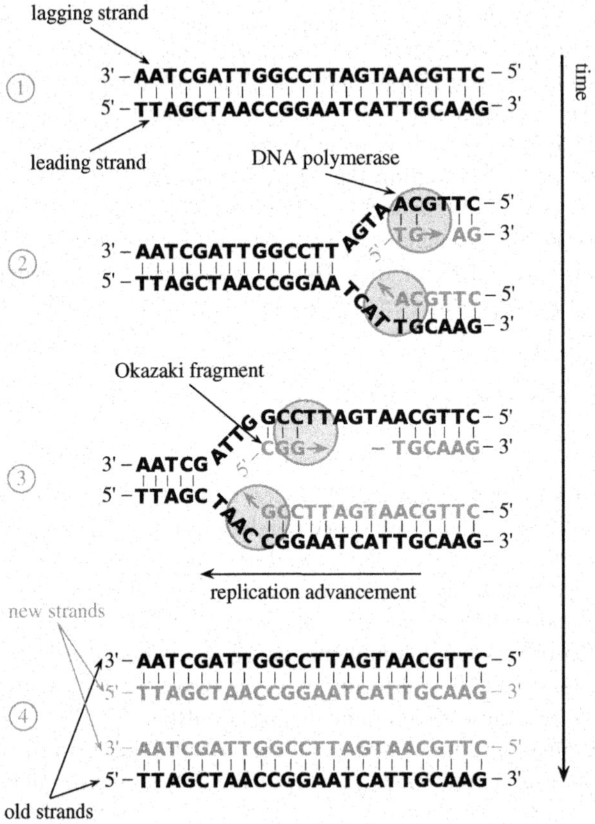

Figure 3.3 A simplified depiction of DNA replication. In black are the old DNA strands and in grey the new ones. The grey circles represent DNA polymerase while grey arrows stand for the direction in which the new DNA strands grow. Time step 1 shows the old DNA double-stranded molecule before replication begins, while time steps 2 and 3 show the advance of the replication fork and the elongation of the leading and lagging strands. The final time step (4) shows the two daughter double-stranded DNA molecules, each composed of one old (black) and one new (grey) DNA strand.

instead of the correct **T**), but the DNA polymerase has very good error-correction (or *proofreading*) capacities, allowing the detection and correction of most of these mistakes. The resulting fidelity is amazingly high, of the order of one erroneous nucleotide per billion basepairs or about 10,000 times the expected fidelity in the absence of proofreading (Snustad and Simmons, 2010, pp. 258–259).

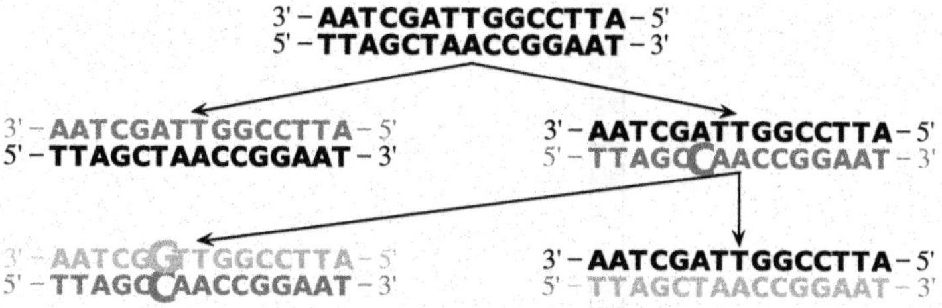

Figure 3.4 A point mutation. Top: the original double-stranded DNA molecule, undergoing semi-conservative replication to produce an identical DNA daughter molecule (left) and another DNA daughter molecule (right) carrying a **T** → **C** mutation (exaggerated size). As before, black, dark and light grey represent the old and new strands. Subsequently, this DNA produces two daughter molecules, one identical to the original (right) and one mutant (left), which has replaced the original pair $\genfrac{}{}{0pt}{}{A}{T}$ by the new $\genfrac{}{}{0pt}{}{G}{C}$ pair.

Such simple errors, resulting in the replacement of one nucleotide by another one, are called **point mutations** and represent one type of **mutation**. Figure 3.4 shows an example of point mutation resulting in the appearance of a new variant or **allele**, so that we now have identical copies of the original double-stranded DNA molecule $\genfrac{}{}{0pt}{}{3' - \text{AATCGATTGGCCTTA} - 5'}{5' - \text{TTAGCTAACCGGAAT} - 3'}$ as well as copies of the new mutant, $\genfrac{}{}{0pt}{}{3' - \text{AATCGGTTGGCCTTA} - 5'}{5' - \text{TTAGCCAACCGGAAT} - 3'}$. Thus, for basepair 6 (counting from the left and referring to the lower $5' - 3'$ strand) we have two possibilities: the original (or *wild-type*) allele **T** and the new (mutant) allele **C**. When frequent enough in a group of organisms (a population), such point mutations are also known as *Single Nucleotide Polymorphisms* (or **SNP**s, pronounced /snɪp/) and are currently very important in modern genetics (these will be discussed later in the book).

3.5 Chromosomes are DNA molecules

In the living cell, the two complementary strands of the double-stranded DNA molecule are coiled around each other, forming the well-known double helix which, in eukaryotic cells, is packed with a special type of proteins called **histones**. Besides their structural role, the histones also play a major part in **gene regulation** but we are only beginning to understand these mechanisms. The complex DNA + histones is called **chromatin** and forms the **chromosomes**, the physical packages of genetic material in a cell. One chromosome contains a single, extremely long molecule of DNA (the longest is about 8.5 cm), and humans have 46 such chromosomes coming in 23 pairs. Twenty-two of these pairs (the *autosomes*) are identified with the numbers 1 to 22 in decreasing order of size, while the 23rd pair represents the *sex chromosomes* and is

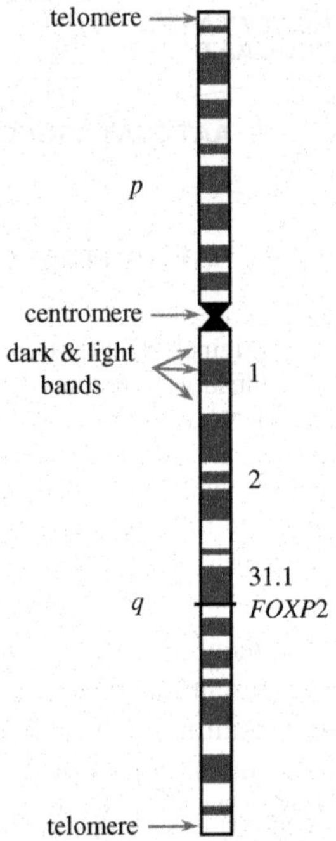

Figure 3.5 A graphical representation of chromosome 7, also showing the position of the *FOXP2* gene denoted as 7q31.1.

identified by the letters X and Y: females have two Xs (thus being XX), while males have an X and a Y (being thus XY). Males have, therefore, the *karyotype* (or full complement of chromosomes) { (1 1) (2 2) ... (21 21) (22 22) (X Y) }, while females have { (1 1) (2 2) ... (21 21) (22 22) (X X) }.

On a chromosome, genes are usually identified by their position, taking the *centromere* (a structure located more or less in the middle of the chromosome and essential in cell division) as the reference point, and specifying on which *arm* and *band* the gene is located. For example, *FOXP2* (a gene involved in speech and language that we will encounter many times in this book) is located on the long arm of chromosome 7, on band 31 and sub-band 1 (see Figure 3.5), an "address" written as 7q31.1 (conventionally, the short arm is denoted as *p* from the French *petit* and the long one as *q* from *queue*). These light and dark bands become visible when staining the chromosome with specific dyes and can be used as landmarks for localizing genes.

The *telomeres* are structures located at the end of the chromosomes and play an important role in regulating the number of replications the chromosomes can undergo, among other things (Matuli et al., 2007; Silvestre and Londoo-Vallejo, 2012). Generally in somatic cells, the DNA replication machinery cannot copy the ends of the chromosomes and each round of replication results in their shortening, by losing pieces of the telomeres instead of actual genetically important information. Thus, the number of replications such a cell can undergo is limited by the length of the telomeres, and consequently they have been implicated in ageing and cancer (Calado and Young, 2012). Interestingly, in germ and cancer cells, telomeres are restored by a specialized enzyme, *telomerase*, ensuring the indefinite transmission of genetic information across generations in the first case, and unregulated tumour growth in the second (Calado and Young, 2012; Matuli et al., 2007). The centromeres are essential to cell division (Brar and Amon, 2008).

The Y chromosome in humans does not contain many genes, but among these is the gene called *SRY*[2] (*S*ex-determining *R*egion *Y*) which is essential for sex determinism: a fetus having no *SRY* activity develops into a female, while one having it initiates the development of testis leading, in the normal case, to the development of a male. Thus, the default mode of development in humans produces females unless *SRY* overrides it towards males. Importantly, *SRY* itself is just a high-level switch, a trigger whose activity (or lack thereof) canalizes the developing embryo towards two possible developmental programs that usually are mutually exclusive, but there are also many "intermediate" cases. Thus, *SRY* regulates the activity of other genes, as we will discuss in more detail later. Of course, this mechanism can (and does) go astray, with, for example, cases of males who are genetically XX (they have a functional copy of *SRY* on one of their X chromosomes triggering the development of testis but lacking other male features whose development requires other genes on the Y chromosome). Moreover, our system of sex determinism is not the only one possible, other systems having independently evolved, with birds, for example, having ZW females and ZZ males (their sex chromosomes are called Z and W), while other animals (such as the Nile crocodile) completely relegate the choice of sex to environmental factors, such as the temperature at which the eggs incubate.

Turning back to humans, sexually adult males and females both produce **gametes**, which are cells specialized for reproduction. The male gametes are called **sperm** and the female ones **ova** and they differ in many respects including size (ova are much larger than sperm) and mobility (sperm are designed to travel long distances, using a long, flexible tail named *flagellum*, towards the largely immobile ova). Nevertheless, both contain *half* the genetic material of the adults that produced them: as discussed above, the adults have 23 pairs of chromosomes, for which reason they are called **diploid**, while the gametes contain just 23 non-paired chromosomes and are called **haploid**. During the complex process of gametogenesis which produces the gametes

[2] By convention, gene names are *italic*.

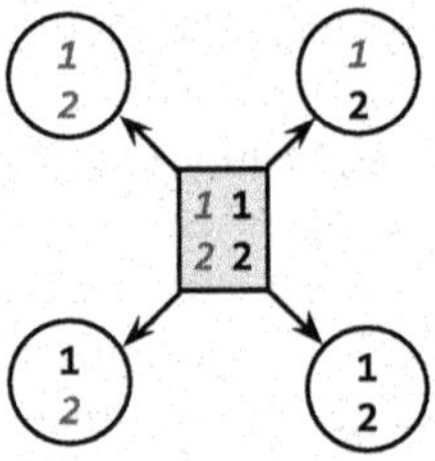

Figure 3.6 Gametogenesis: going from a diploid adult to the haploid gametes. For simplicity, we are focusing on just two chromosomes, 1 and 2, and the two individual members of the chromosome pairs in the adult (square in the middle) are differentiated using *italic* and grey colour. These individual chromosomes are transmitted to the resulting gametes (circles) independently, resulting in four possible such gametes.

specific to each sex, the full *complement* of 23 pairs of chromosomes of the adults is randomly shuffled through a series of cell divisions, such that only one representative of each pair makes it into any given gamete.

To make things clear, let us focus only on two chromosomes, 1 and 2: an adult diploid (male or female) has two 1s and will produce the appropriate haploid gametes as shown in Figure 3.6. Given that there are 23 pairs of chromosomes in an adult human, there are $2^{23} = 8,388,608$ possible different gametes. The process of random allocation of individual chromosomes to gametes is called **segregation** (Mendel's First Law or **Law of Segregation**), while the fact that different chromosome pairs segregate independently (in our example, the actual 2 that gets into a gamete does not depend on what actual 1 gets there) is called *independent assortment* (Mendel's Second Law or **Law of Independent Assortment**).[3]

New humans result from the **fertilization** of an ovum by a sperm, a process whereby these two haploid cells fuse and produce a diploid *zygote*, which then undergoes repeated cell divisions and develops into an embryo, then a fetus and, through birth, a new-born baby. Again for simplicity, let us focus on a single autosome, 1, and on the sex chromosomes, as in Figure 3.7. The two parents have their chromosomes marked as follows: the female parent's chromosomes are **bold** while the male parent's chromosomes are in regular font, and within each parent one member of the pair is *grey italic* and the other black upright. Thus, as per Mendel's first and second laws, there are

[3] Of course, when the monk Gregor Mendel (1822–1884), the father of modern genetics, published his laws in 1866, he did not talk about chromosomes, genes or DNA, but he was concerned with the transmission of discrete characters, such as pea colour, across generations. It turned out much later that these characters are in fact encoded by genes sitting on chromosomes.

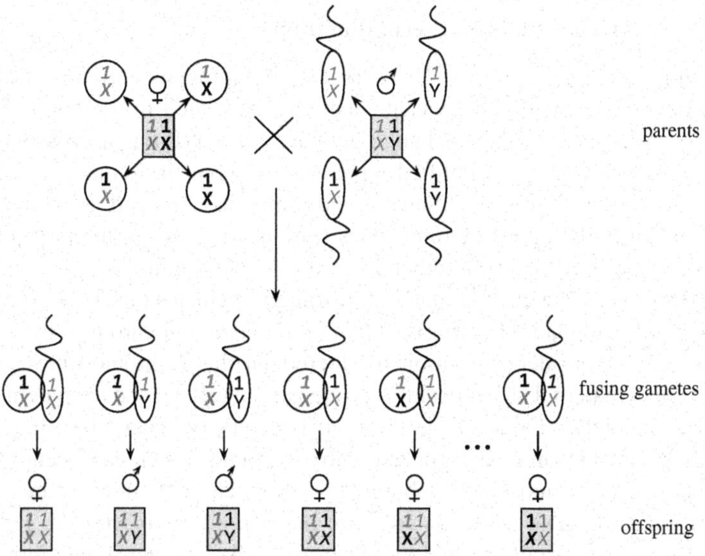

Figure 3.7 The parents (male and female, top) produce gametes which fuse (middle) to produce offspring (bottom). Here we focus on autosome 1 and on the sex chromosomes. For illustration purposes, the female parent's chromosomes are **bold** while the male parent's chromosomes are regular, and within each, one member of the pair is *grey italic* and the other black upright. Only six of the possible 16 zygotes are shown.

four possible female gametes (round) and four possible male gametes (elongated with tails) which can form $4 \times 4 = 16$ possible zygotes, which may, in turn, develop into offspring. Looking at the sex chromosomes, it can be seen that the female gametes are all carrying one X, while the male gametes can carry either an X or a Y and thus the zygote's sex depends on the male gamete (an X will result in a full complement of XX and thus a female, while a Y will give XY and thus a male). All ova carry X, while sperm can be X half of the time and Y the other half, ensuring that, under normal conditions, the possible sex of the zygote is equally split between male and female.

With 23 pairs of chromosomes, there are $2^{23} = 8,388,608$ possible gametes and $2^{23} \times 2 = 16,777,216$ unique zygotes, which might seem a big number, but it would mean that each of the around 7 billion people currently inhabiting Earth would be, on average, genetically identical to approximately another 400 human beings! However, this is patently false as there are several other mechanisms responsible for generating a lot of extra diversity, including point mutation (which we have discussed above) and *recombination*, to which we now turn.

3.6 Genetic loci and recombination

Following Mendel's First Law, the two homologous chromosomes forming a pair in the diploid parent are randomly allocated to the haploid gametes. However, they always exchange genetic material through **crossover**; Figure 3.8 shows a highly simplified representation, but the actual process is more complex (see for example Snustad and Simmons, 2010, pp. 143–146). The two parental homologous chromosomes (one inherited from the parent's mother, shown in black, and the other from the parent's father in grey) pair up and sometimes swap genetic material, leading to gametes (and, if everything goes according to plan, offspring) containing *recombinant* chromosomes.

Let us consider a position (or **locus**) on these chromosomes, where the parent's maternal chromosome has one variant (allele) and the paternal another one, denoted in Figure 3.8 as A and a. This type of generic labelling, denoting alleles of the same gene (or alternative variants which can occupy the same locus on a chromosome) by big and small letters (e.g., A and a are two alleles of the same gene while B and b are the two alleles of another genes) or by numeric subscripts attached to a letter (e.g., A_1, A_2 and A_3 are three possible alleles of one gene), is very common in genetics. However, with the increase in our understanding of the molecular nature of the alleles, more and more often we use actual characteristics of the alleles such as the type of change from the normal, or "wild", type. (As a side note, the big/small letter convention might look nice on paper but is a pain for oral discourse, including lectures!) They could stand for single nucleotide differences (for example, A

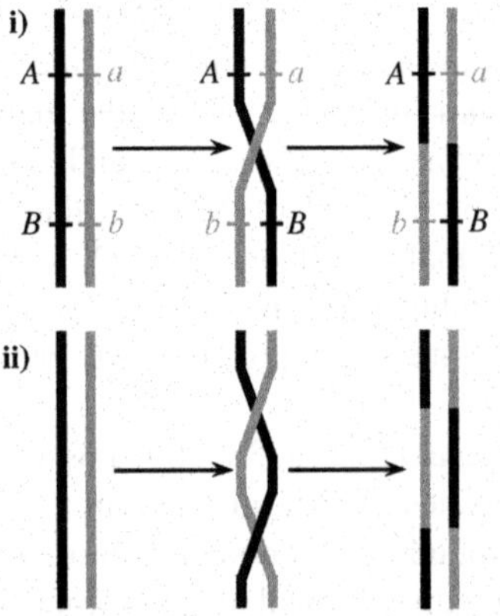

Figure 3.8 A highly simplified representation of chromosomal crossover. Left: the two original homologous parental chromosomes (black and grey) are paired. Middle: they cross over once (top) or twice (bottom) to result in two new recombinant chromosomes (right).

in Figure 3.8 could represent thymine, **T**, while *a* could be cytosine, **C**), whole stretches of DNA spanning several base pairs (bp) up to thousands (kb) or even millions (Mb) of base pairs, or, when the exact nature of the variants is not known, these would denote their observable (or phenotypic) effects (e.g., *A* could stand for "whatever actual DNA variant causes blue eyes in humans" and *a* for "whatever actual DNA variant causes green eyes in humans", both being alleles of the putative locus or gene "eye colour in humans").

Our example in Figure 3.8 panel (i) on top has two loci, one with possible alleles *A* and *a* and the other with alleles *B* and *b*. The parent (left) inherited from its mother (black chromosome and letters) the **haplotype** (or combination of alleles at different loci on the same chromosome) *AB* (from top to bottom) and from its father (grey chromosome and letters) the haplotype *ab*. Without recombination, it should produce two types of gametes (for this particular pair of chromosomes): one carrying the haplotype *AB* and the other the haplotype *ab*, resulting from the semiconservative replication of DNA in its own chromosomes, as discussed in Section 3.4. However, because crossover happens to occur between these two loci and genetic material is exchanged between the homologous chromosomes, the recombinant gametes will inherit instead the *recombinant* haplotypes *Ab* and *aB*.

Thus, in the absence of crossing over, two loci on the same chromosome do *not* assort independently (as per Mendel's Second Law) but tend to "travel together" across generations, their association being broken from time to time by recombination. The probability that this dissociation will happen measures the **linkage disequilibrium** between the two loci, and will turn out to be an extremely important concept in our search for actual genes. But now we need to understand what genes are and how they work at the molecular level.

3.7 What is a gene?

So far we have discussed genes and genetic loci in fairly abstract terms, as pieces of DNA that have a certain location on a chromosome, that are transmitted across generations and that, somehow, manage to influence aspects of the organism's phenotype. But now it is time to get our hands dirty and dig into the fascinating details of gene structure and function.

While it is true that the essence of the genetic information is represented by linear sequences of the four bases A, T, C and G, a "gene" is not a continuous, unitary read of such bases (the way words are usually written on paper), nor is it an abstract digital message floating in a pure void populated by mathematically precise translators and effectors (even if this is how things are usually illustrated in papers and books, including this one). Not a chance! They are composed of different bits spread across the DNA molecule, bits with different functions and properties, all of them wet and real biological molecules floating in complicated environments having water as a main ingredient, and interacting with and manipulated by probabilistic, error-prone molecular complexes open to many signals originating both within and outside the cell. In the following I will try to give a better appreciation of all this biological complexity while still keeping it very abstract and far from the realities of a wetlab. For more information on these aspects one can consult any of the good introductions to biochemistry, molecular biology and molecular genetics such as Wilson and Walker (2010), Alberts (2009), Lewin and Krebs (2011), Voet and Voet (2011) or Lodish et al. (2012).

To get a sense of the scale of the problem, there are about 2 metres of DNA in a normal human cell[4] of several tens of **micrometre** (or **microns**; μm) in size (1 μm = 1×10^{-6} m, so that there are 1 million μm in 1 m); thus, the DNA is about 2 million times longer than the cell it resides in. However, to begin with, this is not a single molecule but 46 of them (2×22 autosomes and 2 sex chromosomes), to which we have to add the multiple copies of the tiny mitochondrial DNA. Each of these molecules is folded and coiled at multiple levels (Figure 3.9): the basic unit is the *nucleosome*, which consists of 146 DNA base pairs (bp) of DNA coiled around a core composed of special

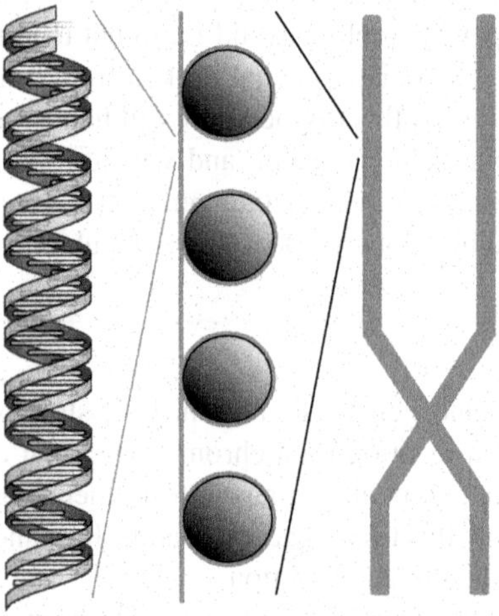

Figure 3.9 Schematic representation of the structure of a chromosome. Leftmost is a fragment of the DNA double helix molecule. This molecule is wrapped around histones (black spheres) forming the nucleosome (middle). Rightmost is the whole chromosome. See text for details; the drawing is not to scale.

[4] There are about 3 billion bases per genome copy, each of 0.34×10^{-9} m long, and there are two copies per diploid cell.

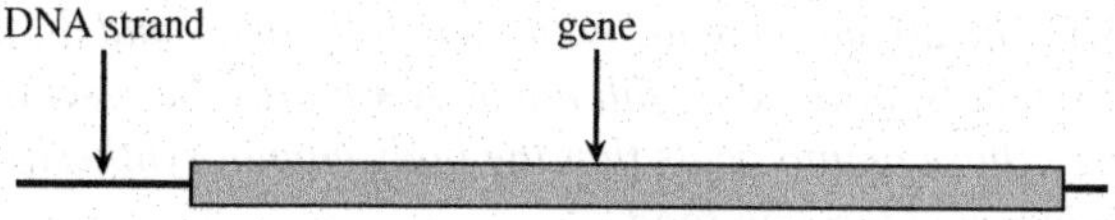

Figure 3.10 Schematic representation of the simplest imaginable gene: just a single continuous block of meaningful letters on a DNA strand.

proteins, the **histones**. Nucleosomes resemble "beads on a string", which are further packed together, reducing the required volume even further. Thus, the DNA is embedded in a tri-dimensional structure which has great influence on how the message is read and interpreted.

Even more, this DNA molecule is not in a sort of "pure" form in this coiled-up and folded structure – like a message safely locked away in an ivory tower where an old wizard may be allowed to enter and read the bits required only when required to do so (a widespread misconception is that this reading happens only once during "development", after which the scroll is folded back and locked away again). In fact, there is a continuous frenetic activity involving the DNA, with most of it probably read (The ENCODE Project Consortium, 2012) and transcribed, with many proteins binding to specific locations on the DNA (based on matching a particular pattern there) and influencing the way other proteins bind (or fail to bind) and read the DNA, attempt to repair it or copy its message into a new DNA molecule. You have to imagine a landscape teeming with life, a true informational ecology living off the DNA molecule and continuously changing the way its messages are read, interpreted and delivered to the world outside.

Imagine a strand of DNA and the letters it contains as a message that needs to be read and expressed; the simplest way a gene could look is a continuous stream of letters that have a clear begining and end and which are reliably read and expressed. Let's say we represent the strand of DNA as a line and this gene as a single block on it as in Figure 3.10: then this block would mean something like "please make me a protein composed of a single chain of amino acids". We saw that amino acids are really the building blocks of life (Section 3.2) and there are 20 of them involved (Table 3.1), all built upon the same basic plan but differing slightly in biochemical properties such as electrical charge (some are negatively charged, some are neutral and others positive) and ability to react with other substances (for details see any introductory texts such as Alberts, 2009, or Lewin and Krebs, 2011).

3.7.1 Decoding genes: transcription and translation

An ideal gene such as the one in Figure 3.10 carries a message represented by a sequence of nucleotides that will ultimately result in a protein. This is a

Table 3.1 *The 20 amino acids with their full names and 3- and 1-letter symbols, ordered by full name. A ◂ marks the* essential *amino acids, i.e., those amino acids that the body cannot synthesize and must be ingested through food.*

Full name	3-letter symbol	1-letter symbol
Alanine	Ala	A
Arginine	Arg	R
Asparagine	Asn	N
Aspartic acid	Asp	D
Cysteine	Cys	C
Glutamic acid	Glu	E
Glutamine	Gln	Q
Glycine	Gly	G
Histidine ◂	His	H
Isoleucine ◂	Ile	I
Leucine ◂	Leu	L
Lysine ◂	Lys	K
Methionine ◂	Met	M
Phenylalanine ◂	Phe	F
Proline	Pro	P
Serine	Ser	S
Threonine ◂	Thr	T
Tryptophan ◂	Trp	W
Tyrosine	Tyr	Y
Valine ◂	Val	V

complex process composed of two major steps: **transcription** and **translation**. Transcription is conceptually simpler and basically consists of copying the DNA sequence into an intermediate form, embodied by an RNA molecule. For our purposes here, RNA is very similar to DNA except it is single-stranded and instead of **T** uses **U** (uracil). Transcription proceeds along the DNA molecule and is initiated by the binding of an enzyme, **RNA polymerase**, to special locations on the DNA known as **promoters**. Transcription involves only one of the two DNA strands (the *template* strand) and produces an RNA molecule complementary to it but identical (except for using **U** instead of **T**) to the other DNA strand (the *coding* strand). Figure 3.11 shows a schematic representation of transcription ending with an RNA molecule – called a *messenger RNA* (or **mRNA**). Transcription takes place in the cell nucleus, but the resulting mRNA is then exported from the nucleus into the cytoplasm, where the next step, translation, takes place.

Now, a different molecular machinery will read the message encoded by the mRNA and translate it into a string of amino acids. As you can easily see, there are 20 amino acids that have to be coded using the four DNA

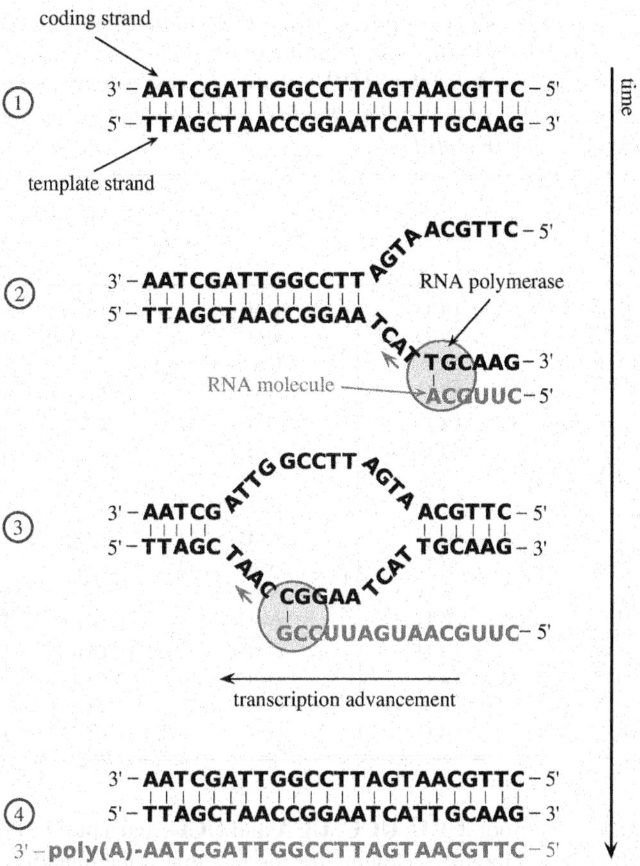

Figure 3.11 Schematic representation transcription.

(or equivalently, RNA) letters – this requires a minimum "word" size of 3, allowing the encoding of $4^3 = 64$ "meanings", much more than the actual 20 needed. This results in the **genetic code** – the mapping between the 3-letter "words" (also known as **codons**) and the corresponding amino acids – being degenerate (or redundant): for some amino acids there's more than one "word" allocated. However, the code is *not* ambiguous, meaning that there's no 3-letter "word" that means more than one amino acid, and the meaning is not context-dependent in the sense that the reading of such a "word" might depend on the neighbouring words. There are multiple ways to visualize this code, a very popular one being reproduced in Table 3.2. It can be immediately seen that the code is indeed degenerate (for example, the amino acid phenylalanine, F, is encoded by the two codons **UUU** and **UUC**, while serine, S, by no less than six

Table 3.2 *The genetic code showing for each possible triplet or codon (tri-nucleotide "word") the corresponding amino acid (represented by its one-letter symbol) or ■ for the STOP codons. The first column gives the first nucleotide ("letter") of the codons, the first row gives the second nucleotide, and the last column the third one.*

1st	2nd								3rd
	U		C		A		G		
U	UUU	F	UCU	S	UAU	Y	UGU	C	U
	UUC		UCC		UAC		UGC		C
	UUA	L	UCA		UAA	■	UGA	■	A
	UUG		UCG		UAG		UGG	W	G
C	CUU	L	CCU	P	CAU	H	CGU	R	U
	CUC		CCC		CAC		CGC		C
	CUA		CCA		CAA	Q	CGA		A
	CUG		CCG		CAG		CGG		G
A	AUU	I	ACU	T	AAU	N	AGU	S	U
	AUC		ACC		AAC		AGC		C
	AUA		ACA		AAA	K	AGA	R	A
	AUG	M	ACG		AAG		AGG		G
G	GUU	V	GCU	A	GAU	D	GGU	G	U
	GUC		GCC		GAC		GGC		C
	GUA		GCA		GAA	E	GGA		A
	GUG		GCG		GAG		GGG		G

codons: a block of four, **UCU, UCC, UCA** and **UCG**, and a block of two, **AGU** and **AGC**). This ambiguity accounts for the fact that some mutations are **synonymous mutations** in the sense that they result in the same amino acid. For example, a mutation changing the last **U** of the **UUU** codon into a **C** (**UUU** → **UUC**) would still result in phenylalanine, resulting in the same protein. This redundancy makes the genetic code more fault-tolerant and is very important in the inference of natural selection as we will see later. Another important aspect to note is that there are three codons (**UAA, UAG** and **UGA**) that do not code for amino acids and instead determine the stopping of the translation process, ending the protein. Another important special codon is **AUG** (coding for methionine, M) and this is where translation usually starts.

The genetic code is almost universal in the living world, but there are several slight variations (NCBI's webpage "The Genetic Codes"[5] was listing 17 variants as of April 2013). For example, our own mitochondria use a slightly

[5] `http://www.ncbi.nlm.nih.gov/Taxonomy/taxonomyhome.html/index.cgi?chapter=tgencodes#SG2.`

different genetic code to interpret their tiny DNA, a code in which **AGA** and **AGG** are stop codons instead of the standard arginine (R) but with the standard stop codon **UGA** coding here for tryptophan (W), **AUA** codes for methionine (M) instead of the standard isoleucine (I), and **AUU** is also a start codon. This raises the intriguing question of its origins and evolution, and the functions that it might have been optimized for (see for example Vetsigian et al., 2006, and Bollenbach et al., 2007, for some proposals and discussion). Moreover, we have recently learned how to extend the genetic code by adding new "meanings", by forcing new associations between codons and amino acids, some of which are outside the standard set of 20 (Liu and Schultz, 2010; Davis and Chin, 2012), and even by creating new 4-letter codons (Neumann et al., 2010; Wang et al., 2012).

Returning to translation, this is the complex process whereby the message encoded by the mRNA is decoded and transformed into a corresponding sequence of amino acids. This is taken care of by complicated cellular machines called **ribosomes**, which read the mRNA molecule in words of three nucleotides (thus, one codon at a time) and add the corresponding amino acid (as given by the genetic code) to the growing string of amino acids (also called a polypeptide), as shown in Figure 3.12. The ribosome is composed of two subunits ("small" and "large", each composed of polypeptides and RNA molecules) that assemble and start translating the mRNA from the first (in the $5' \rightarrow 3'$ direction) start codon **(AUG)**.[6] This start codon is extremely important not only because it initiates the translation process but also because it defines the **reading frame**, the manner in which the codons are read off the mRNA.

The mRNA in Figure 3.12 reads

$$5' - \text{GCCACCAUGGCUGGAUUUUAGACUGAAAAA} - 3'$$

and there are many ways in which the codons could be read: for example, we could start with the very first nucleotide and read **GCC**, **ACC**, **AUG**, ..., or we could start with the second and read instead **CCA**, **CCA**, **UGG**..., or with the third, resulting in **CAC**, **CAU**, **GGC** (Figure 3.13).

Each of these three possible reading frames generates a very different message and the way the ribosome decides which one to pick is dictated by the first start codon; in this case the correct reading frame is thus **GCC**, **ACC**, **AUG**, ... In this reading frame, the message is (using the genetic code, amino acids in parentheses but only between the start and before the stop codons): **GCC**, **ACC**, **AUG** (M, start), **GCU** (A), **GGA** (G), **UUU** (F), **UAG** (■, stop), **ACU**, **GAA**, **AAA**. A sequence of codons in a reading frame that does not

6 The process is in fact very complex and the **AUG** codon is embedded in a so-called *Kozak consensus* (a small sequence of DNA that conforms to a certain pattern of nucleotides) that influences the translation process (Kozak, 1999).

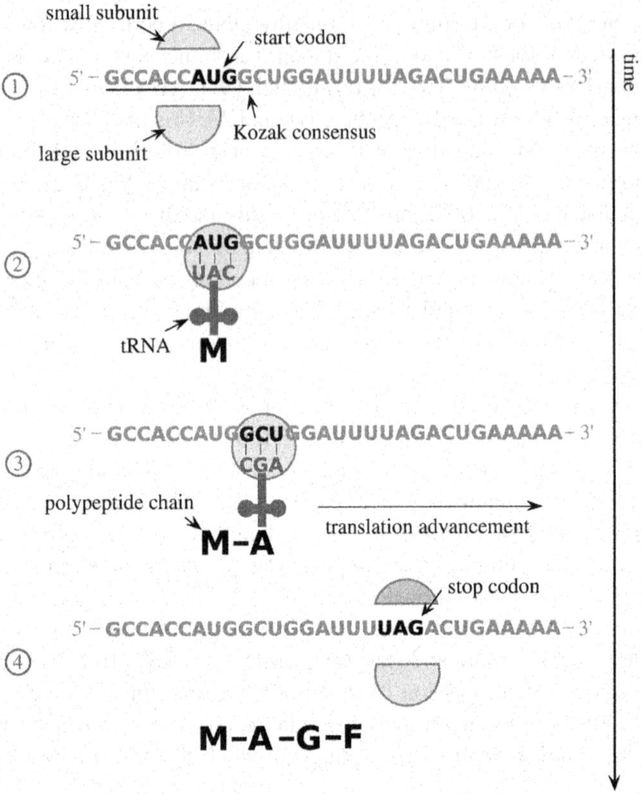

Figure 3.12 Schematic representation of translation. The ribosome is composed of two subunits (large and small) that assemble and begin the translation process with the start codon **AUG**. Translation advances in three-nucleotide units (codons) in the $5' \rightarrow 3'$ direction (here, left → right) and consists in the elongation of the polypeptide chain by adding the amino acid carried by the tRNA corresponding to the current codon.

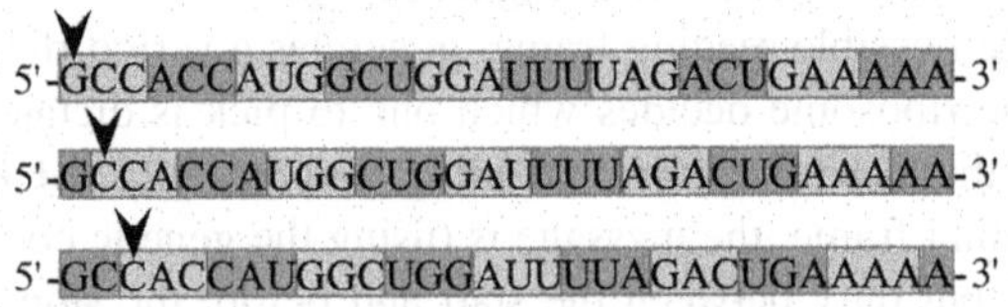

Figure 3.13 Representation of the three possible reading frames in the three rows of the figure. The arrow shows the first letter of a codon and, for visual contrast only, the neighbouring codons are highlighted using light and dark grey.

have a stop codon is called an **open reading frame** (or **ORF**) and its presence signals that this region might function as a protein-coding gene. It is important to realize that mutations that change one nucleotide for another (such as SNPs) will usually affect a single amino acid (synonymous ones will even be "invisible"), but could sometimes prematurely terminate the protein when producing a new stop codon or add new amino acids when destroying an existing one. But there are other types of mutation where nucleotides are inserted (**insertions**) or deleted (**deletions**) – collectively known as **indels** – and these can cause a reading frame shift (or frameshift mutations) when the number of inserted/deleted nucleotides is not divisible by three, producing garbage after the indel's position.

The ribosome advances in jumps of three nucleotides (one codon) from the first start codon onwards and uses yet another molecule, a so-called **transfer RNA** (or **tRNA**, just a folded strand of RNA that looks like a clover leaf), to join the appropriate amino acid to the growing polypeptide chain. tRNA can be thought of as a sort of adapter, associating each codon (in fact, the complement of the codon) with its corresponding (or "cognate") amino acid. In the figure, the ribosome selects first a tRNA for the start codon **AUG** (thus carrying the complementary sequence – or *anticodon* – **UAC**) with a methionine (M) amino acid attached; then it binds to this one the second amino acid (alanine, A) transported by the tRNA corresponding to the **GCU** codon, and so, extending the amino acid chain (the polypeptide) one amino acid at a time. This process continues until the ribosome hits the first stop codon, in this case **UAG**: this results in the termination of translation, dis-assembly of the ribosome and the liberation of the newly produced polypeptide (in this case the short M-A-G-F).

In reality, these amino acid chains are much longer, consisting of hundreds or thousands of amino acids but they are not the final product; while their sequence (so-called *primary structure*) is a faithful representation of the original message encoded in the DNA, the actual function depends on their tri-dimensional shape. The process of *folding* occurs concurrently with the translation, and is driven by thermodynamics (some other proteins might help by offering protection), resulting in their *secondary* and *tertiary* structures, with the latter being the actual tri-dimensional form that can implement the protein's intended function (technically, proteins are composed of one or more such polypeptides). Moreover, most proteins are actually subunits of higher-level complexes, and this represents the so-called *quaternary* structure. Of course, as you might already expect, this is a massive simplification of a complex process, as most proteins cannot act alone but need to combine with others with different functional and structural properties. Moreover, there are other *post-translational modifications* that can affect them, such as changes to some amino acids or cutting of some portions of the protein.

3.7.2 Genes have structure: introns and exons

If this process seems complicated, it is not yet realistic in one important respect: we made the simplification that our gene is composed of a single continuous block of codons encoding a single protein (Figures 3.10 and 3.11). However, in humans and other Eukaryotes, this is not generally the case. To begin with, a gene is not a continuous stream of codons but is composed of alternations of **exons** (from *ex*pressed regi*on*; continuous stretches of nucleotides intended to be read as amino acid-coding codons) and **introns** (from *intr*agenic regi*on*; stretches of nucleotides that do not code for amino acids). Thus, a more realistic depiction of a gene is as in Figure 3.14.

The transcribed mRNA (*primary transcript*, *precursor mRNA* or *pre-mRNA*) becomes *mature* ***mRNA*** by **splicing** away the introns and joining together (some of) the exons, as shown in Figure 3.15. However, splicing is a very complex process controlled by many factors (Nilsen and Graveley, 2010; Chen and Manley, 2009) and is in large part responsible for the existence of many more proteins (the set of all proteins is called the **proteome**) than there are genes (the genome) in an organism (Nilsen and Graveley, 2010; Kelemen et al.,

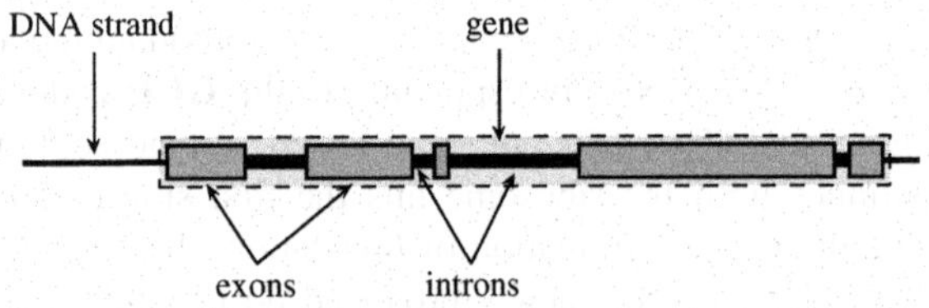

Figure 3.14 Schematic representation of a gene composed of several exons and introns.

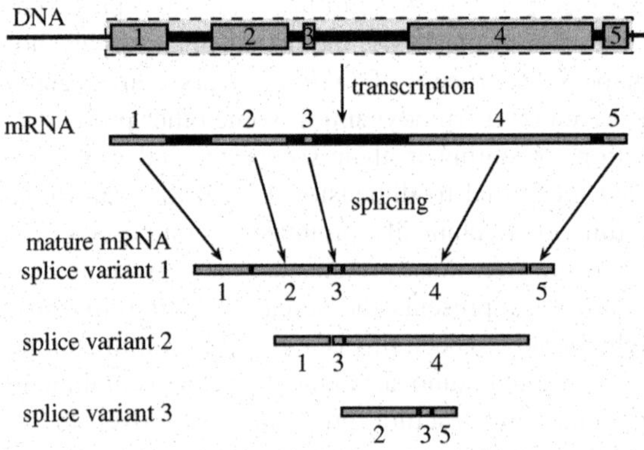

Figure 3.15 Schematic representation of alternative splicing.

2013). The various forms of a protein are called *isoforms* and many isoforms are produced from the same gene by alternative splicing. It has been estimated (Nilsen and Graveley, 2010) that 95–100% of genes with multiple exons produce more than one mature mRNA through alternative splicing (although many of these still encode the same protein as the extra exons are not transcribed). A striking example is the *DSCAM* gene in *Drosophila* (the fruit fly), which can generate about 38,000 mRNAs through alternative splicing (Schmucker et al., 2000), much more than the 14,000 or so genes this animal is currently estimated to have. *FOXP2* in humans is currently estimated to have 19 splice variants[7] ranging in length from no protein product (for two variants), to between 67 amino acids (three exons) and 740 amino acids (18 exons) for the others. In this context, it is important to note that what makes an intron an intron is far from simple and involves several mechanisms: some introns *self-splice* (remove themselves from the pre-mRNA molecule), but most require a separate complex machinery (the *spliceosome*) to actively identify and cut them out (Snustad and Simmons, 2010, pp. 299–307).

All the exons in all the genes in the genome are known as the organism's **exome** and, for humans, this represents about 1% of the whole genome, amounting to about 30 million bp across approximately 180,000 exons (Ng et al., 2009). Even if it represents such a small portion of the whole genome, the exome is very interesting for several reasons. First, while we do not yet really understand what most of our genome does, we do know that exomes result in proteins and it is thus easier to predict what impact changes in the exome might have, and, second, it is estimated that about 85% of disease-causing mutations affect it (Rabbani et al., 2012). Third, while whole genome sequencing keeps getting cheaper,[8] it is still a major investment of time and money, but whole exomes still cost less at about $1,000 per sample,[9] allowing the sequencing of larger samples. Currently, sequencing the whole exome (*Whole Exome Sequencing* or WES) is an area of active research especially for discovering rare and *de novo* (i.e., new mutations that the individual did not inherit from his/her parents) mutations (Veltman and Brunner, 2012; Cirulli and Goldstein, 2010) and as a complement to linkage studies in the analysis of Mendelian (or monogenic) phenotypes (Ku et al., 2011; Rabbani et al., 2012). However, many see WES as a temporary stop before the costs associated with *Whole Genome Sequencing* (WGS) drop to a level where it becomes feasible to routinely perform it for very large samples (e.g., Teer and Mullikin, 2010),

[7] `http://www.ensembl.org/Homo_sapiens/Gene/Summary?g=ENSG00000128573;r=7:113726382-114333827`, April 2013.

[8] The National Human Genome Research Institute (NGRI) data (`http://www.genome.gov/sequencingcosts`, April 2013) lists the cost per genome in January 2013 at $5,671.

[9] For example, as of April 2013, BGI Tech Solutions Co., Ltd. offers $899/sample; `http://bgitechsolutions.com/exome_promotion_2013/`.

and even at the time of writing (April 2013) the difference in cost is not that impressive anymore.

Biological systems are most of the time not elegant in an engineering sense but display unexpected complexities and intricacies that are totally counterintuitive, being the result of the workings of the "blind watchmaker" (Dawkins, 1996) – biological evolution. Thus, it will come as no surprise that this description is even messier in reality with *overlapping* and *nested* genes. Overlapping genes (Ho et al., 2012; Sanna et al., 2008; Makalowska, 2004) share exonic material and can be both on the same DNA strand (read in the same direction), in which case they might be in different reading frames, or on the opposite strands (read in opposite directions). Nested genes (Kumar, 2009) are usually to be found entirely within an intron of another gene; as an example, three genes (*OMG*, *EVI2B* and *EVI2A*) are contained within intron 27 of the *NF1* gene in humans (Makalowska, 2004). This type of non-independence between genes raises interesting questions concerning their origins and evolution given that changes in one will also trigger changes in the other.

Finally, there are various **regulatory elements** such as **promoters**, **enhancers**, *suppressors* and *insulators* (Riethoven, 2010). A schematic representation is given in Figure 3.16. Promoters are regions that initiate the transcription of a gene or set of genes, they are located on the same strand as these genes and influence their expression either by turning on and off the transcription of the genes or by affecting the quantity of transcript a gene produces. Enhancers and silencers are other types of regions that affect the transcription of genes by increasing and decreasing it, respectively. Insulators ensure that the effects of an enhancer do not spill over to the neighbouring genes. The mechanism involves the creation of *loops* in which regulatory proteins, while still bound to their enhancer and suppressor loci, are brought physically close to the gene they regulate (Dean, 2011). Moreover, the histones can be chemically marked (for example, becoming *methylated* or *acetylated*) and these marks can influence the way the message encoded by the DNA around them is interpreted. For example, histone methylation changes the rates at which the corresponding DNA is transcribed into RNA, the first necessary step in expressing the

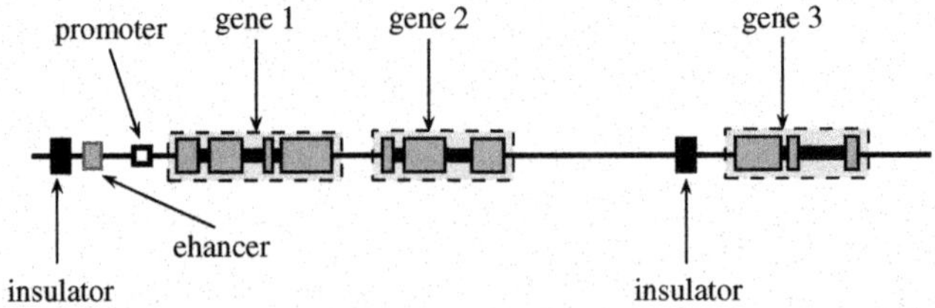

Figure 3.16 Promoters, enhancers, suppressors and insulators.

encoded message, in turn playing essential roles in development and gene regulation (Greer and Shi, 2012).

Gene regulation is extremely important for understanding the evolution, development and maintenance of complex organisms and is currently a very active area of research. We will discuss several concrete examples in Sections 6.7, 6.9 and 6.9.1, and we will encounter its effects in several places throughout this book.

4 Effects of genes on phenotype

Here we encounter the patterns of transmission of genes across generations and the main types of effects they can have on the phenotype. We will discuss autosomal versus sex-linked loci, exemplified using the dominant speech and language pathology apparent in members of the famous British KE family (and due to a mutation in FOXP2, a gene that we will encounter again later in the book), a very interesting case of recessive transmission of hearing loss that ultimately resulted in the emergence of a new sign language in a village on the island of Bali, and, finally, the fascinating transmission of deficits in colour perception (which might turn out to affect language). We will introduce here some concepts of classical genetics and basics of statistical testing (such as the χ^2 test).

In diploid organisms, such as ourselves, the vast majority of genetic loci have two alleles, as discussed above. For example, the two loci in Figure 3.8 part (i) in the parent (left) have alleles *A* and *a* at the first locus, and *B* and *b* at the second, but there are also many loci at which the two alleles are identical. To use a real example studied by William Bateson, Edith Saunders and Reginald Punnett at the beginning of the 20th century (Bateson et al., 1905; Lobo and Shaw, 2008), let's for a moment look not at a human but at a sweet pea (*Lathyrus odoratus*), and more precisely at a locus that influences the flower's colour, with allele *A* resulting in purple flowers while allele *a* results in red flowers.

4.1 Dominance and recessiveness

But how can we talk about the effects of a single allele, such as *A*, in diploid organisms? We can look at **homozygous** individuals that happen to have two copies of the same allele at this locus and observe that *AA* individuals produce purple flowers while *aa* ones have red flowers. But what happens to **heterozygous** individuals that have different alleles at the same locus? In this particular case, the heterozygous *Aa* individuals (the order in which the alleles are written, *Aa* or *aA*, is irrelevant here) produce purple flowers, just like the homozygous *AA* individuals. This shows that the *A* allele is **dominant** over the *a* allele (or, equivalently, that *a* is **recessive** relative to *A*); a common practice is to use capital letters for the dominant alleles and small letters for the recessive ones.

Recessive alleles, such as our red sweet pea flower colour *a*, are effectively hidden in heterozygous bodies by their dominant counterpart, purple colour *A*, and revealed only in recessive homozygous individuals, *aa*. Thus, by looking only at the phenotype (flower colour) of a specific individual (sweet pea plant) one cannot infer the precise **genotype** (the actual combination of alleles) at this locus except when it is red; purple only shows that there is a dominant *A* at work but there's a 50% chance that the second one is also a dominant *A* (as opposed to a recessive *a*).

To learn about the actual genotype one has to either try to infer it from looking at **pedigrees** (*genealogies* or *family trees*) containing the individual of interest, or to directly genotype the individual using one of the several currently available techniques. The possible patterns of inheritance for sweet pea flower colour are represented in Figure 4.1 using the usual conventions for pedigrees: circles represent females and squares males,[1] with colour standing

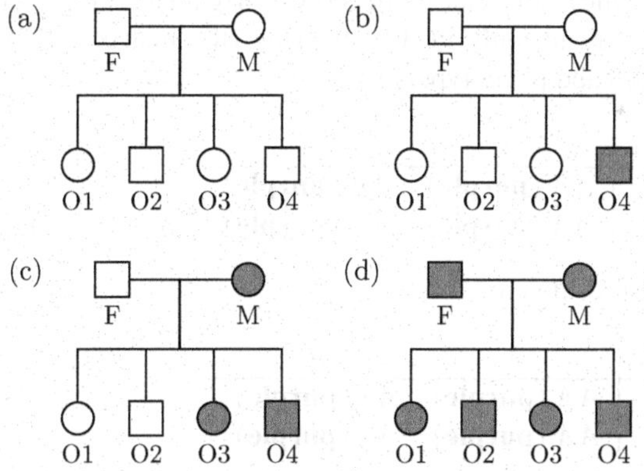

Figure 4.1 The four possible patterns (a) – (b) of transmission of flower colour in sweet peas. Circles ○ represent females and squares □ males; **white** individuals have **purple** flowers and **black** individuals have **red** flowers. "F" represents the father (male parent) and "M" the mother (female parent), while "O1"–"O4" are the four possible types of offspring.

[1] In most animals (humans included) an individual has a single sex, while in plants (sweet peas included) an individual carries both male and female structures (in *angiosperms* – the flowering plants – such as the sweet pea, *stamens* and *ovaries*).

for the different phenotypes (here, **white** individuals have **purple** flowers and **black** individuals have **red** flowers). In human medical genetics, white usually stands for the normal (i.e., disease-free) individuals while black represents affected persons (i.e., suffering from the disease). Horizontal lines connect mating partners and vertical lines show descent by connecting offspring to their parents; time usually flows from the top while the horizontal dimension has no special meaning.

Panels (a) and (b) show that the mating of two **purple**-flowered plants can produce either all **purple** progeny (a) or a ratio of 3:1 **purple**:**red** progeny (b). How can we interpret the pattern in (a), all four possible types of offspring being **purple**? Both homozygous dominant *AA* parents would obviously fit (they would produce only *AA* offspring), as would one homozygous dominant *AA* parent (irrespective which) and one heterozygous *Aa* parent (producing three *AA* and one *Aa* offspring, all phenotypically identical, **purple**). Moreover, one parent being the dominant homozygote *AA* and the other the recessive homozygote *aa* would result in all heterozygous *Aa* offspring, still phenotypically uniform **purple**! Thus, case (a) cannot disambiguate by itself between the matings $AA \times AA$, $AA \times Aa$ and $AA \times aa$.

This verbal description is made much clearer if we use **Punnett squares** showing the possible genotypes (and phenotypes) of the offspring resulting from the combination of certain gametes. If both parents are *AA* (mating shown in the top-left cell), they both can produce only *A* gametes (top column and left row), which can result only in *AA* offspring (the matrix cells show the offsprings' genotypes and phenotypes):

$AA \times AA$	A	A
A	AA (**purple**)	AA (**purple**)
A	AA (**purple**)	AA (**purple**)

The mating $AA \times Aa$ results in:

$AA \times Aa$	A	a
A	AA (**purple**)	Aa (**purple**)
A	AA (**purple**)	Aa (**purple**)

while that between the different homozygous parents *AA* and *aa* results in the same pattern of offspring phenotypes:

$AA \times aa$	a	a
A	Aa (**purple**)	Aa (**purple**)
A	Aa (**purple**)	Aa (**purple**)

Case (b) is much better as the ratio 3:1 **purple**:**red** can be obtained only if both parents are heterozygous *Aa*, which would result in 1 *Aa*, 2 *Aa* (thus 3 **purple** phenotypes) and 1 *aa* (phenotypically **red**):

$Aa \times Aa$	A	a
A	AA (**purple**)	Aa (**purple**)
a	Aa (**purple**)	aa (**red**)

Case (c) is also simple, as the ratio 2:2 **purple:red** can be obtained only if one parent is heterozygous *Aa* (**purple**) and the other is homozygous recessive *aa* (**red**), producing 2 *Aa* (**purple**) and 2 *aa* (**red**):

$Aa \times aa$	a	a
A	Aa (**purple**)	Aa (**purple**)
a	aa (**red**)	aa (**red**)

Likewise, case (d) can be produced only by a single combination of parents (which one?).

Punnett squares are very useful for inspecting the expected types and ratios of offspring genotypes given that we know the parental genotypes. If we also have a good account of the phenotypic effects of all possible genotypes (as in this simple case of the sweet pea flower colour) then they will also reveal the expected distribution of the trait of interest. However, as shown by case (a) discussed above, we might need some very special combinations of parents in order to learn about a genetic system. Today, when we have a better understanding of the molecular bases of such traits, we know that a dominant pathology is usually due to the product of the deleterious allele being either toxic or non-functional and, in this latter case, the phenotype requires the full dose of the normal product for its correct development or maintenance; recessive pathologies usually happen when the product of the normal allele is still sufficient to compensate for the other, non-functional half produced by the deleterious allele.

We will exemplify dominance and recessiveness with two real cases relevant for language and speech: the KE family showing an inherited disorder with a broad spectrum also affecting speech (Developmental Verbal Dyspraxia, OMIM 602081) and the inherited deafness in the Bengkala village in northern Bali. Both are examples of *autosomal* dominant/recessive disorders where the cause is an allele at a locus on one of the non-sex chromosomes (autosomes); the pattern shown by *sex-linked* recessive diseases will be discussed later.

4.2 Autosomal dominance and recessiveness

4.2.1 A dominant speech and language disorder

The first example concerns the "star" gene ***FOXP2*** (Lai et al., 2001; Fisher and Scharff, 2009): everything started with the identification (see Watkins et al., 2002) of a large multi-generational family living in the UK (generally known in the literature as the "KE family") and showing what seemed to be a disorder

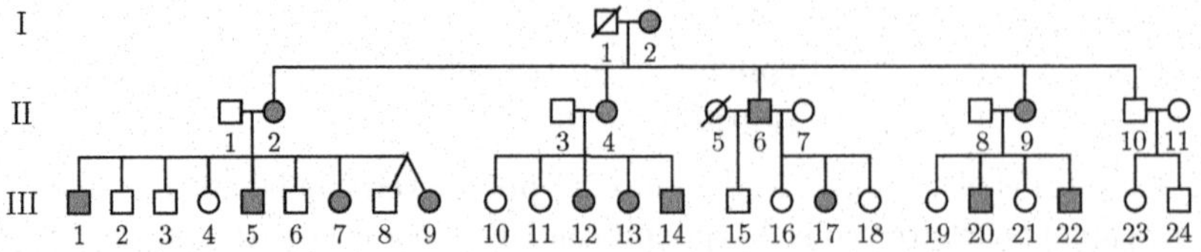

Figure 4.2 The KE family: a three-generation pedigree displaying an autosomal dominant pattern of inheritance. As usual, ○ are females and □ males; **white** individuals are normal and **black** individuals are affected by the pathology (Developmental Verbal Dyspraxia, OMIM 602081). Individuals crossed by an oblique line ⁄ were dead at the time the pedigree was constructed. Individuals 8 and 9 in generation III (connected by $\wedge$) are non-identical twins.

affecting speech and language being transmitted in a simple dominant manner. We will return to the details of this gene later, but for now let us have a look at this pedigree (as reported in Fisher et al., 1998; Lai et al., 2001, and reproduced in Figure 4.2).

It is clear that the disease is dominant because a single affected parent is enough to transmit it to his/her children (generation I, female 2; generation II, females 2, 4 and 9 and male 6). If these affected individuals were homozygous for the **deleterious** (i.e., disease-causing) allele, then all their children should have had the disease, while in the KE family the ratios of affected:non-affected children with an affected parent are 4:1 (generation II representing the children of the first-generation couple), 4:5, 3:2, 0:1, 1:2 and 2:2 (for the third-generation children), which do not deviate too much from the expected ratio of 2:2. This can be formally tested using a **chi-square test** (or χ^2 *test*), which, in this case, cannot reject the hypothesis that the distribution of disease in the KE children follows a 50%:50% probability distribution (see Box 4):

affected:non-affected	χ_1^2	p
4:1	1.80	0.18
4:5	0.11	0.74
3:2	0.20	0.65
0:1	1.00	0.32
1:2	0.33	0.57

Moreover, none of the children resulting from the unaffected couple 10-11 in generation II shows the pathology, further supporting the idea that the disease is dominant autosomal. In fact, after extensive research (discussed later in Section 5.4), the gene responsible was found to be located on chromosome 7 and identified as *FOXP2*, with people carrying a single deleterious allele developing the disease, while carrying two such alleles is probably not compatible with life (such people would fail to be born).

Box 4: The chi-square (χ^2) test

The chi-square (χ^2) test is a very general statistical test that works on discrete variables (i.e., with a limited number of clearly different possible values) that can be used to check whether the actual distribution of such a variable departs from a theoretical expectation or that several such variables are independent. Here we will focus on its first use as a so-called *goodness-of-fit* test and we will consider the variables that determine whether an individual is affected or not (two possible values).

Among the first generation's children (generation II) in the KE pedigree (Figure 4.2) four are affected and one is not. Assuming the pathology is dominant and their affected mother is heterozygous for the disease allele while their father is homozygous for the normal allele, how probable is it to see something like four affected and one non-affected children resulting from such a marriage?

Denoting the normal allele as *a* (recessive) and the disease allele as *A* (dominant), the father is *aa* and the mother *Aa*. The Punnett square for such a mating corresponds to panel (c) in Figure 4.1 (with a different interpretation of the alleles and phenotypes) and is

aa (**normal**) × *Aa* (**affected**)	*A*	*a*
a	*Aa* (**affected**)	*aa* (**normal**)
a	*Aa* (**affected**)	*aa* (**normal**)

suggesting a ratio of 2:2 (50%:50%) for affected:non-affected children.

To compare the observed 4:1 with the expected 2:2 we can apply the χ^2 goodness-of-fit test which quantifies the deviation of the observed from the expected distribution. To this end, we compute the test statistic

$$\chi^2 = \sum_{i=0}^{n} \frac{(O_i - E_i)^2}{E_i}$$

where n is the number of cells (alternatives), in our case $n = 2$ (affected vs non-affected), E_i is the expected number of instances of type i (here, affected or non-affected children) and O_i the actually observed number of such instances. If we take type 1 as the affected, $O_1 = 4$, and type 2 as the non-affected, $O_2 = 1$, giving a total of $4 + 1 = 5$ observations. For a 50%:50% distribution of cases, what is the expected number of types for a total of 5 observations? Well, $E_1 = E_2 = \frac{5}{2} = 2.5$, so, if we were to observe many such matings resulting in 5 children each, we would see on average 2.5 children of each type.

Thus, $\chi^2 = \frac{(O_1-E_1)^2}{E_1} + \frac{(O_2-E_2)^2}{E_2} = \frac{(4-2.5)^2}{2.5} + \frac{(1-2.5)^2}{2.5} = \frac{(1.5)^2}{2.5} + \frac{(-1.5)^2}{2.5} = \frac{2.25}{2.5} + \frac{2.25}{2.5} = 0.9+0.9 = 1.8$. To assess whether this departure of the observed from the expected is reasonable given our hypothesized model, we need to compute the *degrees of freedom* of our test, df, which are equal to the number of alternatives minus 1: $df = n - 1 = 2 - 1 = 1$. Then we compare our value, denoted $\chi_1^2 = 1.8$, with the pre-computed distribution of values of such tests. For this, we can use one of the many statistical packages, such as `R` (R Development Core Team, 2010) and their implementation of the chi-square test (in `R` I typed the command `chisq.test( c(4,1), p=c(0.5,0.5) )`). Doing so will give us a so-called *"p value"*, which in our case here is $p = 0.18$.

The correct interpretation of this p value is one of the trickiest parts in statistics (see Box 5); here, it means that there is an 18% chance of observing 4 affected out of 5 children resulting from an *aa* × *Aa* mating under the model we assumed, which is much bigger than the usual cut-off point of 5% (technically formulated something like "the χ^2 test is not significant at the 0.05 alpha level"). Therefore, we cannot rule out that such a family could have resulted from our assumptions (which is to say that there's no reason to suppose that our model is rejected by these data).

For more about statistics the interested reader should consult any of the dozens of good introductory books such as Field et al. (2012), Field (2013), Dalgaard (2008) or Baayen (2008).

Box 5: What statistical tests are and how to interpret them

In "classical" (or *frequentist*) statistics, we usually want to see if some observed data (let's call it d) depart from the predictions of a *null hypothesis* (usually denoted H_0) strongly enough to allow us to *reject* this null hypothesis in favour of the alternative hypothesis (H_1) with a certain degree of confidence.

To make things clearer, let's say that we are interested in testing the hypothesis (H_0) that the probabilities that an individual is affected (denoted ☺) or not (denoted ☹) are equal. This hypothesis thus makes the prediction that if we count the number of affected ($N_☺$) and unaffected ($N_☹$) in a family, they should be roughly equal. Why "roughly"? Well, the real world is really far from being a clean, deterministic universe and even if H_0 were completely true chance events might still happen that distort the observed counts: either affected or non-affected individuals might fail to be counted (they died, left or simply failed to be born), the diagnostic is not perfect (so that some are wrongly called the opposite), etc. Thus, the real question then becomes: having observed $N_{☺}$ affected and $N_{☹}$ non-affected individuals, how far from the predicted equality $N_{☺} = N_{☹}$ can we deviate before saying that H_0 probably doesn't hold in this case (i.e., reject it)?

Let us imagine a family with 10 children, 5 affected and 5 not affected ($N_{☺} = 5$ and $N_{☹} = 5$): in this case intuition strongly suggests that the hypothesis H_0 is true, and indeed applying a chi-squared test (see Box 4 on page 77) comparing the observed 5:5 $N_{☺}$:$N_{☹}$ to the expected $\frac{10}{2} = 5$:$\frac{10}{2} = 5$ gives a p-value of $p = 1.0$. What this says is that we have almost no grounds (a probability of 1.0 is equivalent to 100%) to reject H_0.

What about observing 4 affected and 6 unaffected (or, equivalently from the point of view of this test here, 6 affected and 4 unaffected)? In this case the chi-squared test gives $p = 0.53$ which, even if less that 1.0, still suggests that just random fluctuations could have produced this distribution from equal probabilities (i.e., if H_0 holds), so no reason to reject it. But 3 versus 7? Now $p = 0.21$, suggesting that there's about 21% chance of seeing this result if H_0 were true, no good grounds for rejecting it. Only 2 versus 8 results in $p = 0.058$, suggesting that there's about 6% chance of seeing this if H_0 were true, a probability that most statisticians would consider insufficient to reject the null hypothesis.

However, 1 versus 9 results in a *low* (i.e., *small*) p-value of $p = 0.011$, saying that there's only 1.1% chance (i.e., about 1 in 90) of observing such a result if H_0 were true. It is worth dwelling on this: if H_0 were true (i.e. equal chances of ☺ and ☹) it could still be possible to see a family with 10 children with 1 ☺ and 9 ☹ *just by chance*. Put differently, if you collect many families with 10 children where the chances of ☺ and ☹ are equal, on average, for about each 90 families 1 will be this skewed. Would you consider this probability too small to be "just" chance and to suggest that something else is at work, pushing the probabilities away from equality? If so, you are in good company: most statisticians would take a p-value smaller than 0.05 (i.e., 5%) as enough reason to reject the null hypothesis. A distribution of 0 versus 10 is even more extreme, with $p = 0.0016$ (i.e., 0.15% or about 1 in 666) suggesting a very low probability that it would have been produced if H_0 were true.

This cut-off point (or *alpha-level*, sometimes written as α) of 0.05 (or 5%) is in many senses arbitrary (and, in fact, sometimes an alpha-level of 0.01 or 1% is used), but it still captures the idea that an observation which is very improbable under the null hypothesis H_0 is good ground for rejecting it. Please observe that the null hypothesis can only be rejected and never confirmed and that rejecting it still leaves the possibility that the extraordinary evidence observed might have been generated, under very unusual circumstances, by the poor null hypothesis. As usual, consulting one or more of the recommended statistics books will help the interested reader delve into these fascinating topics bordering deep questions in the philosophy of science. Likewise, this is not the only approach to statistics there is and an interesting alternative is offered by Bayesian statistics (see, for example, Berry, 1996; Press, 2003, for general introductions, Gelman and Shalizi, 2013, for some philosophical implications, and Beaumont and Rannala, 2004, for an early but readable assessment of its impact on genetics).

4.2.2 Recessive hearing loss

As our second example we will turn to Bengkala, a small village in the northern part of the Indonesian island of Bali. A form of **congenital** (i.e., present at birth) deafness has been around there for a very long time (estimated at 10–20 generations ago or 150–300 years; Winata et al., 1995) at high enough frequencies to result in the emergence, maintenance and evolution of a local sign language, **Kata Kolok**. The deaf are well-integrated in the community and most hearing people are bilingual, and there is even a deaf God in the local pantheon (De Vos, 2011, 2012).

The genealogy of the deaf people in the village has been extensively studied and excellent pedigrees going as far back as seven generations have been reconstructed (Liang et al., 1998), clearly showing that the deafness is due to a single gene acting in a recessive autosomal manner. Part of this extensive genealogy – "kindred K1" from Winata et al. (1995) and Friedman et al. (1995) with added information from Liang et al. (1998) – is reproduced in Figure 4.3.

It can be seen that the genealogy (truncated at individual 60 in generation III) covers five generations from the present (the 1990s, bottom, generation V) well into the past. The first indication that this might be a recessive disease is given by the relatively frequent cases of deaf persons born to hearing parents (individuals 379, 41 and 42 in generation II, individuals 60, 115 and 40 in generation III, and individual 39 in generation IV). Let us denote the disease-causing allele as d (for deafness) and the normal variant as D; with these notations, the two hearing parents with deaf children must be both non-symptomatic carriers of allele d (mating $Dd \times Dd$): such a mating would result in hearing:deaf children in the proportion 3:1 (any other type of mating between hearing parents would not fit the bill – see the Punnett squares above). In Kindred K1, the hearing parents with deaf children are as follows:

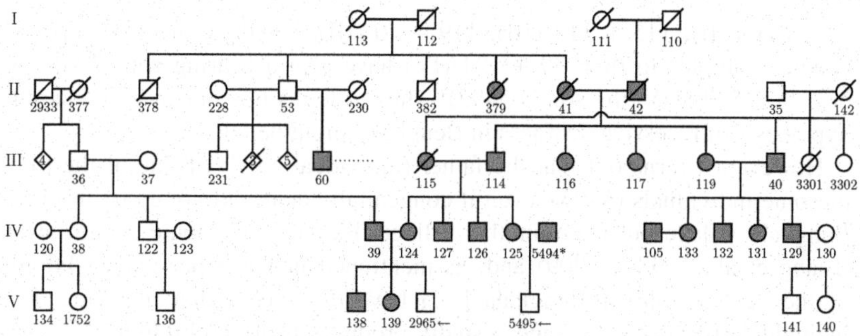

Figure 4.3 Recessive autosomal deafness in Bengkala, Bali: a five-generation pedigree. As usual, ○ are females and □ males; **white** individuals are normal and **black** individuals are congenitally deaf. Individuals crossed by an oblique line / were dead at the time the pedigree was constructed. Diamonds ◇ represent individuals of unspecified sex (and the numbers within specify how many). Individual 60's spouse and children are not represented here. Individual 5494 (marked with *) was not born in the village, while individuals 2965 and 5495 (marked with ⟵) are hearing children born to deaf parents, apparently violating the recessive autosomal model (see text for details). The horizontal distances between siblings and spouses are not important. Please note that individual 115 is a daughter of the couple 35 + 142 and her descent line does not cross the line connecting the parents 41 + 42 to their children. The figure was created by combining and selecting the information in Winata et al. (1995), Friedman et al. (1995) and Liang et al. (1998), and preserves the original individual numeric identifiers.

parents	generation	hearing:deaf	χ^2_1	p
112+113	I	2:3	0.60	0.44
111+110	I	1:0	3.00	0.08
53+230	II	1:0	3.00	0.08
35+142	II	2:2	1.33	0.25
36+37	III	1:2	0.11	0.74

with all families having hearing:deaf children close enough to the expected 3:1 ratio (all χ^2-tests are not significant at the 0.05 level; moreover, given the context in which the genealogy was constructed, there could be a slight bias towards mentioning deaf offspring more than hearing ones, making the p-values smaller than they probably should be).

There is a single mixed, deaf + hearing marriage in our genealogy (129 + 130 in generation IV) and this produced only hearing children, which suggests that the hearing mother (130) was homozygous *DD*. Another hallmark of recessive diseases, as mentioned above, is that children of deaf + deaf marriages (*dd* × *dd*) should all be deaf (*dd*): this is clearly the case for matings 41 + 42 (generation II), 115 + 114 (III), 119 + 40 (III), but there are two apparent counterexamples to this: 39 + 124 (IV) have also a hearing son (2965), as do 125 + 5494 (IV; the son is 5495)! Thus, it looks like our nice and simple hypothesis of recessive deafness in Bengkala might be false!

However, it turns out that the father (individual 5494) of one of the troublesome individuals (5495), even if congenitally deaf, did not come from the Bengkala village but from another village, Banjar Jawa, some 22 km away (Liang et al., 1998, p. 909) and, as mentioned in the paper cited, "[s]ome of the deaf people of Bengkala [...] have discovered that marrying deaf individuals from other villages usually produces hearing children" (p. 909). Far from being an issue for our recessive story, this revelation does, in fact, support it: there are many ways in which congenital deafness can come about and some have genetic causes that disrupt the developmental process in many different ways (for some examples see Sections 6.1, 6.2 and 6.3).[2] To make things bluntly clear, one can hypothetically be congenitally deaf for failing to develop an ear drum, or by having non-functional hair cells (the cells that transform sound wave energy into nerve impulse) or by not developing the right neural connections or the primary hearing areas of the brain, and so on; plus, there can be many genetic reasons for having any one of these failures. The essence, thus, is that two individuals having the same phenotype might not necessarily do so because of the same (genetic) cause.

The deaf mother of 5495 (125) is homozygous for our *d* allele (being, thus, *dd*) as she comes from kindred K1, but her husband (5494) is deaf for other reasons, so that he is most probably homozygous for the *D* allele at this locus (being, thus, *DD* at this locus). Let us suppose that his deafness is caused by a non-genetic factor (such as a prenatal infection with rubella, cytomegalovirus or perinatal anoxia; Fisch, 1969; Fowler and Boppana, 2006; Barbi et al., 2003): then, his son (5495) would not be able to inherit his mother's deafness (as he would be heterozygous at the locus affecting her, *dD*) nor his father's deafness (as this is non-genetic), resulting in a hearing individual.

Or, alternatively it could be that the father's (5494) deafness is indeed genetic but caused by another locus segregating independently from the first one (to keep things simple) with alleles *H* (dominant) and *h* (recessive). Now, things get a bit more complex, depending on how the deafness due this new locus arises: is it determined by the dominant allele *H* (and thus transmitted in a dominant fashion) or is it determined by the recessive allele *h* (and thus transmitted in a recessive manner)? In the dominant case (deafness due to *H*),

[2] This is called *complementation* and points to recessive mutations in different genes as opposed to in the same gene.

the father can be either *HH* or *Hh* at this locus and *DD* at the other, while the mother is presumably *hh* and *dd*; we can write their full genotype at both loci as *hhdd* (mother) and *HHDD* or *HhDD* (father). Thus, their mating can result in all children being *hHdD* (deaf due to father's gene) in the first case or 50%:50% *hHdH* (deaf due to father's gene):*hhdD* (hearing) in the second – clearly the first possibility is ruled out by the son 5495 not being deaf. If the father's deafness is caused by the recessive allele *h*, then he must necessarily be *hhDD*, resulting in all hearing heterozygous children *hHdD*. We will return later to the complexities of interactions between loci and their importance for a proper understanding of genetics.

But what about individual 2965 (in generation V), the hearing son of two deaf parents from Bengkala (39 and 124), thus probably homozygous *dd*? Clearly, he should have also been homozygous *dd* and thus deaf, so what is going on here? Well, some of us might want to believe that humans are strictly monogamous animals, but common sense, TV soap operas and genetic research (e.g., Bellis et al., 2005) suggest otherwise. Thus it could well be that 2965 is not in fact fathered by 39 but by another, hearing man (we do not know if this indeed the case here, but Winata et al., 1995, mention another case from a different kindred where genetic paternity testing of a hearing child of two Bengkala deaf parents revealed a case of extramarital conception).

Further investigations (which we will detail later in Section 6.3) pinpointed the gene causing recessive deafness in Bengkala to chromosome 17, namely the *MYO15A* gene.

4.3 Sex-linked dominance and recessiveness

A special case is represented by the genes on the sex chromosomes. Those genes found only on the Y chromosome, such as the sex-determining region *SRY*, are either completely absent (as in normal females) or in a single copy (as in normal males). Thus, a trait due to such a gene (for example, developing testis) will be transmitted from father to sons, faithfully following the paternal line.

The sex chromosomes X and Y are very different, with X being bigger and much more gene-rich than Y. However, the regions at the tips of the sex chromosomes, called the *pseudoautosomal regions* (*PAR*), are similar and carry the same genes. The genes in these two regions (about 24 in *PAR1* and 4 in *PAR2*), thus, are present in two copies in both females (one on each X chromosome) and males (one on the X and one on the Y) and behave like autosomal genes, showing the "normal" patterns of dominance and recessiveness (Mangs and Morris, 2007).

Finally, there are genes unique to the X chromosome not found on the Y and these show very interesting sex-linked patterns of inheritance.

4.3.1 X chromosome inactivation

Human females have two copies of the more gene-rich X chromosome, while the males have only one. Thus we might naively expect that females express a double dose of the X-linked genes compared to the males. However, this does not happen and most genes on one of the two X chromosomes in females are *inactivated*, ensuring **dosage compensation** (Snustad and Simmons, 2010, pp. 108–110). Which of the two X chromosomes is actually inactivated is determined mostly at random, but this process can also be biased, preferentially inactivating the X chromosome inherited from one of the parents (Wong et al., 2011; Loat et al., 2004).

Interestingly, X-inactivation happens early during development (Loat et al., 2004; Jørgensen et al., 1992) and apparently independently in each cell. Thus, in some cells at this stage the maternal X chromosome will be inactivated while the others will inactivate the paternal X, and each such cell will go on to produce numerous descendant cells which will retain the founder cell's X inactivation decision. These descendant cells will end up in various tissues and organs of the resulting female fetus, child and finally, adult, so that human females are, in fact, mosaics from the point of view of the expressed X chromosome. This mosaicism means that for those X-linked genes where the paternal and maternal alleles differ in their effects, the daughter's body will show a patchy expression of both alleles.

A well-known example of X-chromosome inactivation mosaicism is represented by fur colour in cats, which is determined by a gene on the X chromosome. Thus, female cats carrying the black and orange alleles on their Xs will have a patchy black and orange coat (the tortoiseshell pattern; Snustad and Simmons, 2010, p. 108). Another important consequence of X chromosome mosaicism is that **monozygotic** (or identical) female twins do sometimes differ in the expression of traits influenced by genes on the X chromosome, potentially including behavioral (Loat et al., 2004) and language-related (Stromswold, 2006) phenotypes.

4.3.2 Anomalous colour perception

A very nice example of an X-linked trait is represented by certain types of *anomalous colour perception* (Deeb and Kohl, 2003; Deeb, 2006). The human eye is an exquisite structure adapted for focusing light on the **retina**, the actual light-sensitive layer of tissue found at the back of the eye and capable of transforming incoming photons into nervous impulses sent to the brain for further processing. This capacity is due to specialized cells (the **photoreceptors**), **rods**, which are very sensitive to light, ensuring vision in low-light conditions, and **cones**. The cones are normally of three types (Deeb, 2006): those sensitive to light of short wavelength[3] (≈ 420 nm corresponding to "blue", the so-called *S-cells*), others sensitive to medium wavelength light (≈ 530 nm corresponding to "green", the *M-cells*) and others specialized in detecting long wavelength light (≈ 560 nm corresponding to "red", the *L-cells*). We see colours because these three types of cones are maximally sensitive to light of different wavelengths and, from the integration of the different signals generated by them, the brain can infer the colour of objects in the visual field. Interestingly, the distribution and density of these types of cones in the retina seem to vary a lot across individuals and even to be randomly patterned within a given retina (Deeb, 2006).

The *photopigments* (the molecule that actually reacts to light) are very similar between rods and the three types of cones and consist of the protein *opsin*, bound to *retinal* (derived from vitamin A). The opsin used in rods is called *rhodopsin* and the opsins in cones are collectively known as *iodopsins* and are of three types (S, M and L) absorbing light with short, medium and long wavelengths, respectively. The S opsin is encoded by a gene on chromosome 7 (named *OPN1SW*) while the genes encoding the M and L opsins (*OPN1MW* and *OPN1LW*, respectively) sit on the non-pseudoautosomal region of the X chromosome.

Thus, leaving many of the fascinating details aside, the "green" (M) and "red" (L) photoreceptors in humans are encoded by two genes found on the X chromosome, *OPN1MW* and *OPN1LW*. So, what happens if an individual carries an anomalous allele of one (or both) of these genes? Let us denote the normal allele of *OPN1MW* as *G* and consider, for simplicity, that there is a deleterious allele *g* which results in a completely non-functional "green" pigment (likewise, the alleles of the *OPN1LW* genes are *R* and *r*, respectively).

A male having the *g* allele (please note that this discussion focuses on the "green" gene, but the same applies to the "red" gene as well) on his single X chromosome will completely fail to produce functional "green" photopigments and will, therefore, be unable to perceive the normal colour distinctions; this severe colour deficit is called *deuteranopia* and affects about 1% of all males. Likewise, a homozygous female having two *g* alleles, one on each of her two X chromosomes, will also be a deuteranope. However, a heterozygous female carrying a *g* on one of her X chromosomes and a *G* on the other is a much more interesting case. The vast majority of such females are phenotypically normal,

[3] From a physical point of view, light is an electromagnetic wave characterized by a certain wavelength on the scale of nanometers (symbol *nm*, 1 nm = 10^{-9} m). Different colours correspond to different such wavelengths, going from invisible ultraviolet (UV) to violet, blue, green, yellow, red and to invisible infrared (IR) radiation.

showing trichromatic vision; coupled with the previous discussion of autosomal dominance/recessiveness, this suggests that the functional allele, G, is dominant over the non-functional one, g, as g fails to produce any "green" photopigments but G manages to compensate, so that there are enough functional "green" cones in the retina. However, due to X-chromosome inactivation, would it be possible that X chromosome mosaicism affects colour vision in heterozygous Gg females as well? Apparently, we know of at least one case involving two monozygotic female twins, one of whom is trichromatic normal and the other has abnormal colour vision (Jørgensen et al., 1992), explained by a very skewed X-inactivation of their paternal chromosome carrying the defective allele (their father is also affected).

Thus, the possible matings and resulting offspring are represented by the Punnett squares below, where the possible phenotypes are **T** (trichromatic, normal vision) and **D** (dichromatic, deuteranopic vision). The males are **hemizygous**, having only one copy of the gene (as they have a single copy of the X chromosome and their Y chromosome does not have a locus for the gene), the missing locus being represented by a dash –. Carrier females Gg are mostly normal.

The mating between trichromats results in all normal offspring:

GG (♀, **T**) × $G-$ (♂, **T**)		sperm	
		G	–
ova	G	GG (♀, **T**)	$G-$ (♂, **T**)
	G	GG (♀, **T**)	$G-$ (♂, **T**)

just like the mating between a trichromat homozygous female and a dichromat male:

GG (♀, **T**) × $g-$ (♂, **D**)		sperm	
		g	–
ova	G	Gg (♀, **T**)	$G-$ (♂, **T**)
	G	Gg (♀, **T**)	$G-$ (♂, **T**)

However, the mating between an apparently normal trichromat carrier female Gg and a trichromat male $G-$ will result in half the male children being dichromat:

Gg (♀, **T**) × $G-$ (♂, **T**)		sperm	
		G	–
ova	G	GG (♀, **T**)	$G-$ (♂, **T**)
	g	gG (♀, **T**)	$g-$ (♂, **D**)

while the mating with a dichromat male $g-$ results in half the male and half the female offspring being dichromat:

Gg (♀, **T**) × $g-$ (♂, **D**)		sperm	
		g	$-$
ova	G	Gg (♀, **T**)	$G-$ (♂, **T**)
	g	gg (♀, **D**)	$g-$ (♂, **D**)

Finally, a female dichromat gg gives birth to all male dichromat offspring and either no female dichromats (when the father is trichromat $G-$) or all female dichromats (when the father is dichromat himself, $g-$):

gg (♀, **D**) × $G-$ (♂, **T**)		sperm	
		G	$-$
ova	g	gG (♀, **T**)	$g-$ (♂, **D**)
	g	gG (♀, **T**)	$g-$ (♂, **D**)

gg (♀, **D**) × $g-$ (♂, **D**)		sperm	
		g	$-$
ova	g	gg (♀, **D**)	$g-$ (♂, **D**)
	g	gg (♀, **D**)	$g-$ (♂, **D**)

Thus, this type of sex-linked recessive phenotype affects preponderantly the males and the affected males do not transmit it to their sons but only to their daughters (either "hidden" or manifest, depending on the mother's genotype). However, the reality is a bit more complex and much more interesting (as always): not all carrier heterozygous females have normal trichromatic vision and this is due to X chromosome inactivation.

The discussion above was simplified by considering only the completely non-functional alleles (g and, to a lesser extent, r) but, in reality, there is a wide range of partially deleterious alleles which do produce functional photopigments, but these pigments are sensitive to slightly different wavelengths than the normal ones. The severity of the deficit is related to the difference between the peak sensitivity of the "green" and "red" pigments and the milder forms of *deuteranopia* ("no green") are called *deuteranomaly*, while the milder forms of *protanopia* ("no red") are called *protanomaly* (Deeb, 2005). Let us denote such a deuteranomaly allele as g' (a real-world example of such an allele for the "red" pigment has been described (Deeb, 2006) and involves a change of a single amino acid in the pigment's opsin which alters its peak sensitivity by ≈ 4–7 nm); its inheritance pattern is identical to the g allele described above, but the affected individuals would have a much less pronounced phenotype.

However, this raises an intriguing possibility: a female heterozygous for the normal G and the deuteranomalous g' alleles might, due to X-inactivation, express four types of photopigments (normal "blue", "green" and "red", plus the anomalous – but fully functional – "green′"), making her potentially *tetrachromatic*, capable of seeing colours inaccessible to normal trichromats! A precedent is represented by the New World monkeys, which are normally dichromat but some heterozygous females become effectively trichromat through this X-inactivation mechanism (Deeb, 2004). However, a recent study by Jordan et al. (2010) found that the majority of such heterozygous potentially tetrachromat females are in fact normal trichromats, with a single female participant (out of 24) showing convincing perceptual tetrachromacy. This suggests that tetrachromacy is very rare in the human population, probably due to an incapacity of our brains to make use of this fourth chromatic dimension, while the monkey brain seems perfectly capable of using the fortuitous third dimension available to some heterozygous females.

4.3.3 Variation in colour perception and language

Why would the genetic variation in colour perception be relevant for students of language and speech?[4] The relationship is subtle and related to the process of *biased cultural transmission* (more details in Section 9.4), and concerns the way in which different languages choose to categorize and name the various parts of the colour spectrum.

The universality (or not) of colour naming has been a hotly debated topic for many decades, made very salient by Berlin and Kay's 1969 book. What seems clear from the data collected so far (such as the *World Color Survey* available online at `http://www.icsi.berkeley.edu/wcs/data.html`) is that there is an impressive amount of variation in the way languages classify colours, but that there also seem to be some universal principles at work. Recently, an elegant solution has been proposed (Regier et al., 2007) suggesting that due to the asymmetries inherent in the way humans perceive colours, there are universal constraints resulting from perception, "pulling" languages towards optimal solutions, but this "pull" is not all-powerful and allows languages a certain freedom in their choices (Regier and Kay, 2009).

However, the vast majority of these studies assume (as is usual in psychology and linguistics) that all humans perceive colours fundamentally in the same manner (the same way that linguists usually assume that the core components of the language faculty are the same across the whole species, barring clearly cut "pathologies"). But we just saw that there is a whole range of variation in colour perception, ranging from "normal" trichromatic vision to rare dichromats, even complete colour blindness, and even rarer tetrachromats, with

[4] I thank Asifa Majid for bringing this possibility to my attention and for extensive discussions and collaboration on this topic.

many shades in between (I myself have slightly anomalous trichromatic vision, a feature I share with my brother and which seems to confer on us quite good low-light vision in exchange).

Recent work by Jameson and Komarova (2009a; 2009b) using computer models seems to suggest that dichromats do play a crucial role in allowing mixed populations (normal trichromats and dichromats) to arrive at recurrent colour categories (universal tendencies). Thus, it seems probable that the type and population frequency of anomalous colour vision play a role in both the existence of universal tendencies and the specific patterns of cross-linguistic variation in colour categorization. More empirical work is required to test this hypothesis by comparing the distribution of variation in colour perception among speakers with the colour categories used by the language(s) they speak. Of course, this type of biasing would not be the only factor affecting language, and probably not the most powerful either, and other factors, such as the existence of a material culture where colours are very salient (Casson, 1994), play the major role.

5 Linkage disequilibrium and its role in finding genes

Linkage disequilibrium (which we have encountered earlier in the book) strikes back as it is the basis of very powerful methods for finding genes involved in speech and language. We describe in more detail what linkage disequilibrium is and how it is used. We begin by looking at association studies, which use unrelated participants to find genetic loci correlated with variation in the phenotype, as well as a variant that uses familial groups (for example, the parents and an affected child); we will survey some recent examples of using association studies to discover genes relevant for language and speech. We then discuss linkage studies that use large families to pinpoint genes, focusing on the example of FOXP2. In this chapter we will encounter more statistics as we need to deal with many complications such as what happens when we test hundreds of thousands of genetic markers, and on how to interpret the results of association and linkage studies.

The concept of **linkage disequilibrium** (or the tendency for some loci to be transmitted non-independently) is extremely important in our hunt for the genes influencing phenotypes. We will discuss here the two main methods that use it, *association* and *linkage studies*, introducing also some statistical concepts fundamental to properly conducting such studies and interpreting their results.

5.1 What is linkage disequilibrium?

As we saw in Section 4, the sweet pea flower colour is controlled by a locus with alleles *A* (dominant, purple flowers) and *a* (recessive, red flowers). Another locus studied by Bateson, Saunders and Punnett in sweet pea controls the shape of pollen grains and has two alleles: *B* (dominant, long grains) and *b* (recessive, round grains). By crossing *pure lines* of homozygous plants with purple flowers and long grains, *AABB* plants, with homozygous plants with red flowers and round grains, *aabb*, they obtained, in the first generation (usually denoted F_1), only heterozygous plants with purple flowers and long grains *AaBb*, as expected given the dominance relationships at these two loci.

But how would one make sure, in practice, that indeed the parental plants are truly homozygous at the desired loci? One common technique is to produce **inbred lines** by repeatedly mating related individuals (brothers and sisters in the case of animals and even the same individual in the case of plants where *selfing* – pollination using the plant's own pollen – is possible). A very important example of inbred lines (or strains) is represented by the over 450 (as of 2000) inbred strains of mouse, some of which are an essential tool in biological and biomedical research (Beck et al., 2000).[1] It is worth quoting part of the guidelines for producing such strains proposed in 1952 (and revised several times since; there is even an International Committee on Standardised Nomenclature for Mice whose current members can be viewed on the Mouse Genome Informatics website: `http://www.informatics.jax.org/mgihome/nomen/inc.shtml`.): "A strain shall be regarded as inbred when it has been mated brother×sister [...] for twenty or more consecutive generations (F20), and can be traced to a single ancestral breeding pair in the 20th or a subsequent generation [...]" (Beck et al., 2000, p. 23), at which point more than 98% of the loci are homozygous.

However, for Bateson, Saunders and Punnett, the surprise came when they further selfed F_1 plants to obtain the second generation, F_2. Based on Mendel's Second Law of Independent Assortment, one would expect a ratio of phenotypes 9:3:3:1 as shown by the Punnett square below (black and grey mark the two parents; **p** means pink flower, **r** stands for red flower, **l** for long grains and **s** for round/short grains):

$AaBb$[2] × $AaBb$	AB	Ab	aB	ab
AB	$AABB$ (**pl**)	$AABb$ (**pl**)	$AaBB$ (**pl**)	$AaBb$ (**pl**)
Ab	$AAbB$ (**pl**)	$AAbb$ (**ps**)	$AabB$ (**pl**)	$Aabb$ (**ps**)
aB	$aABB$ (**pl**)	$aABb$ (**pl**)	$aaBB$ (**rl**)	$aaBb$ (**rl**)
ab	$aAbB$ (**pl**)	$aAbb$ (**ps**)	$aabB$ (**rl**)	$aabb$ (**rs**)

1 Besides these inbred strains there are also **transgenic** (or genetically modified) mice which have had their genome altered using genetic engineering; there are currently many more such lines (maybe more than 10,000) as they are very important tools for biomedical research as for example models for various diseases (see for example `http://www.findmice.org/` and `https://www.komp.org/index.php`).

2 Please note that for a single locus the notation *Aa* refers to the full genotype at this locus (here heterozygous with allele *A* on one chromosome and *a* on the other), while for two (or more) loci physically on the same DNA molecule (chromosome), the notation *AaBbCc*... refers to the genotype at all loci simultaneously but without specifying which alleles are on which of the two homologous chromosomes, one inherited from the mother (maternal) and one from the father (paternal). Thus, we know that allele *A* is on one chromosome at the first locus (say,the maternal one) and allele *a* on the other chromosome (paternal), and allele *B* on one

Thus, there should be 9 **pl**, 3 **ps**, 3 **rl** and 1 **rs**, but when counting their actual F_2 phenotypes (Lobo and Shaw, 2008) they observed 1528 **pl**, 106 **ps**, 117 **rl** and 381 **rs** or a ratio 1528:106:117:381 = 14.42:1:1.10:3.59, which seems very far from the expected 9:3:3:1! Indeed, a χ^2 test (Box 4) is very highly significant, $\chi_3^2 = 966.61$ with the p-value $p < 2.2 \cdot 10^{-16}$ much smaller than the standard 0.05 significance level.

What is the explanation for the fact that they found many more **pl** and **rs** plants than expected and fewer **ps** and **rl**? Clearly, something links together the two loci controlling the flower colour and pollen grain shape, hindering their independent assortment and producing many more grandparental phenotypes (**pl** and **rs**) and fewer *recombinant* phenotypes (**ps** and **rl**) than expected.

This is due to the linear structure of the chromosomes (Section 3.6): in Figure 3.8 locus A could be taken as controlling the flower colour and locus B as controlling the pollen grain shape. When there is no crossover, the current individual's parental combination of alleles at these loci (or **haplotype**) will be faithfully transmitted to the individual's children as such, violating the law of independent assortment. To use our example, a plant from the first generation F_1 has the genotype *AaBb* (with haplotypes *AB* and *ab* at these loci, one on each of the two homologous chromosomes, where black and grey are used to distinguish the two parents). One extreme possibility is that these two loci are so tightly linked as to never recombine (there will never be any crossover occurring between them): in this case, they are effectively transmitted as a unit, as a single locus, and all offspring will inherit the unchanged parental haplotypes (here, *AB* and *ab*), with no offspring showing the recombinant haplotypes *Ab* and *aB*. In this case, knowing an individual's alleles at one locus will offer perfect information about the alleles at the other.

Another extreme is when the two loci are so far apart on the chromosome that when a sex cell is produced, recombination can take place or not between these loci with equal probability. In this case, Mendel's independent assortment law holds and the two loci are perfectly independent: knowing the alleles at one locus does not provide any bit of information about the alleles at the other.

In between these two extremes, perfect correlation (or no recombination) and perfect independence (or 50% chance of recombination), there is a continuum of recombination probability. Two loci which have a 1% chance of recombination in a single generation are said to be separated by one **centimorgan**, 1 cM: thus, independent loci are separated by 50 or more centimorgans, and perfectly linked loci by 0 cM.

chromosome for the second locus and allele b on the other, but not necessarily that alleles A and B are on the same chromosome and a and b on the other (i.e., the two-locus genotype $AaBb$ can correspond to $\frac{AB}{ab}$ or $\frac{Ab}{aB}$) where first row denotes the maternal and the second the paternal chromosomes.

By studying the probability of recombination between pairs of loci it is possible to build *genetic maps* representing the linear relationships between loci on the same chromosome. Moreover, by using physical landmarks on the chromosome (such as bands visible under a microscope), one can place loci in actual physical space, ultimately pinpointing the precise location of a gene on the chromosome in terms of the number of nucleotides (or basepairs, bp, and multiples such as thousand basepairs, kb, and millions of basepairs, Mb) from a conventional landmark (such as one end of the chromosome; for details see, for example, Snustad and Simmons, 2010, pp. 146–156).

From this discussion it should be clear that there is a close relationship between linkage (probability of recombination) and physical distance between two loci on the same chromosome. However, this relationship is far from simple and linear: on average, 1 cM corresponds to about 1 Mb (one million basepairs) in humans, but there is huge variation in the actual relationship along the chromosomes. There are "hot" and "cold" spots which represent regions on the chromosomes where recombination happens orders of magnitude more (or less) often than the average (Hey, 2004). Interestingly, the location of these hotspots is partly under genetic control and a recently identified gene, *PRDM9*, seems to be involved in the sense that the protein it produces binds to patterns of DNA nucleotides and marks that location as a hotspot (Berg et al., 2010; Parvanov et al., 2010; Baudat et al., 2010). Even more interesting, mutations in this gene change the patterns of DNA nucleotides to which it binds and thus the locations it marks as hotspots, and it has been proposed that these changes play a role in **speciation** (the process whereby a species splits into two) by making the incipient species' chromosomes incompatible (Mihola et al., 2009; Baudat et al., 2013; see also the cover story in *New Scientist*, 12 February 2011, pp. 33–25).

This non-independence of transmitted loci, the **linkage disequilibrium** (or *LD*), is extremely important not only for locating and identifying genes but also for understanding and studying evolutionary phenomena such as natural selection.

5.2 Using linkage disequilibrium

The linear structure of the chromosomes and the ensuing non-independence of the linked loci (Sections 3.5, 3.6 and 5.1) are great tools in our hunt for genes using a very simple idea. If we could somehow find some conspicuous "landmarks" on the genome that co-vary with the phenotype (or trait) of interest, then we should be able to approximate the position of the actual gene(s) involved with a precision related to the properties of these landmarks. Please note that finding a correlation between such a landmark (or **marker**) and our phenotype of interest by no means entails that the marker itself has a causal

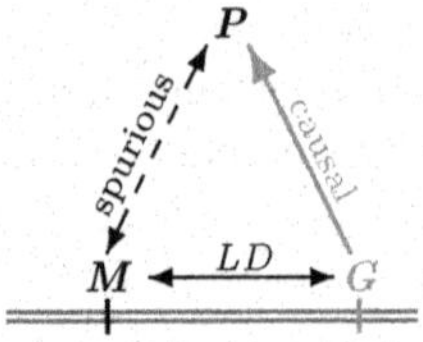

Figure 5.1 The observable relationship (←--→) between the phenotype ***P*** and the marker ***M*** is indirect (spurious), being the result of (i) the causal relationship flowing from the hidden genetic variant ***G*** towards ***P*** (→) and (ii) the linkage disequilibrium (LD) between ***G*** and ***M*** (⟷) due to their physical closeness on the same chromosome (the long parallel lines).

connection to the phenotype. Instead, it is usually just a visible signpost signalling that somewhere in its vicinity one of the true causal factors is hiding. More precisely, the observable correlation between the marker (let's call it ***M***) and the phenotype (***P***) is *indirect* (or *spurious*), being due to this hidden third factor (***G***) which is causally related to ***P*** on one hand, and in LD with ***M***, on the other (Figure 5.1).

Thus, if we could find a dense set of such markers, $\boldsymbol{M}_1, \boldsymbol{M}_2, \ldots$, covering the whole genome, then we should be able to look for correlations between these markers and our phenotype of interest, ***P***. Luckily, there are several such sets available, ranging from the ones that are truly observable under the microscope (such as the light and dark bands appearing after staining the chromosomes with specific dyes; see Figure 3.5) but having a very low coverage, to high-density *SNPs* (single nucleotide polymorphisms) that can be cheaply and quickly identified across the genome.

There are two basic ways that we can use these ideas to hunt for genes: one is to look for correlations between markers and phenotypes in large families (**linkage studies**) and the other is to look for correlations between markers and phenotypes in very large samples of unrelated individuals (**association studies**). We will now turn to each of these two types using examples relevant to speech and language and, given that association studies are conceptually simpler, we will begin by looking at them.

5.3 Association studies

The basic idea behind an association study is to discover a correlation (association) between the genotype and the phenotype by comparing how the two co-vary across a large sample of individuals (for good reviews see for example Hirschhorn and Daly, 2005, or Lewis and Knight, 2012). Sometimes the phenotype ***P*** is *dichotomous* (or can be treated as such by dichotomization) as is the case, for example, in some diseases where people can be classified as either affected (also called **cases**) or not (normal, non-affected or **controls**). In our earlier examples (Section 4.2), a Bengkala villager can be either deaf (a case) or with normal hearing (a control), while a member of the KE family can either present the symptoms of DVD (Developmental Verbal Dyspraxia; a case) or not (a control). For such dichotomous phenotypes, we can conduct an association study by comparing the genotypes of the cases to those of the controls in a *case-control genetic study*.

More precisely, we would compare the alleles at a set of genetic markers $\boldsymbol{M}_1, \boldsymbol{M}_2, \ldots$ in the case and the control groups, hoping that some of them would show different distributions across these two groups. Let us consider for now only biallelic markers (such as SNPs), such that each marker $\boldsymbol{M}_i$ will have only two possible alleles, let us call them A_i (the most frequent allele, also called the major allele) and a_i (the rarer, or **minor**, allele; please note that in this case the capital and small letter notation does not imply anything related to dominance/recessiveness). Thus, for each of the autosomal markers each participant will have a genotype composed of the two alleles (one on each homologous chromosome carrying the marker), which can be $A_i A_i$, $A_i a_i$ or $a_i a_i$, while the genotypes for the sex chromosome markers will depend on the participant's gender. Thus, the control group is composed of n non-affected participants (controls) while the case group is composed of k affected participants (cases), each participant with their own genotype for each of the markers $\boldsymbol{M}_i$. By simply counting the number of participants with a given genotype for each marker in each group, we can compare the frequencies of these genotypes across groups and test these group differences for statistical significance.

Table 5.1 shows a hypothetical distribution of genotype and allele frequencies at one genetic marker across the two groups. It can be seen that the major allele A is slightly more frequent in the controls, while the minor allele a is more frequent among the cases, and we can conduct a chi-squared test[3] and check the statistical significance of these differences (top part of Table 5.1): $\chi_1^2 = 36.78$, $p = 1.32 \cdot 10^{-9}$, with a p-value much smaller than the standard threshold of 0.05 (and even than the more stringent one of 0.01). Thus, it seems that this marker is associated with the phenotype of interest (more specifically, having major allele A is associated with being affected). Likewise, we can test the association between the genotype and the phenotype (bottom part of Table 5.1), $\chi_2^2 = 37.58$, $p = 6.92 \cdot 10^{-9}$, suggesting

[3] The chi-square test (χ^2) has several limitations, such as the requirement that each cell in the contingency table has at least five observations. When these assumptions are violated it is better to use other tests such as *Fisher's exact test* (implemented for example in `R` by the function `fisher.test()`). I will keep this example simple by applying the χ^2 test, but Fisher's gives very similar results here.

Table 5.1 *The distribution (counts) of each allele (top) and genotype (bottom) for the case and control groups, with their totals. Hypothetical data from Li et al. (2008).*

Allele	Cases	Controls	Total
A	1790 (= $2 \times AA + Aa$)	1900	*3690*
a	210 (= $2 \times aa + Aa$)	106	*316*
Total	*2000* (= 2×1000)	*2006* (= 2×1003)	$\chi_1^2 = 36.78$, $p = 1.32 \cdot 10^{-9}$
Genotype	Cases	Controls	Total
AA	800	900	*1700*
Aa	190	100	*290*
aa	10	3	*13*
Total	*1000*	*1003*	$\chi_2^2 = 37.58$, $p = 6.92 \cdot 10^{-9}$

Table 5.2 *The dominance model: genotypes AA and Aa have similar effects (and are thus counted together) but different from aa (counted separately).*

Genotype groups	Cases	Controls	Total
AA and Aa	990 (= $AA + Aa$)	1000	*1990*
aa	10	3	*13*
Total	*1000*	*1003*	$\chi_1^2 = 2.81$, $p = 0.094$

that indeed different genotypes are distributed differentially among the two groups.

However, this test cannot give more information about the type of mechanism involved: maybe A is dominant (or recessive) in relation to a, or maybe they act additively (with the heterozygote intermediate between the two homozygotes), or there might be a more complex model of how the two alleles influence the resulting phenotype. We can test the dominance/recessive model by deriving the appropriate contingency table from the genotype counts. A being dominant versus a would mean that the effects of the genotypes AA and Aa are similar to each other and different from the effects of aa (Table 5.2), meaning that the appropriate approach is to test whether the pooled genotypes AA and Aa versus genotype aa are differentially distributed across the two groups: $\chi_1^2 = 2.81$, $p = 0.094$. Similarily, to test the recessive model we pool together Aa and aa versus AA: $\chi_1^2 = 36.18$, $p = 1.80 \cdot 10^{-9}$. The additive model requires a slightly different statistical test (the *Cochran-Armitage trend test*[4]) which performs a chi-square test while taking into account the order of the genotypes; for our data, $\chi_1^2 = 37.36$, $p = 9.82 \cdot 10^{-10}$.

In conclusion, we have quite strong support for the association of this marker with the phenotype of interest, with its major allele A over-represented among the cases, with it being recessive over (or acting additively with) a.

5.3.1 *Statistical concepts: power, multiple testing correction and effect size*

Before we continue, it is necessary to clarify some essential statistical concepts that will greatly help understanding not only the main issues in conducting association studies but also the proper interpretation of their results. In fact, statistical power and effect size are fundamental concepts that pop up in many other fields such as psychology, psycho-linguistics, sociology, medical science and economics, while multiple testing has tended to be relatively neglected but more and more researchers realize that it is essential to tackle the undesirable effects of not accounting for it.

Outcomes of tests of statistical hypotheses

Let us focus again on the case of an association between a single marker $\boldsymbol{M}$ and a trait of interest P; here the **null hypothesis** H_0 that we want to test is that $\boldsymbol{M}$ is not associated with P, and we collect some data d (for example doing a genetic association study) that should help us make a decision. We now have the following four possibilities shown in Table 5.3.

It might be the case that really the null hypothesis H_0 is true, but our test using the data d incorrectly rejects it, producing what is called a *type I (one) error* or a **false positive**; this is what the alpha-level α (0.05 or 0.01)

Table 5.3 *Possible outcomes of a statistical test and their probabilities.*

	The null hypothesis (H_0) is:	
Decision	True	False
Reject H_0	Type I error *False positive* Probability α	Correct decision *True positive* (*Power*) Probability $1 - \beta$
Fail to reject H_0	Correct decision *True negative* Probability $1 - \alpha$	Type II error *False negative* Probability β

[4] Implemented in R by `prop.trend.test()`.

is supposed to guard against and which, in our case, would amount to concluding that $\boldsymbol{M}$ is associated with P even if it really isn't (unfortunately it turns out that many of the early reported such associations are in fact false positives!). Another outcome could be that the test and data d fail to reject H_0, producing the correct decision to conclude that $\boldsymbol{M}$ and P are not related; this *true negative* happens with probability $1 - \alpha$ and we should seek to maximize it. Now, if out there the null hypothesis H_0 is indeed false and our test on the data d manages to reject it, then we make a correct decision and report that $\boldsymbol{M}$ is associated with P (potentially resulting in a nice paper that also happens to be correct); the probability of these *true positives* is given by $1 - \beta$ (β will be discussed below) and represents the **statistical power**, something that we would also like to maximize. Finally, our test on the data d fails to reject the false H_0, resulting in a *type II (two) error* or a **false negative** whereby we miss the association between $\boldsymbol{M}$ and P with probability β. Ideally, we would like to have as many correct decisions as possible, namely a high $1 - \alpha$ and $1 - \beta$, but there are trade-offs between them that mean we have to make some tough choices: do we prefer to miss some true positives (low power, small $1 - \beta$) or would we rather miss some true negatives (small $1 - \alpha$)? The first is preferable if the cost of a false negative (or type II error) is very high such as when failing to diagnose a fatal but treatable disease or to find a technical failure that can crash a plane; the second when the cost of a false positive (type I error) is higher, such as when falsely diagnosing a healthy person results in unnecessary interventions with life-altering consequences or when fear of minor technical failures disrupts busy flights.

Why is it necessary to worry about multiple tests?

When testing simultaneously multiple markers $\boldsymbol{M}_1, \boldsymbol{M}_2, \ldots \boldsymbol{M}_N$, the simplest method is to perform these association tests for each marker $\boldsymbol{M}_i$ individually. However, modern **Genome-Wide Association Studies** (GWAS) use very large sets of SNPs (single nucleotide polymorphisms) numbering in the large hundreds of thousands or low millions and covering the whole genome. These SNPs are cheap and easy to genotype (making them observable) using a variety of methods, probably the best known being the standardized and comercially available *DNA chip* or *microarray* platforms such as those offered by Affymetrix and Illumina (for a gentle but somewhat old overview see, for example, Perkel, 2008). This means that we would have to perform hundreds of thousands of separate tests of association, one for each SNP, but the cost we have to pay for such a high density coverage of the genome is that we astronomically increase the chance of finding **false positives**. A false positive in this particular context means that we would reject the hypothesis of no association between a certain SNP and our phenotype when in fact there is *no* association between them. But how is this possible? Don't statistical testing and the requirement that a p-value is smaller than 0.05 (or 0.01) guard us against such mistakes?

Well, not really: in our context, the *alpha-level* (or α-level) of 0.05 (or 0.01) really means that there is a 5 in 100 (or 1 in 100) chance of obtaining a distribution of alleles which looks as if it signals an association with the phenotype when, in fact, the **null hypothesis** (or H_0) that there is no association between the two holds. Our fictional tests exemplified in the previous section produced, however, much smaller p-values, of the order of 10^{-9}, which means that there is a very tiny chance of 1 in 10^9 (one in a billion) that even these clear signals of association are in fact false.

To see why, I generated 1000 random samples with a similar composition to our hypothetical example, namely having about 1000 cases and 1000 controls and about 11.7 alleles A for each allele a (i.e., 3690:316 = 11.7:1.0 A:a) in the whole population. Crucially, when generating these random samples, I assumed that there is no difference between the frequencies of the two alleles among the two groups, which means that the contingency tables derived from these samples should show no association between the alleles and the groups (as there is none)! However, in one particular run of this simulation, I got 52 out of 1000 (5.2%) samples for which the contingency tables suggested a significant association at alpha-level 0.05 (i.e., their χ^2 test's p-value was smaller than 0.05), and 7 (0.7%) significant at alpha-level 0.01 (see Table 5.4 for some examples and the Appendix for the actual `R` code, which you are encouraged to run and understand). It should be clear, thus, that for an alpha-level of 0.05, about one in 20 (1/20 = 0.05) tests will turn out significant for a population in which there is no association between the genetic marker and the phenotype (**false positives** or type I errors).

What this practically means for genetic association studies is that if we were to use 1000 independent genetic markers each one of them *not* associated with the phenotype (i.e., independent of it), then, by pure chance, about 50 of them will appear significant at a "standard" alpha-level of 0.05 and 10 will do so for the much more stringent level of 0.01. Thus, before drawing any conclusions from such results, we need a way to guard ourselves against such false positives. An obvious approach is to lower the alpha-level depending on the number of independent tests performed, such that the *overall* probability of a false positive (the so-called *family-wise error rate* or FWER) is still the nominal 0.05 (or 0.01). Intuitively, given that we perform many tests we require each one of them to be much more "convincing" than would normally be the case. But how much more convincing? A simple method of correction used in many scientific disciplines is represented by the **Bonferroni correction** whereby the individual test alpha-level is lowered to the FWER (the nominal 0.05 or 0.01) divided by the number of independent tests: in our example this results in an

Table 5.4 *Simulations of extracting samples from a population in which there is no association between a biallelic genetic marker and a binary phenotype. Shown are the contingency tables of three such samples that are significant at the standard alpha-level 0.05.*

Allele	Cases	Controls	Total
Sample 1			
A	1865	1824	*3698*
a	136	181	*317*
Total	*2001*	*2005*	$\chi^2_1 = 6.54$, $p = 0.011$
Sample 2			
A	1845	1867	*3712*
a	119	175	*294*
Total	*1964*	*2042*	$\chi^2_1 = 8.92$, $p = 0.003$
Sample 3			
A	1900	1801	*3701*
a	175	130	*305*
Total	*2075*	*1931*	$\chi^2_1 = 3.88$, $p = 0.049$

alpha-level of $0.05/1000 = 0.00005 = 5 \times 10^{-5}$, a pretty stringent requirement. With this correction, none of our simulated samples is significant any more.

However, while very effective, the Bonferroni correction has a very nasty downside: it also reduces drastically our chances of finding *true* associations that are actually present in the data. To see why, imagine we were testing 999 genetic markers not associated with the phenotype, but the 1000th one is statistically weakly (but causally importantly) associated with the phenotype, with a *p*-value of "only" 0.0003, which, alas, is above the corrected alpha-level of 0.00005. This is a real concern in many branches of science where multiple hypotheses must be tested simultaneously, but it reaches cosmological magnitudes in GWAS where hundreds of thousands of markers might be simultaneously tested for association with more than one phenotype. Thus, if we were to use a standard SNP array platform such as Affymetrix 6.0,[5] which at the time of writing can genotype more than 906,600 SNPs, to test for association with various phenotypic measures related to language and speech (say, vocabulary size, rapid naming, grammaticality judgments, phonological fluency and non-word repetition), then we would have to correct for more than

[5] `http://www.affymetrix.com/estore/browse/products.jsp?productId=131533&categoryId=35642&productName=Genome-Wide-Human-SNP-Array-6.0#1_1`, retrieved on the 18th of August, 2012.

$906,600 \times 5 = 4,533,000$ tests, resulting in a requirement that individual association (i.e., SNP × measure) have a p-value of less than $1.1 \cdot 10^{-8}$! And, indeed, guidelines for GWAS do recommend an alpha-level of about 10^{-8} (Clarke et al., 2011) for an individual association to be deemed statistically significant. How can we deal with this issue? There are several (not mutually exclusive) possibilities.

Methods for multiple testing correction

The Bonferroni multiple testing correction method assumes that all tests are independent, which is not generally true for a **GWAS**: the neighbouring **SNPs** on the same chromosome are in LD (**linkage disequilibrium**) due to the physical linkage between them (see Sections 3.6 and 5.1), and other evolutionary forces (covered later) ensuring that there are correlations between SNPs such that knowing the alleles for some allows us to predict with various degrees of confidence the alleles for others. Moreover, when using multiple phenotypic measures, these measures are rarely independent as they reflect the study's general concern, such as language and speech. Thus, the independence assumptions behind Bonferroni's method are too strong and provide a *conservative* lower bound for the alpha-level.

Other methods for controlling for multiple testing[6] are more liberal but still strict enough to guard against accepting false positives as real, such as *Holm*'s (Holm, 1979) method, which does not assume that there must be a unique lowered alpha-level for all individual tests to ensure an appropriate FWER (as done by Bonferroni) but instead adjusts each individual alpha-level depending on the tests. Also popular are methods for controlling the proportion of *false discoveries* (i.e., incorrectly rejecting the null hypothesis of no association) known generally as **False Discovery Rate** methods (**FDR**), which produce q-values (somewhat equivalent to p-values).

Recently, **permutation** methods have gained a lot of popularity and some regard them as a "gold standard", as they do not require many assumptions and use the data itself to test various hypotheses. This allows them to take into account various dependencies and types of structure in the data when estimating the statistical significance of a genetic association. As a simple example, we can randomly shuffle the participants' labels (i.e., being in the "control" or the "treatment" groups) and recompute the associations each time, finally comparing the real association with the distribution of these permuted associations:

6 Several such methods are implemented in `R` by the `p.adjust()` function (the `method` argument selects, for example `"bonferroni"` or `"holm"`). The approach is a bit different from what we describe here (but equivalent to it), in that this function requires the list of the individual tests' p-values and returns the list with the adjusted p-values after the application of the multiple correction method, which can then be directly compared with the FWER standard alpha-level (e.g., 0.05 or 0.01).

the p-value is in this case the proportion of permuted tests more extreme than the real one. The fundamental idea is that the permutations should destroy the pattern of interest in the data (but not other patterns) and would show how different from randomness that pattern actually is. There are several good general introductions to this vast and expanding methodology (for example, Edgington and Onghena, 2007; Dugard et al., 2011) and its applications to genetics are an active research field (see Li et al., 2008, Clarke et al., 2011, for brief overviews and pointers to the literature); resampling approaches to multiple-testing are described in Dudoit et al. (2003) and Westfall and Troendle (2008). However, the price for this power and flexibility is the much bigger computational costs and the careful attention to be paid to the actual implementation of the permutation technique and the computation of the statistical summary of interest.

Increasing statistical power

The power of a test is the probability that the test will actually reject the null hypothesis when this is false. In our context, this means that it will correctly identify the association between a marker and a phenotypic measure when this association is, in fact, true. Thus, given a null hypothesis (usually denoted as H_0; here, that a marker M is not associated with a phenotypic measure P) which can be true or false, and the two decisions we might make using a test (to reject it or to fail to reject it[7]), we have the four possibilities of a false positive, a true positive, a true negative and a false negative, as shown in Table 5.3.

We have already met the alpha-level (or significance level) α as being the probability of making a type I error (a false positive): this represents the probability of falsely rejecting the null hypothesis when it is in fact true (i.e., falsely "finding" an association between M and P when there's none) and usually is set at 0.05 (5% chance of a false positive) or 0.01 (1% chance). Another type of error happens when we wrongly fail to reject a false null hypothesis (i.e., there really is an association between M and P but our test is unable to find it by producing a p-value larger than the alpha-level); this is called a type II error and represents the probability, denoted β, of producing a false negative.

On the other hand, failing to reject the null hypothesis when it is in fact true (i.e., deciding that M and P are associated when they aren't) is a correct decision which can happen with probability $1-\alpha$, and represents the *specificity* of the test. Another correct decision is that of rejecting a false null hypothesis (i.e., correctly finding an association between M and P when they are indeed

[7] Please note the actual formulation of "failing to reject H_0" as opposed to "accepting H_0" – all you can hope for in a classical statistical testing framework is to reject the null hypothesis; not being able to do so means only that and not its "acceptance". Put differently, you can *never* prove the null hypothesis but you might be able to reject (falsify) it.

associated), which will hapen with probability $1 - \beta$, also known as **statistical power**.

The relationship between type I and type II errors (or equivalently between the alpha-level α and the power $1-\beta$) is complex but they are inversely related: increasing one decreases the other and vice versa. To see why, imagine you want to lower the probability of a type I error α (thus the probability of a false positive rejection of a true null hypothesis) and to achieve this you will want really strong evidence in the data that H_0 is false, thus a more extreme deviation of your data from what would be expected by chance if H_0 were true. In our example, you would really want the allele frequencies in the contingency table's cells to be very different from what you would expect if there was no association between M and P, as measured by a more extreme value of the chi-squared statistic χ^2. However, this burden of proof will make it very probable that you will miss the slightly less extreme contingency tables resulting from a weaker (but no less real) association between M and P. Thus, you will commit more type II errors (or false negatives) by failing to detect these associations, effectively increasing their probability β (and decreasing the power $1 - \beta$). Alternatively, if you decide not to miss these weak associations by decreasing the probability of a false negative β (and increasing the power $1 - \beta$), you will inevitably have to accept much weaker evidence for an association, thus less extreme contingency tables (summarized by smaller χ^2 values), which could have resulted simply by chance (with probability α) in sampling from a distribution where M and P are not associated. This effectively results in much more frequent type I errors (false positives), and thus an increase in α.

Given this tug of war between type I (α) and type II (β) errors, one has to strike a balance between the two opposing requirements of avoiding false positives and false negatives. Sometimes, avoiding false positives is percevied as more important, such as when the risk of failure has extreme consequences, as happens in medical screening for conditions with lethal consequences. However, at times, avoiding a false negative is seen as more important, as for example in court cases where the null hypothesis of innocence is rejected only by very strong evidence to the contrary (very low α and high β), and this also seems to be the implicit attitude in most of scientific literature, by requiring a relatively low threshold α for accepting the claim that there *is* an effect (i.e., that the null hypothesis is false).

The statistical power $1 - \beta$ is positively influenced by the sample size N, meaning that, everything else being kept constant, large samples have more power to detect small effects. To illustrate, I simply reduced the sample size in our hypothetical example (Table 5.1) by a factor of 10 while keeping the distribution of the alleles in the two groups the same (Table 5.5): this resulted in a spectacular drop in the value of the chi-squared test to a non-significant $\chi_1^2 = 2.80$, $p = 0.94$. So, if we were unlucky enough to collect only 401 participants from the population, we would be unable to detect the true association between M and P and we would have produced a false negative (type II error). Alternatively, had we had the resources and foresight to keep collecting 4006 participants, we would have successfully rejected the null hypothesis of no association even at a very demanding whole-genome alpha-level of 10^{-8} (our p-value would have been an unbelievable $1.32 \cdot 10^{-9}$). Thus, the sheer number of participants made the whole difference, in this case, between a bitter (and largely non-reportable) null result, and a potentially high-impact *Nature* paper[8] reporting the first ever association between M and P.

Intuitively, however, not all distributions of alleles across groups (quantified by their contingency tables) are equally "extreme": some, such as the one in our hypothetical example (Tables 5.1 and 5.5), are relatively mild, while some are really not that different from what would be expected by chance (Table 5.6), and others are really extremely different from this expectancy (Table 5.7). How can we quantify these intuitions?

Table 5.5 *The effects of sample size on statistical power: same distribution as in Table 5.1 in a ten-times smaller sample.*

Allele	Cases	Controls	Total
Sample 1			
A	179	190	*369*
a	21	11	*32*
Total	*200*	*201*	$\chi^2_1 = 2.80$, $p = 0.094$

Table 5.6 *Contingency table not that different from what would be expected by chance.*

Allele	Cases	Controls	Total
Sample 1			
A	100	100	*200*
a	100	100	*200*
Total	*200*	*200*	$\chi^2(1) = 0.01$, $p = 0.92$

[8] The standards for accepting an association study have become very demanding, as discussed below, but in the early years of GWAS a single highly significant hit like this one might have been enough.

Table 5.7 *A very extreme contingency table.*

Allele	Cases	Controls	Total
Sample 1			
A	5	195	*200*
a	195	5	*200*
Total	*200*	*200*	$\chi^2(1) = 357.21$, $p < 2.2 \cdot 10^{-16}$

The very important concept of **effect size** does just that: in our case, it measures the strength of the association expressed by a contingency table and it is equal to the *phi coefficient*, $\phi = \sqrt{\frac{\chi^2}{N}}$, where χ^2 is the value of the chi-squared test and N is the sample size. ϕ can vary between 0.0 (extremely small effect size) to 1.0 (extremely strong effect size), and is a *correlation coefficient* for dichotomous variables, meaning that its square, ϕ^2, represents the proportion of variance shared by these two variables. Thus, the case in Table 5.6 has an effect size $\phi = \sqrt{\frac{0.01}{400}} = 0.005$, our hypothetical example has $\phi = 0.08$, and the extreme Table 5.7 has $\phi = 0.95$, supporting our intuitions.

To summarize, we have four related parameters that are relevant to conducting a statistical test: the sample size N, the effect size ϕ, the alpha level α, and the power $1 - \beta$. Given any three of them, we can deduce the fourth either manually, by applying the appropriate formulas, or by using specialized software such as `G*Power` (Faul et al., 2007) or `R`. As an example, what is the minimum sample size N required to have a power of $1 - \beta = 95\%$ to detect the associations expressed by our three contingency tables at a standard level $\alpha = 0.05$? Using `G*Power 3.1.2`[9] we find that we would need only 15 participants to detect an association as strong as in Table 5.7, about 2000 for our example, and a whopping half a million for Table 5.6! Relaxing the power to $1 - \beta = 80\%$ (thus accepting a 20% risk of not being able to detect a real association of this strength), we require only 9, about 1200 and a little over 310,000 participants, respectively. Keeping the high power $1 - \beta = 95\%$ but being much stricter (suppose this association is only one of 500,000 SNPs tested in a GWAS) with[10] $\alpha = 10^{-7}$, we will require only 54 participants for

[9] `G*Power` is freely available for download at `http://www.psycho.uni-duesseldorf.de/abteilungen/aap/gpower3/download-and-register` as of 24 August 2012. Please note that the "Effect size w" used by `G*Power` is numerically identical to ϕ (Cohen, 1988, p. 223). The complex issues behind effect sizes and power analysis are discussed in various places such as Cohen's (1988) reference book, and specifically for dichotomous variables one can see, for example, Sánchez-Meca et al. (2003).

[10] G*Power is currently limited to a minimum α of 10^{-7}.

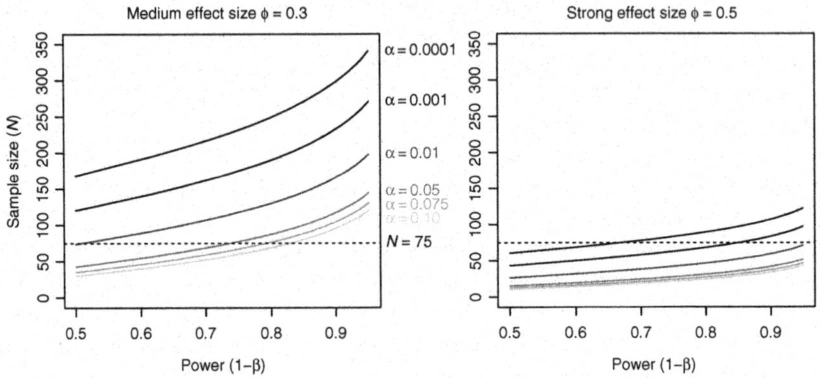

Figure 5.2 Statistical power of the χ^2 test. Each panel shows the sample size required (vertical axis) for a given power ($1-\beta$, horizontal axis) and α-level (the curves); the left-hand panel represents a medium effect size ($\phi = 0.3$) and the right-hand one a strong one ($\phi = 0.5$). The dashed line represents a sample size of $N = 75$: it can be seen that it intersects multiple curves for various powers, effect sizes and α-levels.

the strongest association, about 8000 for our example, and two million willing participants for the weakest association (see also Figure 5.2). Often there is no way to compute exact effect sizes, especially when conducting a new type of study, but there are general guidelines which suggest that effect sizes around 0.1 are small, close to 0.3 are medium, and about 0.5 and above are large.

Generally in genetics the effect sizes are very small, which explains why so many participants are required, and studies enrolling thousands or even tens of thousands of participants are becoming the norm. Of course, this type of "big science" requires a very different approach, funding, project structure and skills from the small group of researchers and assistants designing and conducting an experiment with 20 participants, or, more dramatically, from the lone theoretical linguist working on the grammaticality of self-generated constructions. But with such big samples, anything would become significant, wouldn't it? Put differently, when the sample size N is so large, there are two potential issues to consider: (i) any random fluctuations might be inadvertently interpreted as significant results, and (ii) even if the associations found are real, they are so small as not to really matter.

The first potential issue assumes that there really is no association (so the null hypothesis H_0 is actually true) but that using large samples might result in rejecting it: this is a type I error (a false positive) and is governed by the significance level, α, which is usually fixed before the experiment at a conventional value and is *not* affected by the sample size.

Statistical and real-world significance: effect sizes

The second issue identified above, however, has merit, in the sense that, as we just saw, larger samples make the detection of small effects (i.e., weaker associations) more probable.[11] But do these weak associations actually matter? It is sometimes said, especially when a weak effect is found to be statistically significant using large samples, that statistical significance *does not* equal real-world significance, usually implying something on the lines of "yeah, sure, the p-value might be less than 0.05, but so what?" Calling a statistical test whose p-value is less than the agreed α-level "significant" was probably not the best choice (even when specifying that it only concerns statistical significance). More appropriate names should have reflected better the actual meaning of $p < \alpha$, namely that there's a small chance (less than α, actually given by p) of obtaining such a result if there is no real effect. (Maybe "probableness" would have worked better?)

Given this technical meaning of "statistical significance", it is clear that a "statistically significant" result is not automatically also relevant or significant in the real world, especially given that this type of significance is highly dependent on a complex array of factors, some of them very subjective in nature. Thus, the weak association in our example (Table 5.1 and repeated here as Table 5.8) is statistically significant in a large sample of 4006 participants, but its effect size is only $\phi = 0.08$. Another way to interpret such a contingency table is in terms of the *risk* an individual has of being affected (i.e., belonging in the cases group) as a function of their marker's alleles. The risk of being a case imparted by allele A is the probability of carrying this allele and being a case, namely $R_A = \frac{1790}{1790+1900} = 0.49$; conversely, the risk of being a case when carrying allele a is $R_a = \frac{210}{210+106} = 0.67$. The *risk ratio* compares these two risks and is computed as the ratio of two risks: $RR = \frac{R_A}{R_a} = 0.73$. Thus, the risk of being a case when carrying an A is only about three-quarters (0.73) of that

Table 5.8 *Our example repeated here.*

Allele	Cases	Controls	Total
Sample 1			
A	1790	1900	*3690*
a	210	106	*316*
Total	*2000*	*2006*	$\chi_1^2 = 36.78$, $p < 1.32 \cdot 10^{-9}$

[11] To be sure, these issues are much more general than genetics and are extensively covered in many good introductions to statistics; probably one of the best short discussions of such issues and a must read for any scientist is Ioannidis (2005).

of being a case when carrying an a. A risk ratio of 1.0 would mean that the two alleles are equivalent from this point of view (i.e., no association between M and P).

Another intuitive way to see this contingency table is in terms of the *odds* of being a case given a certain allele: the odds of being a case when carrying A are $O_A = \frac{1790}{1900} = 0.94$, while the odds of being a case when carrying an a are $O_a = \frac{210}{106} = 1.98$. Odds around 1.0 represent a 50:50 chance of being a case, with odds less (or more) than 1.0 standing for events that happen less (or more) than half the time. As for risks, it is informative to compare the two odds using the *odds ratio*, defined as $OR = \frac{O_A}{O_a} = 0.48$, and showing again that the odds of being a case when carrying A are about half of the odds when carrying an a. (Again, an OR of 1.0 means that there is no association between the marker and the phenotype, i.e., they are independent.) An important property of the odds ratio is that it does not depend on the sample size: the OR for the smaller sample in Table 5.5 is $OR = \frac{179/190}{21/11} = \frac{0.94}{1.91} = 0.49$ (due to rounding and the requirement that the number of participants in each cell is an integer, these values are not exactly equal for our examples).

Usually, odds ratios are converted into **log odds ratios**,[12] denoted LOR, by taking the logarithm[13] of the OR: $LOR = log(OR)$. Thus, for our example, $LOR = log(0.48) = -0.73$, which is negative (because the $OR < 1$) and shows that the odds of being a case for A are smaller than those for a. While not as easily interpretable, LORs have other useful properties, such as the ease of computing **confidence intervals**. Briefly, a confidence interval, or CI, for a statistic such as the LOR, represents the range of plausible values that the LOR might have given the observed sample. It is closely related to the significance level α and the type I error, but simultaneously also gives information about the probable effect size, carrying much more useful information. For example, in our case (Table 5.8) the $LOR = -0.74$ with a 95% confidence interval (usually denoted 95%CI) of $(-0.99, -0.50)$: this can be interpreted in the sense that the real LOR value (estimated based on this particular sample as -0.74) falls between -0.99 and -0.50 with a 95% probability (and, as it should be, $-0.99 \leq -0.74 \leq -0.50$). If the 95%CI does not contain the value that the statistic would have if the null hypothesis were true (here, an LOR of 0.0 if there is no association between the marker and the phenotype), then the null hypothesis is rejected at an α-level of $0.05 = \frac{100-95}{100}$. A 99%CI corresponds to

[12] Log odds ratios, even if relatively hard to understand, are very important as they are often used to report the strength of association in the genetic literature.

[13] The logarithm $log(x)$ is a function that returns the power z, at which Euler's number $e \approx 2.7182\ldots$, must be raised to give x: $log(x) = z$ if and only if $e^z = x$. Please note that x must be positive. Thus, $log(e) = 1$, $log(e^2) = 2$, $log(1) = 0$, $log(0) = -\infty$ and $log(\infty) = \infty$. The log is very good at transforming exponential processes into linear ones, which are much easier to grasp.

an $\alpha = 0.01$ and amounts in our case to the interval $(-1.06, -0.43)$, which is wider than and contains the 95% one, as it should be, given that the more strict we want to be about ruling out false negatives, the less precise we can be about the actual values. In extremis, the 0%CI (corresponding to $\alpha = 1.0$) is exactly the estimated value $(-0.74. -0.74)$, meaning that we can never know the actual value of a population parameter (such as the real LOR), while the 100%CI (for $\alpha = 0.0$) is the whole set of real numbers $\mathbb{R} = (-\infty, +\infty)$, meaning that we can never be 100% sure that the null hypothesis H_0 is false.[14]

During this rather long but very important digression, we saw that we can encapsulate the strength of association expressed by a 2×2 contingency table using a chi-squared effect size ϕ, or we can use the log odds ratio LOR for comparing the odds of being a case when carrying one or the other of the alleles. Both ϕ and LOR do not depend on the sample size N, or the significance level α, or the power $1 - \beta$, and reflect an intrinsic property of the association present in the sample as captured by the contingency table. Therefore, these effect size estimates are much better than p-values at judging the strength of association and its possible real-world relevance. Plus, as we saw, we can estimate confidence intervals for the LOR, which come with both a *point estimate* of the association (the actual value of the LOR in the sample) and the limits within which its real value would be found with a certain probability, α, allowing us to simultaneously judge the statistical significance of this particular result in this particular sample. Thus, for our example in Table 5.8, $\phi = 0.096$ and $LOR = -0.74$ with a 95%CI of $(-0.99, -0.50)$. The effect size estimates are not particularly big and suggest that, even if highly statistically significant in our 4000-strong sample, the marker M has a relatively modest practical impact on this particular phenotype P: carrying an *a* roughly doubles the odds of being a case.

So how important is it, in real life, to know somebody's genotype at M given that carrying *a* doubles the odds of being a case? Well, that depends on many other factors. First, what are the practical implications of being a case: does P represent a major deleterious phenotype profoundly affecting the quality of life, such as, for example,[15] developing *Primary Progressive Aphasia* (Rohrer and Schott, 2011; Rohrer et al., 2010; see also OMIM 607485) some forms of which have been related to mutations in the *GRN* (*Granulin Precursor*) gene, or *Developmental Verbal Dyspraxia* due to mutations in *FOXP2* (see Section 4.2)? This information could lead to major decisions concerning treatment and life-style changes for the carriers. And second, what are the

[14] For more about risk and odds ratios, their advantages and pitfalls, please see Bland and Altman (2000), Montreuil et al. (2005) or Simon (2001) among others; for the actual R code used to compute odds ratios and their confidence intervals, please see Appendix A.2.

[15] Please note that in these real cases the genetic effects and effect sizes are different and are used here just for illustration purposes.

theoretical implications this discovery might have? Sometimes finding that a specific gene influences a phenotype, even if this influence is relatively weak, might open totally unexpected and extremely valuable new avenues. It might turn out, following further investigation, that the gene in question is part of a much larger network of genes which influences the development of the phenotype, its day-to-day functioning or its resilience to negative environmental impacts such as infection or injury. We will see later in the book (Chapter 6, especially Section 6.7) some such examples of specific genes that turned out to be "molecular windows" (Fisher and Scharff, 2009) into speech and language.

5.3.2 *Controlling for other factors using regression*

While testing the association of a single genetic marker with a single phenotypic measure using χ^2 or *LOR* tests represents the fundamental approach, usually we want a more powerful approach that would allow us to make use of more data than just a single particular marker in explaining binary (such as disease status) as well as continuous or **quantitative phenotypes**. The **regression** framework (see Section 2.3.5 for an introduction) allows us to do more than just that. Fundamentally (see any good introduction to regression such as Montgomery et al., 2007, or Gelman and Hill, 2006, for more details), for a set of N participants (denoted $1 \leq i \leq N$), we have their measurements for the phenotype of interest (denoted y_i), the information on a set of M genetic markers (denoted m_{ij} with i being the participant and $1 \leq j \leq M$ the marker) as well as other information such as their sex, age, country of origin, ethnicity, medical history (so-called *covariates* and denoted here as x_{ik} where as usual i is the participant and $1 \leq k \leq K$ is a particular covariate out of all K covariates). We would like to explain the phenotype y in terms of the genetic information m and the covariates x. In its simplest form, the phenotype measure y_i will be described as a weighted sum of all the genetic markers and covariates:

$$\underbrace{y_i}_{phenotype} = \underbrace{a}_{intercept} + \underbrace{\left(\sum_{j=1}^{M} b_j \cdot m_{ij}\right)}_{genetic\ markers} + \underbrace{\left(\sum_{k=1}^{K} c_k \cdot x_{ik}\right)}_{covariates} + \underbrace{\epsilon_i}_{error}$$

where a is a constant called the *intercept*, b_j and c_k are coefficients,[16] and ϵ_i is the *error* with which the participant's phenotype y_i is explained by this regression model. The intercept and coefficients are estimated in such a way as to minimize the unexplained variance encapsulated by the errors ϵ_i which

16 Please note that usually the genetic markers and covariates are included in the same sum but we separate them here for clarity.

Table 5.9 *Performance in an artificial language learning task y (0=worst, 100=best), the genotype (and its coding as discussed in the text) at a single marker m, sex as a covariate x (1=female, 0=male), and the predicted performance in a sample of 100 participants (showing the first 10 only).*

		Genotype and coding, m					
ID	Learning, y		(a)	(b)	(c)	Sex, x	Predicted y
1	61	aA	1	1	0	1	59.35
2	27	AA	0	0	0	0	29.63
3	54	aa	2	1	1	1	64.44
4	30	aA	1	1	0	0	34.72
5	38	aa	2	1	1	0	39.80
6	65	aa	2	1	1	1	64.44
7	39	AA	0	0	0	0	29.63
8	34	aA	1	1	0	0	34.72
9	36	aA	1	1	0	0	34.72
10	27	aA	1	1	0	0	34.72
...							

combine together all the factors that affect the phenotype not included in this model.

As an example, consider a single genetic marker m with alleles a and A, a single covariate x representing the gender, and a continuous measurement representing the success in an artificial language learning task, y, in 100 participants (Table 5.9). The actual genotype can be coded in multiple ways reflecting different genetic models, such as:

(a) additive: we code the number of a (or, equivalently, A) alleles: $aa = 2$, $aA/Aa = 1$, $AA = 0$;
(b) a is dominant: we code the presence of at least one a allele: $aa = aA/Aa = 1$, $AA = 0$;
(c) a is recessive: we code the presence of two a alleles: $aa = 1$, $aA/Aa = AA = 0$;

The relationship between y, m and x is shown in Figure 5.3 and the actual regression coefficients are:

$$y = 29.63 + 5.09 \cdot m + 24.64 \cdot x$$

all of them highly significantly different from 0 (all p-values are smaller than 10^{-10}). This regression model explains quite well the data as shown by an adjusted R^2 of 0.87 with only about 13% of the variance in the phenotype

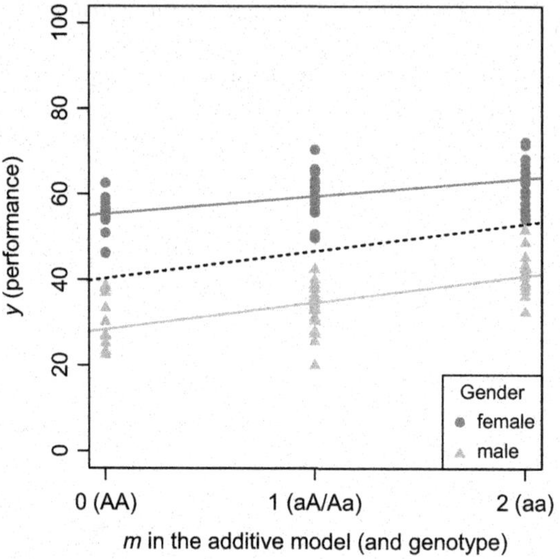

Figure 5.3 Regression of artificial language learning performance y on a single genetic marker m using the additive model (denoted (a) in Table 5.9 and in the text) also showing the influence of sex (x). The black dotted line shows the regression across all participants, while the solid light and dark grey lines show the regressions for males and females only.

not explained by sex and the genetic marker. The interpretation of this regression model is straightforward: the intercept $a = 29.63$ gives the expected test performance for a homozygous AA ($m = 0$) male ($x = 0$), with each a allele adding on average $b = 5.09$ points, and being a female adds an impressive $c = 24.64$ points. It must be highlighted that these are *predicted* (or *expected*) scores given the considered information (the marker and sex) and these will differ from the actual scores by the error ϵ as shown in Table 5.9; it can be seen that for some participants this model is very good (e.g., participant 6 with $\epsilon = 0.56$) but for others less so (e.g., participant 3 with $\epsilon = 10.44$). Importantly, such a regression model is useful beyond this particular sample by predicting the performance of not-yet-measured participants based on their genotype m and sex x, as well as giving the confidence we should have in these predictions.

When the phenotype y is binary (such as in the case-control studies discussed above), we can use **logistic regression**, while the **GLM** (*Generalized Linear Model*) can deal with much more general cases. Moreover, **multi-level regression** (or hierarchical models) can accommodate hierarchically structured data, but these go beyond our goals here. Of course, using regression does not somehow magically alleviate the need for multiple testing correction, but the flexibility of this framework is very well suited for complex designs where multiple factors beyond the individual's genetic makeup might be relevant.

In conducting genetic association studies we can use general statistical packages such as `SPSS` or `R` (R Development Core Team, 2010), but there are many good specialized software packages such as `PLINK` (Purcell et al., 2007) or `Bioconductor` (`http://www.bioconductor.org/`) which are much better suited for most analyses.

5.3.3 Accounting for population stratification

As we just saw, genetic association studies are built on the assumption that the correlation between the phenotype and the genetic marker is due to the genetic marker being ultimately causally involved in the phenotype, either directly or because it is very close to another marker that is directly involved. Focusing on a binary phenotype (case-control) for simplicity, there are, however, other reasons why the genetic makeup of the cases and controls might differ that has nothing to do with their phenotypic status. This is a familiar problem in matching cases and controls in general, where it is extremely important to make sure that the two groups are as equivalent as possible.

For example, let's say we collected data on the ability to imitate retroflex stops (such as /ʈ/ and /ɖ/) in a mixed student population in London and, after running our GWAS, we find several good hits: does that mean we have found the gene(s) for retroflex stops? Leaving aside the complex problem of "what a gene is for" (which we will discuss later in the book), these results are most probably biased by an obvious confound: most of the best performers are of relatively recent Pakistani, Indian or Bangladeshi origins who still to a certain degree speak their heritage languages, languages that are famous for their rich inventory of retroflex stops. Meanwhile, the poorest performers will be mainly of British origins speaking languages such as English, Welsh or Irish Gaelic lacking retroflex stops. Thus, a vast amount of the differences in phenotypic performance will be ultimately explained by the participants' population of origin, manifested as relevant cultural and/or environmental differences, in this case their native languages. However, if these different populations were genetically identical, then there would be no trouble: all putative genetic associations would be obviously false positives and should go away with a good multiple testing correction, while the lucky ones will fail to be replicated (see Section 5.3.4 below).

But this is obviously not the case: as detailed in Section 8.6.4, even if we are a relatively genetically homogeneous species when compared with other mammals, we are far from being clones. These genetic differences are mostly due to inter-individual variation (about 80–85%) but there are also differences between populations as well (the remaining 15–20%). Importantly, these inter-population differences are not categorical but are distributed as more or less smooth gradients, and they reflect geography (the populations separated by longer distances or more effective barriers such as mountains or oceans tend to be more different), demographic history (such as conquests and migrations) and different selective pressures (such as darker skin closer to the equator to protect from harmful UV light, the capacity to digest milk or immune systems better adapted to different pathogens). Thus, individuals from two populations will harbour genetic differences reflecting these factors and these genetic differences are not necessarily correlated with differences in the phenotype of interest. In our case, the best and worst retroflex stop imitators will differ in their genes mostly because they come from different populations with different histories and adapted to different environments and diseases. The best hits in our association study will thus be reflective of these differences and will probably include immune system genes, skin pigmentation and several **AIMs** (**Ancestry Informative Markers**, genetic loci whose allele frequencies differentiate populations).

This source of misleading genetic associations is known as **population stratification** (Astle and Balding, 2009; Freedman et al., 2004; Cardon and Palmer, 2003) and is a real concern when participants might come from different (or admixed) populations which simultaneously show differences in the genetic markers and phenotypes of interest. A classic example of detected population stratification is presented by Knowler et al. (1988), who found an apparently significant association between type II diabetes and an immune system gene, but showed that controlling for their participants' European Caucasian or North American ancestry removed this association. In effect, they note (p. 523) that this genetic locus is a "very sensitive marker for Caucasian admixture in American Indians" being virtually absent from North Americans and at relatively high frequencies among Europeans.

Given the potential biasing effects of undetected and unaccounted for population stratification, compounded by the fuzzy, continuous and complex nature of human genetic diversity, many methods have been used (and new ones are being proposed) to deal with it. One approach is to use family-based designs such as the **Transmission Disequilibrium Test** (TDT; Spielman et al., 1993; see also Section 5.3.6) which considers trios consisting of the two parents and their affected child. *Genomic Control* (Devlin and Roeder, 1999) uses a relatively small set of markers that are (probably) not linked with the trait in question to estimate a correction coefficient due to population stratification, a coefficient that is used to adjust the association test. Other methods (such as **`STRUCTURE`**; Falush et al., 2003) use a set of genetic loci to try to allocate individual participants to a number of discrete populations while simultaneously estimating the structure of these "original" populations as well as their relative contributions to each participant's genetic makeup, thus effectively seeing all participants as admixed to a lower or higher degree. Then these admixture proportions are used as covariates in a regression framework, controlling for population stratification. Yet others conduct **Principal Component Analysis** (**PCA**; Jolliffe, 2002) or **Multidimensional Scaling** (**MDS**; Cox and Cox, 1994) on the full genetic dataset to condense the genetic variation between the participants into a small number of components (`EIGENSTRAT`; Price et al., 2006) or dimensions explaining most of this variance. Then these components/dimensions are used as covariates in a regression framework, being able to account for the continuous nature of genetic variation. It has recently been suggested that adding either clusters derived from MDS (Li and Yu, 2008) or phylogenetic information derived from inter-participant genetic distances (Li et al., 2010) would also account for the relatively discrete aspects of inter-population structure.

5.3.4 Replication as the gold standard

Running a genetic association study and finding a statistically significant association between a marker M and a phenotypic measure P is but the first step in a long process. Even if the definition and measurement of the phenotype P are as good as possible, and even if the genotyping is perfect, and the study controls for all sources of false positives such as population stratification and multiple testing, there remains the possibility that such a significant hit is still not "true". There are several reasons for this, including biases in the study design, sampling and measurement, fiddling with the data and biased reporting, as well as good old false positives inherent in any statistical test. That this is not a theoretical discussion is amply shown by the vast number of published genetic associations involving multiple phenotypes such as diabetes and cancer that failed to be replicated and are now regarded as false positives (e.g., Ioannidis et al., 2001).

There is a widely shared consensus that genetic associations should be replicated, and there are proposed guidelines for conducting replications (Chanock et al., 2007) and systematic reviews of replicated studies (Sagoo et al., 2009), but there are still some issues that must be carefully considered (Kraft et al., 2009; Liu et al., 2008). To begin with, a claimed (and published) replication does not automatically make an association true. While replicating a result greatly reduces the probability that it is a false positive, it does not make it zero: there is still a chance that both results are false positives. Moreover, to be a true replication, the same marker and the same phenotype must be associated in a similar (but different) sample, and the direction of the association must be the same. Thus, if the original study found that allele A of a biallelic marker M is associated with an increase in the phenotype of interest P (or with being a case), then a replication must find the same and not, for example, that carrying allele a increases P (or the probability of being a case).

Second, failing to replicate a finding does not automatically make it false. It is inherently difficult to replicate genetic association studies (Liu et al., 2008) due to many factors, including differences between the samples in their genetic and phenotypic structure as well as the environmental influences on their genes. In this context, it is very interesting to consider the issue of statistical power in replication: generally, new genetic associations are "at the limits" of the studies that discovered them. Given that most interesting traits have a complex genetic structure, involving several markers of small effect, fishing expeditions will, given some luck, return a very few of these. And luck is quite important when statistical power is low: sometimes random fluctuations will conspire to increase the effect of a marker, making it barely cross the significance threshold and pushing it into the select set of "candidate genes". But the replication study, being focused on these lucky markers, will most probably suffer from the inverse effect: luck will conspire to make those same markers dip below the significance threshold. Surely, it might reveal other markers but not the ones it set out to replicate (even if they truly are associated with the phenotype under study). Thus, to beat this "winner's curse", replication studies must take into account the fact that the real effect sizes are most probably much lower than reported by the initial studies to be replicated, by increasing the replication sample's size over that of the initial study.

It is apparent, thus, that replication, while a gold standard, is far from perfect in ensuring that false positives are cast aside and only true associations are accepted as such. But there are other ways of probing an interesting association. While genetic association and linkage studies (and their replications) are, in the end, purely statistical exercises oblivious to the biological, neuro-cognitive and linguistic substance of the relationship between genes and phenotypes, one can delve into these as well. For example, as we will see later, *FOXP2* was identified by conducting a linkage study (also an exercise in statistics similar to associations) in a large family (Lai et al., 2001), but one of the decisive pieces of evidence was its identification as a gene controlling other genes (a **transcription factor**) and, later, the discovery of its involvement in brain development, structure and functioning in mice, humans and birds, and its effects not only on speech but also on birdsong and mouse vocalizations. Thus, convergent evidence from other types of inquiry involving functional experiments in the wet lab using cells or animal models might help transform a statistical association into an accepted real relationship between a gene and a phenotype.

5.3.5 Combining multiple markers

Testing the association of a phenotype with each one of hundreds of thousands or millions of SNPs individually might strike one as a lot of effort, but having to correct for this enormous number of simultaneous tests afterwards definitely doesn't feel quite right. Requiring the significance threshold to be of the order of 10^{-8} does guard against a lot of false positives but it also means that many true associations that are too weak to produce such incredibly low *p*-values will be missed, resulting in a very high rate of false negatives. Moreover, given that we are using these SNPs as landmarks hopefully in linkage disequilibrium (correlated) with the true (but most probably not directly visible) causal genetic loci, we should somehow strive to combine the imperfect information given by the nearby SNPs instead of pitting them against each other in a multiple testing correction competition.

Another issue is that almost certainly when looking into complex phenotypes such as aspects of language and speech, we should definitely expect that SNPs (and, by proxy, their hidden causal loci) do not act in isolation, producing main effects in the absence of interactions with other loci (what is usually called **epistasis**) and environmental factors, during development or afterwards. In fact (as we will see later), we have quite strong indications that this is indeed the case, with genes participating in complex pathways and networks, with their products being but one step in a long chain, acting as regulators on the activity of other genes (or even of themselves), or interacting with quite complex environmental signals such as psychological stress.

Given all these, there are methods of combining markers into sets that are taken as entities in themselves when looking for associations with the phenotype of interest. For example, we can combine all the individual SNPs that fall within a gene[17] and ask the big question "is the gene as a unit associated with the phenotype" instead of asking many smaller questions of the type "is SNP rs#[18] (probably as a proxy for a hidden locus in LD) associated with the phenotype?" The results of the big question will probably make much more biological sense, would be easier to test using different methods such as animal models, human pathologies or cellular techniques, and, importantly, would be easier to replicate. (Remember that replication is a key requirement in accepting an association as potentially real, but also that replicating associations between particular SNPs and phenotypes is very difficult, but using higher-level units such as genes should result in better replication rates.) However, not all is rosy and we should not all jump into doing whole-gene analysis: for a starter, we now have a better understanding of the complexities of what a

17 For example, using dbSNP at `http://www.ncbi.nlm.nih.gov/SNP`, we can see that there are literally thousands of SNPs within the *FOXP2* gene, but the actual number genotyped and tested will usually be much smaller depending on the genotyping platform used.

18 SNPs are often identified by their so-called reference SNP ID number of the form rs# where # is a unique number.

gene is and what it does. We know that genes are not monolithic entities invariably producing a single protein that does a single thing in the organism. Quite the opposite, as we will see later, genes often produce alternative transcripts, essentially meaning that a single gene results in multiple products sometimes with very different functions. Moreover, specific positions on the DNA are involved in different regulatory processes, and a good understanding of what a particular mutation does involves less general statements than "gene X is associated with phenotype Y": we need to know what part exactly of gene X does what to what aspect of phenotype Y and in what context.

Of course, we can define sets of SNPs based on other criteria than spanning a single gene: we can use smaller sets of SNPs in LD trying to better pinpoint the effects of the hidden marker that they all correlate with (and "stand for") or we can use much larger sets encompassing several genes involved in a biologically meaningful **pathway** or process, such as aspects of neural development. There are many methods proposed in this evolving literature, some combining the results from atomic tests of individual SNPs (e.g., Holden et al., 2008) while others consider the joint effects of multiple SNPs by, for example, extracting their first principal components and using them as co-variates (Gauderman et al., 2007), or by using more complex methods such as Logistic Kernel Machines (Wu et al., 2010). However, it is important to realize that SNP-based methods and approaches that combine individual SNPs are not exclusive but complementary and try to answer slightly different questions.

5.3.6 The Transmission Disequilibrium Test

One potentially important issue in "simple" association studies is represented by true associations between a genetic marker M and a phenotype P that are due to population stratification (see also Section 5.3.3): in this case we would have association but not linkage between M and the true (but hidden) causative genetic locus. The **Transmission Disequilibrium Test** (or **TDT**) is a way to overcome this by making sure that we simultaneously test for association and linkage so that the marker M and the hidden genetic cause of the phenotype P are indeed linked. But why would this linkage requirement be so important? As we saw, we very rarely happen to be so lucky that one of the true causative loci for the phenotype of interest is in our set of available genetic markers, so that we are usually reduced to looking for landmarks nearby the true but hidden cause. Being nearby, these landmarks are linked to the causal locus, meaning that due to the mechanics of DNA replication they tend to be inherited together, which also means that we know in what regions of the DNA molecule we have to look for the cause. If this assumption is broken and there is no linkage between our landmark and the cause, we will end up looking in the wrong place, wasting time and resources and inducing false hope.

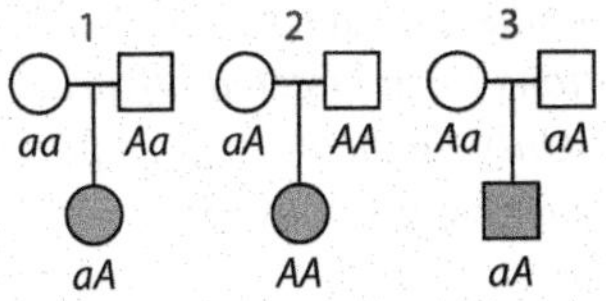

Figure 5.4 A set of three family trios composed of the affected child (dark symbols) and their biological parents (females are circles, males are squares), also showing the genotype at the locus of interest. To facilitate comprehension, we follow the convention that, in the children, the allele on the left is inherited from the mother and the one on the right from the father.

The TDT (Spielman et al., 1993; Ewens and Spielman, 1995) is based on the insight that using families ensures linkage while comparing cases and controls across families ensures association. Thus it is a hybrid between the "pure" association studies that we have just studied and "pure" family linkage studies that we will detail in the next section. In the classic TDT we need to collect trios composed of one affected child and his/her two biological parents, usually recruited through the affected child (also known as the **proband**). Let us, as usual, assume a biallelic landmark locus of interest M with alleles A and a and we want to test its association with a binary phenotype P with two states: affected and non-affected. Figure 5.4 shows a set of three such trios composed of one affected child and his/her biological parents, as well as their genotypes at locus M, with the convention that the allele to the left comes from the mother and the one to the right from the father. Thus, in trio 1 the child's a comes from the mother and the A comes from the father. For this test, the parents' phenotype values (i.e., affected or not) are irrelevant.

The basic insight behind TDT is that if the locus M is really associated and linked with P, then the responsible allele will be overtransmitted to the affected children. More precisely, the two alleles at locus M have a priori equal chances of being transmitted to a given child due to Mendel's first law (Section 3.5) of segregation. However, if one of them, say A, is associated and linked with the disease, then the probands will disproportionately inherit it relative to the alternative allele, a. We should note that only heterozygous parents are informative here.

Using the original notations in Spielman et al. (1993), we can count the number of times each allele was transmitted from the parents to the affected offspring (the probands) and then build a table showing the number of combinations of transmitted and non-transmitted alleles among these parents. If there are n probands, then there will be $2n$ parents, and this table will look as in Table 5.10. Given that only heterozygous (aA) parents are informative, we use only b (the number of heterozygous parents transmitting allele a but

Table 5.10 *Table showing the number of transmitted and non-transmitted allele combinations for n affected offspring (the probands) and their 2n parents. Only the heterozygous parents are informative (b and c), the homozygous ones being not considered (greyed out, a and d).*

	Non-transmitted allele		
Transmitted allele	a	A	Total
a	a	b	$a + b$
A	c	d	$c + d$
Total	$a + c$	$b + d$	$a + b + c + d = 2n$

not A to their respective affected offspring) and c (the number of heterozygous parents transmitting allele A but not a to their respective affected offspring); in our example, $b = 1$ and $c = 3$ for $n = 3$.

With these notations, if neither a nor A is associated and linked with the disease, then we would expect them to be equally transmitted to the probands, which means that $\frac{b}{b+c}$ and $\frac{c}{b+c}$ should be approximately equal. For our example, these ratios are $\frac{b}{b+c} = \frac{1}{4} = 0.25$ and $\frac{c}{b+c} = \frac{3}{4} = 0.75$. We can statistically test this null hypothesis using a chi-squared test computed as $\chi^2_{tdt} = \frac{(b-c)^2}{b+c}$ with one degree of freedom. For our example, $\chi^2_{tdt} = \frac{(1-3)^2}{1+3} = 1.0$, corresponding to a p-value of 0.32, far from significant.

Various extensions to this classic TDT have been developed, for example for continuous traits (Allison, 1997), and the test is included in popular and easy to use software packages such as PLINK (Purcell et al., 2007).

5.3.7 *Examples of association studies for language and speech*

At the time of writing (February 2013), there are literally thousands of published genome-wide or candidate gene association studies and dedicated repositories which allow users not only to browse them, but more importantly to conduct targeted searches using for example phenotype names, genes or individual SNPs. The best-known such repositories are *dbGAP* (`http://www.ncbi.nlm.nih.gov/gap?db=gap`), the *Catalog of Published Genome-Wide Association Studies* (`http://www.genome.gov/gwastudies`; Hindorff et al., 2009), *GWAS DB* (`http://gwas.biosciencedbc.jp/cgi-bin/gwasdb/gwas_top.cgi`), and *GWAS Central* (`https://www.gwascentral.org/index`). Some such association studies are gigantic, the results of enormous consortia involving hundreds of individual researchers from many labs in multiple institutions (both academic and private) across several countries. As an illustration, a recent study into the genetic foundations of height (Allen et al., 2010) needs about two journal pages to list the authors and their affiliations, and investigated no fewer than 183,727 individuals, while efforts looking into the genetic architecture of the body mass index (BMI; a measure combining body height and weight) used 249,796 individuals (Speliotes et al., 2010).

Unfortunately, when it comes to language and speech such massive efforts are still in the relatively distant future, mostly due to the difficulty of providing phenotypic measures that are quick, reliable, cheap and cross-linguistically valid to allow such large sample sizes. Nevertheless, there are already some good association studies aiming at deciphering the complex genetic architecture of these uniquely human traits. I will focus here on a small set of such studies selected not only for their scientific importance but also for their methodological clarity, but I hope that many more will follow in the near future.

As we will discuss in much detail later, the genetic architecture underpinning language and speech is very complex. A "molecular window" (Fisher and Scharff, 2009) into this is offered by the *FOXP2* gene, which, as discussed in the next chapter, was discovered due to the massive effects some of its deleterious mutations have on speech and language. Nevertheless, further studies using techniques such as **cell lines** and **animal models** have helped identify other genes that interact with *FOXP2* and compose the complex genetic networks required for normal speech and language development. One such gene is ***CNTNAP2*** (contactin-associated protein-like 2; OMIM 604569, usually pronounced "cat nap two"), which is a member of the **neurexin** family of genes which play important roles in the nervous systems. This gene has been involved in various pathologies such as Cortical Dysplasia Focal Epilepsy Syndrome (OMIM 610042; Strauss et al., 2006) and, interestingly, in the susceptibility to **autism**, more specifically being associated with the "age at first word" (Alarcon et al., 2008). Vernes et al. (2008) have used cell lines and human fetal brain tissue to show that *FOXP2* down-regulates *CNTNAP2* (i.e., the FOXP2 protein inhibits the expression of the *CNTNAP2* gene). To further test whether *CNTNAP2* actually affects language and speech, the authors conducted a TDT test looking for association between SNPs within the gene and various language measures within the *SLIC* sample[19] and found that, indeed, SNPs within the *CNTNAP2* gene are associated with language performance in this sample. Thus, *CNTNAP2* definitely seems a very interesting gene from the point of view of the genetic architecture of speech and language, and the study we

[19] The Specific Language Impairment Consortium (SLIC) sample (The SLI Consortium, 2002) consists of UK families selected through a family member diagnosed with **Specific Language Impairment** (SLI), discussed in more detail later.

will now focus on (Whitehouse et al., 2011) must be understood in this larger context.

Whitehouse et al. (2011) used 1149 children (606 males) from the Western Australian Pregnancy Cohort (Raine) Study – a longitudinal sample of 2900 mothers and their children recruited between 1989 and 1991. The genetic data was represented by 30 SNPs within the *CNTNAP2* gene, selected to match those used by Vernes et al. (2008). The phenotypic data had to be adapted to the characteristics of the sample (composed of very young children) and consisted of the parents' reports of their children's early communicative behaviour using the "Communication subscale" of the *Infant Monitoring Questionnaire* (IMQ) (Bricker and Squires, 1989). This consists of seven items scored on a three-point scale representing the frequency with which the infant shows the corresponding behaviour ("always" = 2 points, "sometimes" = 1 point, and "never" = 0 points); the scores were summed (resulting in a total score between 0 and 14) and further normalized by taking the z-score (the z-score is used to transform a set of measurements into a normal distribution with mean 0 and standard deviation 1; we will denote this phenotype by P). The authors conducted a quantitative association study of this phenotype P with each of the 30 markers $M_1, \ldots M_{30}$, looking for a statistically significant association between each marker's genotype and the phenotype across all participants. They further assumed that the **risk allele** of each marker was dominant, meaning that an individual carrying the risk allele in one or two copies would manifest the same deleterious effect as quantified by a tendency to have a lower total score relative to somebody carrying no copy of the risk allele. The results of these SNP-level tests of association (their Table 1 on page 453) show three SNPs that are significantly associated with the phenotype P, and three more that show p-values relatively close to the standard α-level of 0.05.[20]

Strikingly, these six SNPs are in the same region of the *CNTNAP2* gene as the significant SNPs found previously by Vernes et al. (2008), which is even more interesting given the differences between these two studies: while Vernes et al. (2008) investigated language performance in an SLI sample, Whitehouse et al. (2011) looked at parental reports for much younger children from the normal population. Moreover, some of these SNPs also overlap with the original findings of Alarcon et al. (2008) investigating language delay in autism. Such convergence of findings using different measures in different populations is a clear indication that *CNTNAP2* indeed affects language and speech, and is strengthened by studies using very different methodologies, such as cell lines and animal models, which help shed light on the actual mechanisms involved.

[20] It is not clearly specified in the paper whether these p-values are corrected for multiple testing, but they probably were.

Still in the domain of language and speech pathologies, much work concerns the genetic bases of SLI and **dyslexia**. While SLI can be broadly defined as "[...] significant language deficits despite adequate educational opportunity and normal nonverbal intelligence [in the absence] of other conditions [...] that may give rise to language impairments" (The SLI Consortium, 2002, p. 384), dyslexia (or *reading disability*) involves "poor literacy skills despite adequate intelligence and opportunity to learn [despite] adequate hearing and no major handicapping condition that might interfere with learning" (Bishop and Snowling, 2004, p. 858), and it is unclear if they are different pathologies or only different aspects of the same underlying problems (Bishop and Snowling, 2004; Newbury et al., 2011). Newbury et al. (2011) conducted association studies in SLI and dyslexia samples between a set of candidate genes and a set of measures of language performance. More precisely, they conducted a comprehensive review of the literature resulting in a set of 31 SNPs from seven genes (*DYX1C1*, *DCDC2*, *KIAA0319*, *MRPL19/C2ORF3*, *CNTNAP2*, *CMIP* and *ATP2C2*). The phenotypic measures covered a broad range of aspects of the oral and written language such as the Non-Word Repetition (NWR; Gathercole et al., 1994), single word reading and spelling, and reading comprehension. The SLI sample is represented by the Specific Language Impairment Consortium (SLIC), while the dyslexia sample is composed of 264 unrelated families from Oxford previously used in studies of the genetic bases of dyslexia (Francks et al., 2004) and a new set of 331 UK cases. The authors used a version of the TDT that allows quantitative phenotypic measures (QTDT; Allison, 1997) to test the association between the selected SNPs and the phenotypic measures separately in the two samples. They also conducted a classic case-control association study by comparing SLI and dyslexia cases from these samples to controls from an independent database (the Human Random Control panel of the European Collection of Cell Cultures). The results of this study (their Tables 2, 3 and 4 on pages 96, 97 and 99 respectively) confirm various associations previously found in the literature as well as providing new findings, but overall they support only weakly a shared genetic architecture between SLI and dyslexia.

The last study we use to exemplify the use of genetic association concerns the genetic bases of normal variation in language and speech. **Non-Word Repetition (NWR)** is a very simple task that involves the repetition of spoken meaningless sequences of syllables of varying length and complexity, such as "glistow", "perplisteronk" or "versatrationist" (Gathercole et al., 1994); performance can be measured in various ways but it essentially quantifies the number of correctly repeated syllables. Bates et al. (2011) focus on ***ROBO1***, a gene previously involved in dyslexia (Hannula-Jouppi et al., 2005; Paracchini et al., 2007), in a cohort of 538 normal families with twins. The phenotypes of interest comprise quantitative measures of reading and spelling, NWR, and measures of working memory. The study found significant associations between several SNPs within *ROBO1* and NWR after controlling for several factors such as age and intelligence, with two of them surviving multiple testing correction. The authors interpret this finding as showing that *ROBO1* is "a strong candidate for involvement in normal variation in language acquisition [being] functionally specialized to support the phonological buffer component [...] of a language acquisition system" Bates et al. (2011, p. 54). However, this might be too strong an interpretation, requiring independent replication of these findings, as well as more work on understanding both the nature of the NWR task and the mechanisms through which *ROBO1* could affect performance on this task.

As we can see from these selected examples, genetic association approaches are an active and promising approach in our quest to understand the genetic architecture of language and speech, both in the normal population and in pathologies, but much more work needs to be done. Several issues need to be solved before more progress is made, the most important, in my view, being the development of fast, cheap and reliable phenotypic measures that have theoretical import as well as good cross-linguistic equivalence. Recent developments in large-scale structural and functional brain imaging are very promising, as are advances in other techniques such as eye-tracking, but we will discuss these later in the book.

5.4 Linkage studies

The basic idea behind linkage studies is very simple: if a variant of a gene is involved in producing a certain phenotype then the two should be *co-transmitted* across generations within families. More precisely, parents should transmit both the variant and the phenotype to some of their children and the two transmission processes should be highly correlated. One of the simplest cases is that of an *autosomal dominant* and fully *penetrant* mutation, meaning that it is located on one of the non-sex chromosomes (or *autosomes*), and having even a single copy of the variant in question (being heterozygous) is enough (and equivalent to having two of them – being homozygous) to show the phenotype. Such a case, which we encountered before in Section 4.2, is represented by a certain mutation in the *FOXP2* gene discovered through performing linkage analysis in the British KE family (Fisher et al., 1998; Lai et al., 2001). The family's pedigree is reproduced here (Figure 5.5), also showing the individuals for whom no genetic information was available at the time of the linkage analysis (Fisher et al., 1998) as well as the inheritance of the *FOXP2* mutation. It can be seen that the mutation (as marked by the lightning symbol) and the affected status (dark symbols) seem to correlate perfectly (assuming the non-genotyped but non-effected individuals also do not carry the mutation),

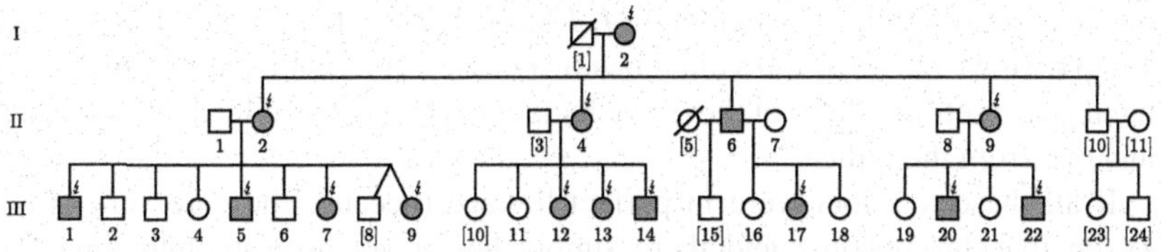

Figure 5.5 The KE family again: ○ are females and □ males; **white** are non-affected and **black** are affected; ⁄ were dead at the time the pedigree was constructed. For individuals between brackets no genetic information was available for genotyping. A lightning symbol marks the carriers of the *FOXP2* mutation.

being co-transmitted. This mutation turned out to be a single letter (nucleotide) change from a G to A, which disrupted a region of the resulting protein critically important for its function (more on this later). But how did they discover that this is the gene responsible in the first place?

As a note before we begin, genes are usually identified by their position on the chromosome (Section 3.5), and we now know that *FOXP2* is located on the long arm of chromosome 7 at 7q31.1 (see Figure 3.5). As we briefly discussed before, the various loci on the same chromosome are not necessarily transmitted independently from parent to offspring. In fact, if chromosomal cross-over (Section 3.6) did not exist, whole chromosomes would be transmitted as indivisible units across generations, all the loci on the same chromosome being perfectly linked. For illustration purposes, let us consider two loci on chromosome 7: the *FOXP2* gene[21] and four nearby *markers* flanking it, M_1, M_2, M_3 and M_4; thus, this portion of chromosome 7 looks like:

$$-M_1 - M_2 - FOXP2 - M_3 - M_4-$$

While the *FOXP2* gene was unobservable (hidden) for us in the mid 1990s, as we didn't then know it to be important for the strange phenotype shown by some members of the KE family, the markers M_i were observable. For these reasons, we will denote this hidden locus as *SPCH1* (the first locus involved in speech) just as its discoverers did at that time (Fisher et al., 1998). For simplicity, let us denote the normal and deleterious (i.e., disease-causing) alleles of *SPCH1* as A and G (based on our current knowledge that it is a single change from a G to an A that caused the disease in the KE family). Likewise, the observable markers have two variants (or alleles) each (a marker with a single variant is completely useless), denoted here using italic and bold numbers: marker M_1 has alleles *1* and **1**, marker M_2 has alleles *2* and **2**, and so on. Thus we can have $2^5 = 32$ possible combinations of alleles for these five loci,

21 For now we'll gloss over the structure of this gene and consider it as a unit.

denoted (ignoring the dashes): *12*A*34*, *12*G*34*, **12**A*34*, **12**G*34*, ..., **12**A**34**, and **12**G**34**. However, each individual has *two* copies of chromosome 7, therefore having two sets of alleles at these five loci: one allele inherited from the mother and one from the father.

If all these five loci were in perfect linkage (because they are very close by on the chromosome and there simply hasn't yet been enough time for recombination to break it), the parental combination of five alleles at these loci would be transmitted as a single unit to offspring, forming what is known as a **haplotype**. In this case, observing which allele is present at any of the four markers $M_1 \ldots M_4$ would give full information about the *FOXP2* allele. However, when these loci are not in perfect linkage, recombination breaks the correlation between them, ensuring that children sometimes inherit new combinations of alleles. The probability of recombination depends on many factors, one of the most important being the physical distance on the chromosome separating the loci: the closer they are, the lower the chance of a recombination occurring between them in any given generation. When the probability of recombination between say M_1 and *SPCH1* is too high (they are too far apart on the chromosome), then knowing M_1's allele gives too little information about *SPCH1*'s allele, dropping to no information whatsoever when the two are independent (the chance of recombination is 50% in each generation; see also Section 5.1).

For example, let us assume that M_1 and M_4 are so far apart from *SPCH1* that they are independent, but that M_2 and M_3 are so close to it that they are perfectly correlated in our family (of course, as long as they are not the same as *SPCH1*, given enough generations recombination between them and *SPCH1* will undoubtedly occur no matter how low its probability in each generation is). In this case, the unobserved *SPCH1* determines the disease phenotype in the family, but some of the markers we can actually observe, namely M_1 and M_4, will show perfect *decorrelation* from the disease, while the others, M_2 and M_3, will be perfectly correlated with it (by being perfectly linked with the unobserved but causative gene *SPCH1*). This would clearly indicate that the unobserved causative locus is somewhere quite close to M_2 and M_3 but far away from M_1 and M_4.

An even better case would be represented by finding that the recombination probabilities between the disease and the four markers are as follows: large with M_1, small with M_2, even smaller with M_3 and large again with M_4. Figure 5.6 illustrates such a hypothetical case using parts of the KE pedigree. For example, individual 2 in generation II inherits the deleterious allele G from her mother but there was a recombination event affecting the M_1 locus, replacing the expected **1** allele on the G-containing maternal chromosome with the *1* allele from the other maternal chromosome. This pattern of recombination would make us think that the causative locus (*FOXP2*) is quite far from M_1,

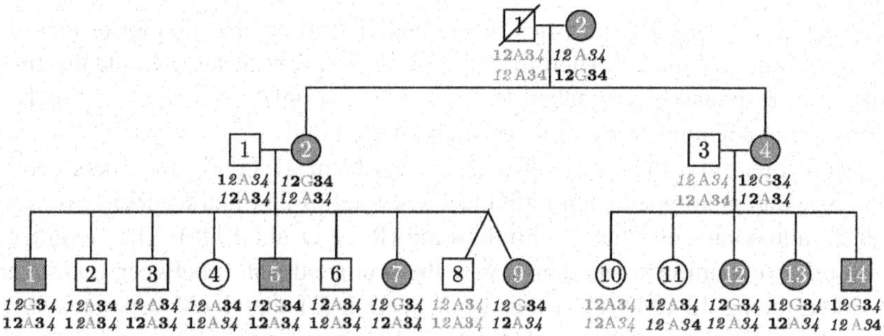

Figure 5.6 Hypothetical example of linkage using parts of the KE pedigree. Shown are the four markers $M_1 \ldots M_4$ discussed in the text, each with two possible alleles (*italic* and **bold**) as well as the unobserved *SPCH1* gene with two alleles (A and G). Each individual has its full genotype at the five loci shown (the upper chromosome is inherited from the mother while the lower from the father); grey stands for unobserved genotypes, either due to the unavailability of the individual or because it is at the *SPCH1* locus.

close to M_2, even closer to M_3, but far from M_4, suggesting the order and relative distances:

$$-M_1 - - - - - M_2 - -SPCH1 - M_3 - - - - - - M_4-$$

Thus, it is clear that our capacity to localize the hidden causative *SPCH1* depends on the density of the observable markers (the more and the closer together the more finely we can pinpoint *SPCH1* between two of them), but markers that are too close might not recombine in shallow pedigrees (the number of generations is too small for that to have happened yet), suggesting that bigger families are better.

To summarize, by tracing the co-transmission of various markers with the phenotype of interest across many generations within a family, it is possible to localize the causative (but usually unobserved) locus within a certain region of the genome. However, the precision of this localization is quite low and other techniques are usually required for the identification of the actual locus involved.

The identification of *SPCH1* itself is a very good and clear example of such an approach with important results for the understanding of the genetic architecture of language and speech. As described in Section 4.2, the British KE family had already been identified and intensively studied by the early 1990s (Gopnik, 1990) and it was clear that the complex disorder involving speech and language affecting about half its members was transmitted in a simple dominant manner (Figure 5.5). This pattern suggested that a single gene might be involved, and the characteristics of the pedigree made the attempt at identifying it look promising. Therefore, the first step towards actually identifying the gene responsible was taken by Fisher et al. (1998) in the late 1990s by conducting a genome-wide linkage study of the family.

More precisely, they secured genetic data from 27 family members (Figure 5.5) and genotyped each individual at a total of 254 marker loci covering all 22 autosomes and the X chromosome (Reed et al., 1994). The coverage was relatively uniform and dense, with the average distance between two such markers being 13 cM (remember from Section 5.1 that 1 cM separates two loci that recombine with a probability of 1% in one generation) and 96% of all loci in the genome are within 20 cM from one such marker (Reed et al., 1994). These markers are not the SNPs we have already encountered but a different type of extremely useful genetic polymorphism, namely *microsatellites*. In general, a microsatellite is a short sequence of nucleotides that is repeated a different number of times in the genomes of different individuals. For example, the microsatellite denoted[22] *D1S228* is located on chromosome 1 and is a repeat of the CA sequence; an allele of this locus might have four repeats (CACACACA) and another one six (CACACACACACA), and people can be, for example homozygous for the 6-repeat allele or heterozygous for the 4- and 6-repeat alleles. Microsatellites are such good markers because their alleles are relatively easily and reliably detected, their mutation rate is relatively high but not too high to blur their inheritance in families, and they are highly polymorphic (i.e., they have many alleles – repeat numbers in this case – that differ among individuals).

Fisher et al. (1998) found that among all these markers, the strongest linkage with the phenotype was found with markers on the long arm (q) of chromosome 7. The strength of evidence for linkage is usually measured by the *Logarithm Of Odds* (LOD score) introduced by Morton (1955) and computed as follows.[23] Given two loci M_1 and M_2 and an observed pedigree, we can estimate the probability of actually obtaining the genotypes in the pedigree assuming the loci are linked with strength θ relative to the probability of observing these if the loci are unlinked (independently recombining, $\theta = 0.5$); let us denote this as $L(\theta)$. For simplicity assume we focus on a single parent who has genotype **12**/*12* at these markers (please note that we assumed the marker loci to be biallelic with *italic* and **bold** denoting the two alleles

[22] The nomenclature rules for the human genome are complex and are detailed in the Guidelines for Human Gene Nomenclature (Wain et al., 2002; see also `http://www.genenames.org/guidelines.html`). Microsatellite names usually start with *D* for DNA, followed by the chromosome symbol, e.g. *1*, *S* to mark a unique DNA segment, and finally a unique sequence number.

[23] For a more detailed explanation and context see Chapter 11 in Strachan and Read, 1999, also available online at `http://www.ncbi.nlm.nih.gov/books/NBK7560`.

as before; the "/" symbol separates the two chromosomes). If the two loci independently recombine, then the probability that a child will inherit the original (or non-recombinant) combination of alleles from this parent (i.e., **12** or *12*) is equal to inheriting a shuffled (or *recombinant*) version (i.e., **1**2 or 1**2**) and is 50% (remember that independent genes assort independently following Mendel's Second Law introduced in Section 3.5, just as loci on different chromosomes do). However, if the two markers are not independent but linked in such a way that they recombine with probability θ in any one generation (you will recognize that θ is measured in centimorgans, cM), then the probability that a child will inherit non-recombinant alleles from the father is $p_{N,\theta} = \frac{1-\theta}{2}$ and that of inheriting recombinant alleles is $p_{R,\theta} = \frac{\theta}{2}$ (the division by 2 is required given that there are two chromosomes in each individual, each independently capable of recombination). Let us also count the number of non-recombinant children (denoted N) and recombinant ones (R); with these, the ratio of the two probabilities is:

$$L_\theta = \frac{(p_{N,\theta})^N \times (p_{R,\theta})^R}{0.5^N \times 0.5^R}$$

Then pick the value of θ that maximizes L_θ (call it θ_{max}), and the logarithm in base 10 of this is the LOD score (also sometimes denoted Z) of the two markers, LOD $= Z = log_{10}(L_{\theta_{max}})$. Usually, a LOD score of 3 (which is equivalent to $10^3 = 1000{:}1$ odds for linkage and approximately corresponds to the usual α-level of 0.05) is taken as evidence of linkage between the two markers. When multiple markers are available (such as in this search for *SPCH1*), the LOD score for each of them in relation to the hidden causative locus is computed and then mapping proceeds depending on these scores and the actual position of the markers on the chromosome. Of course, these computations are done by specialized software packages such as `MERLIN` (Abecasis et al., 2002; available online at `http://www.sph.umich.edu/csg/abecasis/Merlin`) or `GENEHUNTER` (Kruglyak et al., 1996; `http://www.broadinstitute.org/ftp/distribution/software/genehunter/`).

Specifically, Fisher et al. (1998) found very convincing linkages between the hidden locus and several markers in the same region of chromosome 7's long arm (such as a LOD score of 6.22 for *D7S486* and of 5.46 for *CFTR*). Using the pattern of these scores combined with the pattern of recombination between the observed phenotype and these markers in some members of the KE family (both affected and unaffected), they were able to narrow down the actual location of the hidden causative locus to a region of about 27 cM between *D7S527* (LOD score 1.82) and *D7S530* (LOD 2.76). To better localize this locus, they further used an even finer set of 20 markers spanning this region and managed to narrow down the region to only 5.6 cM in the 7q31 band of chromosome 7, between *D7S2459* and *D7S643*.

However, this is a pretty big chunk of DNA (further work investigated about 8 million base pairs from this region; Lai et al., 2000) that contained about 20 known genes at that date, some of them with seemingly decent chances of being the one. But basically, the researchers were stuck: there was no way of deciding which one of these 20 or so genes (if any) is *SPCH1*. This is how far even a good family, such as the KE, and a good phenotype, such as Developmental Verbal Dyspraxia (DVD), can bring you. A different approach was needed to progress and in this case (Lai et al., 2001) it was represented by a new individual (so-called CS), unrelated to the KE family, who showed a very similar phenotype. CS had a translocation between the long arms of chromosomes 5 and 7, denoted t(5;7)(q22;q31.2), meaning that genetic material was exchanged between these two chromosomes right within the region on chromosome 7 where *SPCHI* had been located using linkage in the KE family. This allowed the precise identification of a previously unknown gene belonging to the FOX gene family, the *FOXP2* gene.

6 What do genes actually do?

The roles that genes play are extremely varied; more often than not, the same genes can play multiple roles at different times in the organism's life cycle or in different places throughout the body. To better understand this complexity, we will encounter in this chapter several concrete examples of genes relevant for language and speech, ranging from the mechanical properties of structures of the inner ear necessary for hearing, to energy production, and to neural development and functioning. Here we will discuss many types of gene regulation and its paramount role in the existence of complex biological organisms such as us. The main aim of this chapter is not only to offer actual cases of genetic influences on speech and languages, but also to showcase the incredible complexity and beauty of these actual mechanisms, their counter-intuitiveness and power, the messiness but also elegance that are expected of products of biological evolution. A proper appreciation of these should make clear that abstract, mathematically elegant but simplistic proposals concerning the genetic bases of language have a very low chance of being real.

6.1 Structural proteins: *TECTA* in the inner ear

Some genes produce *structural proteins* that are components of our bones, skin, muscles, or parts of our hearing system such as the **tectorial membrane**. This is a structure of the inner ear (Gavara et al., 2011; Richardson et al., 2008; see Figure 6.1) with specific mechanical properties and which plays an important role in hearing through its interactions with the **hair cells** – the actual receptors converting the mechanical energy of the sound waves into neural impulses to be processed by the brain. The tectorial membrane is composed of collagen (a common family of proteins found in connective tissues such as tendons and skin) as well as non-collagenous proteins,

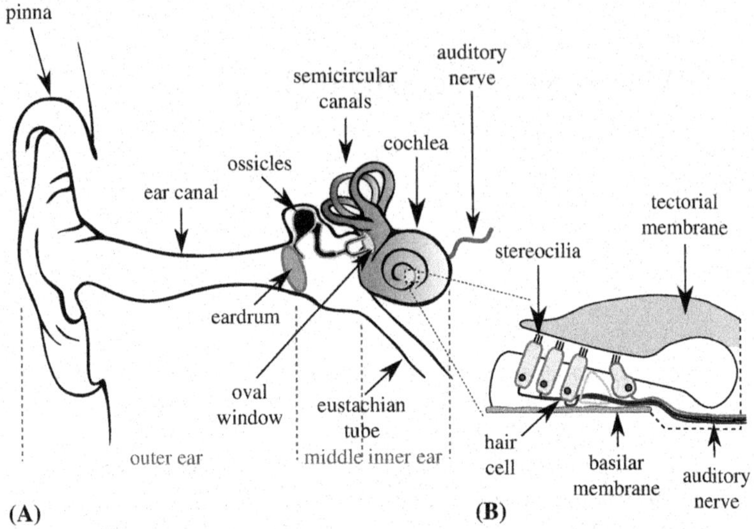

Figure 6.1 The human ear. Panel **A**: The sound waves arrive at the *pinna* (the visible part of the ear), travel through the *ear canal* and are transmitted by the *eardrum* (the *tympanic membrane*) and the three ossicles (the *malleus*, the *incus* and the *stapes*) to the fluid in the spiral-shaped cochlea (the auditory part of the inner ear, which also contains the vestibular system with its three semicircular canals, important for balance). Panel **B**: a zoomed section through the *Organ of Corti*, showing the basilar and tectorial membranes and the hair cells with their stereocilia.

among them the so-called *α-tectorin*, encoded by the ***TECTA*** gene on chromosome 11. Mutations in *TECTA* that produce abnormal forms of α-tectorin result in various types of deafness such as *DFNA12* (OMIM 601543; Verhoeven et al., 1998) and *DFNB21* (OMIM 603629; Mustapha et al., 1999). Interestingly DFNA12 is a dominant pathology (see Section 4.1) meaning that a single disrupted copy of the gene (for example, the allele TYR1870CYS where a tyrosine amino acid was replaced by a cysteine at position 1870) is enough to result in deafness, probably by altering (interfering with) the entire structure of the tectorial membrane. In contrast, DFNB21, caused by different alleles of the same *TECTA* gene, is recessive, allowing the normal copy of the gene to still produce a functional tectorial membrane (thus the one normal allele can *compensate* for the non-functional protein produced by the mutated allele).

Thus, mutations in the same gene (here, *TECTA*) might result in the "same" phenotype (deafness) but through slightly different mechanisms and with different inheritance patterns (recessive versus dominant).

6.2 Energy production: mitochondrial gene *MTRNR1*

Cells are really complicated things (see Figure 3.1) and there are many roles to be fulfilled by the products of the genes. One essential role is to produce energy in a usable form – a complex process in itself taken care of by the highly specialized cell organelles called **mitochondria** (Xing et al., 2007; Vafai and Mootha, 2012). Mitochondria have their own small genomes governed by their slightly different genetic code, and this genome encodes various components essential to the respiratory chain, the process whereby oxygen is used to "burn" food in a carefully controlled manner that stores the released energy in the chemical bonds of *ATP* (adenosine triphosphate), the molecule that transports energy around the cell. The respiratory chain is a complicated and essential process, and its disruption can have dire consequences, from vision loss to systemic disorders (Vafai and Mootha, 2012). However, certain mutations in mitochondrial genes can cause *non-syndromic hearing loss*, i.e. hearing loss that is not accompanied by other abnormalities (Kokotas et al., 2007; Xing et al., 2007), one very interesting example being the ***MTRNR1*** gene (OMIM 561000).

This gene (also known as the *12S rRNA* gene) does not encode a protein, but an RNA molecule – this means that there is no translation after transcription for this gene but instead the transcribed RNA is integrated into the small subunit of the mitochondria-specific **ribosome** (this is where the rRNA notation comes from: ribosomal RNA). Reflecting their deep evolutionary origins as once independently living bacteria that somehow got integrated about 2 billion years ago into the ancestor of the Eukaryotes (the **endosymbiotic theory**; Sagan, 1967), the 12S rRNA molecule in human mitochondria is still very similar to the bacterial 16S rRNA molecule, an essential component of the bacterial ribosome.

Some mutations in the *MTRNR1* gene, such as A1555G substituting a G for an A at position 1555, may result in non-syndromic deafness. Focusing on the A1555G, the first thing to note is that the genotype-phenotype relationship is far from perfect: the **penetrance** is quite low (i.e. not everybody carrying the mutation develops hearing loss) and seems to require the presence of other modifier factors. Moreover, when hearing loss does happen it varies a lot between carriers in both severity and age of onset (Bindu and Reddy, 2008). Thus, A1555G is only a *risk factor* that by itself does not seem able to produce deafness in every carrier of the mutation.

Possible modifiers are mutations in other genes and environmental factors such as treatment with an aminoglycoside antibiotic such as gentamycin or streptomycin (Bindu and Reddy, 2008). This class of antibiotics is used to fight specific types of infections (usually not related to the ear) and is known to damage hearing at high doses or after prolonged application (Vafai and

Mootha, 2012). However, in carriers of the A1555G mutation in *MTRNR1*, even very low doses of the antibiotic may result in deafness (Kokotas et al., 2007; Bindu and Reddy, 2008). The reason seems to be (Ballana et al., 2006) that the A1555G substitution changes the way the resulting 12S rRNA folds into the necessary shape required for its function within the ribosome. This change allows the antibiotic to bind the mitochondrial 12S rRNA with higher affinity than normal, as if it were the bacterial 16S rRNA, and to therefore affect the functioning of the mitochondrial ribosome as it does the bacterial ribosome. As a consequence, mitochondrial activity is impaired and results in less effective energy production, which probably affects the receptor hair cells of the inner ear, leading to deafness.

This example raises several interesting issues. First, it is clear that an allele such as A1555G does not have a clear-cut and deterministic effect on the phenotype: it can be better understood as a probabilistic process which depends on other modulating factors, some internal to the organism (other genetic variants in the person's genome) and some external (antibiotic treatment). Second, mitochondria are essential components of all living cells as they are the cellular "power plants", with these alleles such as A1555G being present in all copies of the mitochondrial DNA in all mitochondria in all of the individual's cells, and yet apparently their only discernible effect is deafness – while the exact mechanism is still not fully clear, presumably the cellular environment specific to the hair cells makes them more sensitive to the damage produced by antibiotics on *MTRNR1* carrying the A1555G mutation. Thus, this is but one example of a widespread mutation (present in all cells) of an essential component (energy production is required by all cells) that nevertheless has a very specific (and apparently counter-intuitive) effect; it would be madness to call *MTRNR1* "the gene for deafness"!

6.3 The transportation system: *MYO15A*

The autosomal recessive non-syndromic deafness resulting in the emergence of **Kata Kolok** in the Benkala village on the island of Bali in Indonesia (Section 4.2) is caused by a mutation in the ***MYO15A*** gene at the *DFNB3* locus. This gene belongs to the *myosin superfamily* – a group of related genes that encode similar proteins, in this case the myosins. During evolutionary history sometimes whole stretches of DNA are duplicated, producing one or more copies of the enclosed genetic information within the organism's genome. Such duplicated genes initially have the same function (as they encode identical genetic information) but with time different paths are possible (see, for example, Conrad and Antonarakis, 2007; Roth et al., 2007; or Lynch, 2007). One copy might start accumulating deleterious mutations and, in the end, stop functioning, becoming a **pseudogene** (the path to so-called **non-functionalization**). Or it can accumulate mutations and hit upon a new, usually related function, producing a new gene evolving under its own selective pressures (**neo-functionalization**), this being possible as the other copy still fulfills the original function. Finally, the two copies might both change such that they implement two different but complementary aspects of the original function (**sub-functionalization**). While the first path – gene "death" by non-functionalization and the production of a pseudogene "ghost" – is by far the most probable, sometimes neo- and sub-functionalization happen, resulting in **gene families** or even *superfamilies* (we have already encountered such examples in the form of the *FOXP* family of genes and the opsin genes involved in visual perception).

The myosin superfamily (Hartman and Spudich, 2012) is an ancient and large group of genes that encode motor proteins (the myosins). These proteins convert the chemical energy packed in ATP into mechanical movement such as that produced by muscles or that required for the ferrying of various molecules within the cell. *MYO15A* is a so-called "unconventional myosin" (Mooseker and Cheney, 1995) – named so simply because they were discovered after the "conventional" ones – and its function in hearing is complex but very instructive. We already saw that the hair cells in the inner ear are the actual sensors transforming the mechanical sound energy into neural activity sent for processing to the brain. They do this using specialized organelles, the *stereocilia*, tubular structures a few μm (micrometres, 10^{-6} metres) in length that give the cells their name. When deformed by the sound waves, stereocilia initiate a chain of events that result in the generation of the electrical nerve impulse.

Myosin XVa (or MYO15A), the protein produced by the *MYO15A* gene, is essential to the structure and function of the stereocilia, but it is not currently completely clear how exactly it plays this role. It seems (Manor et al., 2011; Yang et al., 2012; Stover and Diensthuber, 2012) that MYO15A must transport another protein, whirlin, to the tip of the stereocilia, where they must interact with yet another protein, EPS8, in order to produce normal, functional stereocilia. The mutation affecting the people in Benkala changes a single A into a T at position 2674 (A2674T) in exon 28 of the *MYO15A* gene (Wang et al., 1998), and is one of the several mutations (OMIM 602666) that result in non-functional proteins, incapable of the normal interaction with whirlin and EPS8 and leading to abnormal stereocilia, and ultimately deafness. Thus, a gene (in this case *MYO15A*) might result in a phenotype by being simultaneously involved in two processes necessary for this phenotype (here, transporting another protein to its destination and interacting with other proteins).

6.4 Unexpected processes: stuttering and the lysosome

Yet another function that cells must do is to digest (i.e., break down) stuff, such as waste products and ingested particles, or even invading microbes and viruses; the results of this process might then be recycled into new cellular components. The **lysosome** is a specialized cellular organelle containing a set of digestive enzymes. These enzymes are produced through translation by ribosomes (just as any other protein; Section 3.7) and transported into another specialized cell compartment (the **endoplasmic reticulum**), from which vesicles containing these enzymes bud off and are shipped to the lysosome, where they will perform their proper function (Fisher, 2010; Kang and Drayna, 2012). In order for the transport system to recognize them as destined for the lysosome, they need to have the proper "tag"(a specific molecule) attached to them. Attaching this tag is a two-step process implemented by two enzymes, GNPT and NAGPA, acting in sequence. GNPT itself is composed of three subunits (proteins), denoted α, β and γ; while subunits α and β are encoded by a single gene on chromosome 12 (***GNPTAB***; OMIM 607840), the γ subunit is encoded by a separate gene on chromosome 16 (***GNPTG***; OMIM 607838). In turn, NAGPA – also known as the *uncovering enzyme* (UCE) – is encoded by the ***NAGPA*** gene (OMIM 607985) on chromosome 16. This process is represented in Figure 6.2.

Stuttering is a disturbance of speech fluency characterized by involuntary repetitions or prolongations of syllables or larger units, or by interruptions known as blocks (Fisher, 2010; Kang and Drayna, 2012). It is present in about 5% of children, but in most cases (~80%) it goes away, leaving only ~ 1% of adults affected. It is a heterogeneous disease, with heritability studies suggesting a relatively strong genetic component, but linkage studies have painted the picture of a complex genetic structure with several loci potentially involved. In a small number of stutterers (< 10%; Kang and Drayna, 2012), mutations in *GNPTAB*, *GNPTG* and *NAGPA* have been identified, suggesting the involvement of the lysosomal transport system. However, this is an interesting

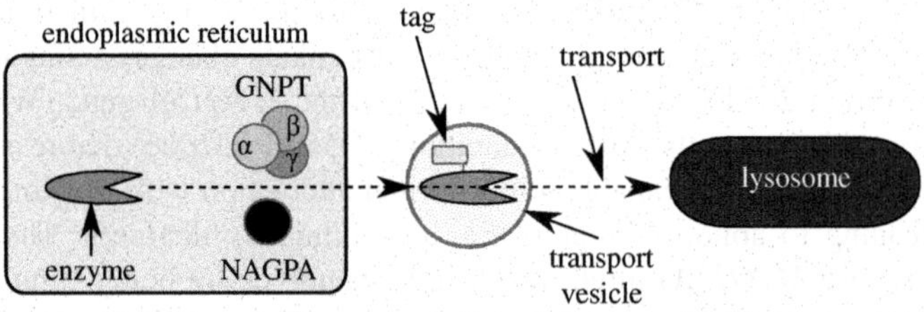

Figure 6.2 In order to be transported from the endoplasmic reticulum to the lysosome, enzymes must be appropriately tagged. This tagging process is controlled by two enzymes: the 3-subunit (α, β and γ) GNPT and the single-subunit NAGPA. Subunits α and β of GNPT are encoded by the *GNPTAB* gene on chromosome 12, subunit β of GNPT is encoded by the *GNPTG* gene on chromosome 16, and NAGPA is encoded by the *NAGPA* gene on chromosome 16

story and illustrates how discovering one gene involved in a phenotype can open the window on a larger set of genes (we will see another example in *FOXP2*).

A genome-wide linkage study conducted on a large number of inbred Pakistani families with stuttering (Riaz et al., 2005) has identified a strong signal on chromosome 12, later (Kang et al., 2010) refined to the G3598A missense mutation in the *GNPTAB* gene. Knowing that GNPT is composed of three subunits and that subunit γ is encoded by gene *GNPTG*, and that NAGPA, encoded by the *NAGPA* gene, is necessary for the same process, Kang et al. (2010) reasoned that mutations in these two genes might also be involved in stuttering. Indeed, they identified mutations in these genes in other people affected by stuttering (not coming from the original Pakistani families), showing that starting from one gene (*GNPTAB* in this case) and knowing the biological mechanisms the gene is involved in (here, tagging for lysosomal transport) allows hypotheses to be generated in a principled way and tested.

However, it is currently unclear why these mutations in *GNPTAB*, *GNPTG* and *NAGPA* lead to such a specific phenotype, given that the lysosome is such a ubiquitous organelle with a generally important function. Furthermore, other mutations in these genes do result in very severe diseases with fatal outcomes (mucolipidoses types II and III; Kang and Drayna, 2012). One proposal is that the mutations resulting in stuttering are much less severe, affecting only slightly the enzymes' function, but this is most probably not the full story (Kang and Drayna, 2012), and it will be fascinating to understand the exact mechanisms involved. Thus we saw, again, that a set of genes involved in a ubiquitous cellular function can result in a very specific phenotype, with no a priori reason to suspect that this pathway would have anything to do with this specific phenotype, highlighting again the suprising and complex nature of the genetic architecture required for language and speech.

6.5 Guiding axons: *ROBO1* in dyslexia and normal variation

Dyslexia or *Reading Disorder* (RD) is a relatively common developmental condition characterized by impaired reading in the absence of educational, cognitive, receptive or neurological causes. Nopola-Hemmi et al. (2001) conducted linkage analysis in a large four-generation Finnish family with severe dyslexia, identifying a locus on chromosome 3 (*DYX5*, OMIM 606896), later (Hannula-Jouppi et al., 2005) refined using a non-related patient with a **translocation** that disrupted the *ROBO1* gene. Translocations represent exchanges of genetic material between different (i.e., non-paired) chromosomes; in this case, it involved chromosomes 3 and 8 and specifically exchanged the short arm of 3 from band 12 (p12) with the long arm of 8 from band 11 (q11), denoted as t(3;8)(p12;q11). Translocations (see, for example,

O'Connor, 2008) may result in gene fusion (when two genes on different chromosomes are brought together and combined) or gene truncation (when part of a gene is replaced by non-coding material), either way disrupting the gene at the break point. This mutation in *ROBO1* was present in heterozygous form and resulted in reduced or absent expression of the gene, leading Hannula-Jouppi et al. (2005) to suggest that this form of dyslexia may result from (partial) **haploinsufficiency** – a condition in which one copy of a gene is (partially) rendered inactive (due to several causes such as failure to transcribe or translate, or lack of function of the resulting protein) and the other, normal copy cannot produce enough of its product (usually protein but could also be an RNA molecule) by itself to ensure a normal phenotype.

Interestingly, Bates et al. (2011) found that normal variants of *ROBO1* seem to be associated with normal variation in the **Non-Word Repetition** task (NWR), whereby participants must repeat back legal but meaningless sequences of syllable (i.e., non-words) of varying difficulty and length scored depending on the number and type of errors committed (Gathercole et al., 1994). However, it is currently unclear what exactly NWR measures, but it is probable that it focuses on phonological working memory (Gathercole, 2006).

But what exactly does *ROBO1* do? In the *Drosophila* (fruitfly), the *roundabout* (*robo*) gene controls the axonal crossing of the midline of the central nervous system (Kidd et al., 1998). As in mammals, the fruitfly central nervous system is bilaterally symmetric, being composed of relatively similar left and right halves, and some neurons' axons must cross this midline in order to reach their appropriate targets (remember that one hemisphere is generally responsible for the contralateral part of the body, one of the reasons for this midline crossing). This process is complex and involves several mechanisms and genes conserved across invertebrates and vertebrates (Evans and Bashaw, 2010), with *robo* being one of the genes responsible. Interestingly, there are three homologue *robo* genes in mammals forming a subfamily of genes, and indeed mutations of *ROBO3* (the most divergent of them) disrupt the crossing of sensory and motor pathways in humans. The evidence for *ROBO1* is less clear-cut but a recent study (Lamminmäki et al., 2012) showed, using **magnetoencephalography** (**MEG**; a brain imaging technique recording the tiny magnetic fields generated by brain activity with extremely good temporal resolution), that in dyslexic members of the same Finnish family studied by Nopola-Hemmi et al. (2001) the normal crossing of auditory pathways was impaired. This impairment was linearly dependent on the amount of mRNA *ROBO1* expressed by lymphocytes (a type of white blood cell used in this study as they are so much more easily accessible in living humans than brain tissue).

Thus, while much more work is needed before we have a complete understanding, it seems that *ROBO1* probably influences the guidance of the growing axons and their crossing of the midline in ways that ultimately result in specific forms of dyslexia.

6.6 Brain growth and development: *ASPM* and *MCHP1*

The human brain is one of the most complex objects in the known universe, its size and complexity have increased during human evolution (see, for example, Shultz et al., 2012), and it has recently been suggested that this is due to an increase in overall brain size (Barton and Venditti, 2013; Herculano-Houzel, 2012). It is widely believed that language and speech are dependent upon our big brains, making the genes behind brain growth and development important targets of research.

Brain growth and development is a complex process, with its size being principally determined by *neurogenesis* during embryonic development. Neurons are born (Götz and Huttner, 2005; Huttner and Kosodo, 2005) from the *neural progenitor cells* (NPC) that exist in the epithelium lining the *brain ventricles* (the brain's main cavities) and then migrate to their final position in the cortex. NPCs can undergo two types of cell division: initially they divide *symmetrically* (through *proliferative divisions*), each NPC giving rise to two identical daughter NPCs and increasing the pool of NPCs in the developing brain. These are followed by *asymmetric* (or *neurogenic*) divisions, whereby each NCP gives rise to one neuron (which then migrates) and one daughter NPC; these asymmetric divisions give birth to roughly the same number of neurons per initial NPC, neurons that migrate to seemingly form one single *cortical column* (Montgomery and Mundy, 2010). Therefore, changing the number of asymmetric divisions would result in a thinner or thicker cortex (and an aberrant cortical structure), but changing the number of symmetric divisions would mainly result in more or less cortex (of approximately normal structure) in terms of cortical area. The orientation of the *cleavage plane* (or *division plane*, separating the two daughter cells) seems to determine whether the division will be symmetric or asymmetric (Götz and Huttner, 2005; Huttner and Kosodo, 2005). More precisely, if the two daughter cells both inherit the apical region of the parent cell (i.e., the cleavage plane passes *through* this region), then they will both be NCPs (symmetric division), but if the cleavage plane misses this apical region then only one daughter cell will inherit it, becoming an NCP, the other differentiating into a neuron (asymmetric division). See also Figure 6.3 for a schematic representation.

Thus, the genes controlling the number of symmetric divisions will play an important role in influencing cortical (and brain) size. In this respect, there are rare natural experiments represented by **Primary Autosomal Recessive Microcephaly** (**MCPH** from *M*icro*C*ephaly *P*rimary *H*ereditary), which is characterized (Passemard et al., 1993; Mahmood et al., 2011) by a very small

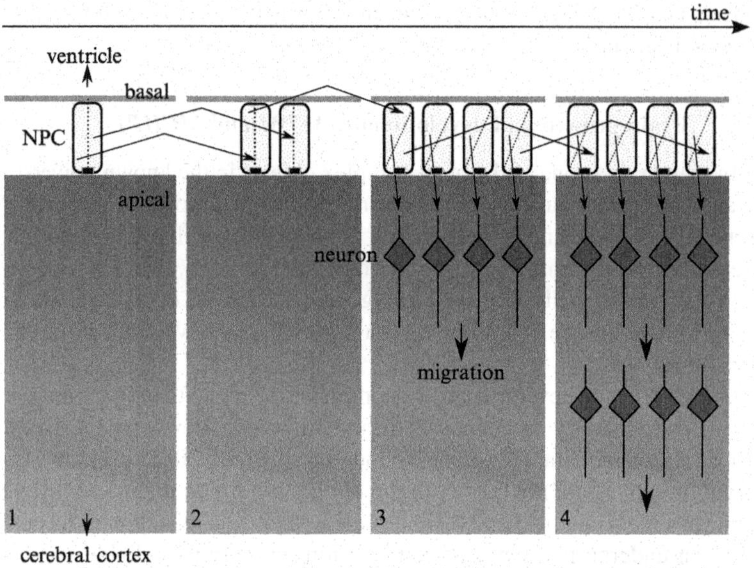

Figure 6.3 Neurogenesis and symmetric and asymmetric divisions. **Step 1**: a sigle NPC (neural progenitor cell) is shown, with its apical end (darker square) pointing towards the cerebral cortex (outside the brain) and its basal end towards the ventricle (inside the brain), also showing the vertical cleavage plane (dotted line) resulting in a symmetric division. **Step 2**: the two NPC daugther cells resulting from the two halves of the original NPC now undergo a second round of symmetric divisions (vertical cleavage plane), resulting in four NPCs in **step 3** that switch to an oblique cleavage plane leading to asymmetric divisions and four neurons (migrating towards their final destination in the cortex) and four new NPCs. **Step 4**: a new round of asymmetric divisions produces four new neurons and four potentially dividing NPCs.

head (occipito-frontal circumference at least two standard deviations below the mean for matched age, sex and ethnicity), associated mild to severe mental retardation and absence of other malformations or neurological problems (except sometimes for mild seizures). The brain volume is correspondingly reduced, with reduction especially marked for the cerebral cortex, but with largely normal gross brain anatomy. To date, several loci are known to be involved in MCPH, but we will focus on two of them: ***MCPH1*** (represented by the gene *Microcephalin*) and *MCPH5* (the ***ASPM*** gene). Deleterious mutations in both these genes result in primary microcephaly by affecting the number of symmetric divisions, but the actual mechanisms are different and instructive.

Microcephalin (*MCPH1*; OMIM 607117) is located on chromosome 8 and seems to be involved in cell cycle checkpoint control and DNA repair (Kaindl et al., 2009; Mahmood et al., 2011). Basically, the life cycle of a cell is a complex dance involving many steps that must be closely coordinated to ensure their proper timing and sequence, and this depends, among other things, on mechanisms for DNA damage repair – if there is DNA damage detected, the cell cycle must wait for it to be repaired, if possible, before continuing. *Microcephalin*, thus, seems to affect the number of symmetric divisions mainly through these roles, and the deleterious mutations resulting in MCPH may disrupt the ability of NPCs to undergo the required number of symmetric divisions.

*A*bnormal *SP*indle-like *M*icrocephaly-Associated (*ASPM*; OMIM 605481) on chromosome 1 is involved in determining the orientation of the cleavage (division) plane, helping thus to maintain the symmetrical cell divisions, and is down-regulated (i.e., its expression is reduced) when the cell switches to asymmetric divisions (Higgins et al., 2010; Kaindl et al., 2009). The deleterious mutations resulting in MCPH are proposed to alter the proper orientation of the cleavage plane, resulting in a reduced number of symmetric divisions. It is interesting to speculate on why mutations in such a gene involved in an essential process throughout the organism result in such a circumscribed phenotype (MCPH): one suggestion (Higgins et al., 2010) is that this specificity might be due to the extremely elongated shape of the NPCs requiring a very accurate control of the cleavage plane direction in order to produce a symmetric division.

Besides their potential role in explaining human brain evolution and development, *ASPM* and *Microcephalin* seem to play a role in the normal variation in brain size and anatomy as well, with an intriguing twist. Wang et al. (2008) have recently reported that an exonic SNP (rs1057090) within *Microcephalin*, resulting in a valine to alanine change at position 761 in a region of the protein seemingly important for its function in DNA repair and cell cycle checkpoint control, is associated with *cranial volume* in a sample of 867 unrelated Chinese individuals, but only in males (it has no effect in females)! This pattern of sex-specific association seems supported by a different study (Rimol et al., 2010) looking at variation in several measures of brain anatomy using MRI (such as mean cortical thickness and total cortical area) in a discovery sample of 287 Norwegian participants and a replication sample of 656 North American participants. Among other results, they found that three SNPs (rs2816514, rs2816517 and rs11779303) in *Microcephalin* and one SNP (rs10922168) in *ASPM* are associated with variation in brain volume, cortical area, and intracranial volume, but only in females. The authors also present brain-area-specific associations with cortical area (e.g., their Figures 2 and S1), clearly showing that these SNPs have not only global brain effects but also local, area-specific influences. Interestingly, these SNPs are not in coding regions of the genes but supposedly affect the regulation of these genes. As Montgomery and

Mundy (2010) point out, these studies could have several very important consequences. First, they are in line with the mechanism by which these genes are proposed to influence neurogenesis in that most associations in Rimol et al. (2010) concern cortical area and none cortical thickness. Second, they suggest that sex-specific factors might influence gene expression and normal variation, but also that different variants in the same gene (*Microcephalin*) might influence one aspect of a phenotype (cranial volume) in one gender (males) and others (brain volume, cortical area, and intracranial volume) in the other[1] (females). Moreover, and important for our discussion here, Rimol et al. (2010) showed that genes involved in such general, brain-wide phenomena can nevertheless have brain-area-specific effects. This could be down to area-specific interactions with other genetic and non-genetic factors. Of course, pending further replication and independent uncovering of molecular mechanisms using experimental techniques, these findings must be taken as suggestive only.

These two genes, *ASPM* and *Microcephalin*, might be relevant for language and speech, as was proposed some years ago by myself and D. Robert Ladd (Dediu and Ladd, 2007). Our proposal is that two alleles of these genes are associated with the use of linguistic tone (i.e., linguistic distinctions carried by voice pitch such as famously done in Chinese or Yoruba) by populations having different frequencies of these alleles (see Section 9.4.1). This proposal, based on a population-level correlation, has recently received support from an association study (Wong et al., 2012) showing that *ASPM* is indeed associated with tone perception and brain response to lexical tone in the auditory cortex. However, the sample size is extremely small and the detected effect is apparently in the opposite direction to the one suggested by Dediu and Ladd, 2007, and further work is needed before accepting that *ASPM* (and possibly *Microcephalin*) have anything to do with tone. Needless to say, even if this relationship were secure, it is currently unclear how these alleles might have such specific effects on pitch processing.

6.7 Regulating the expression of other genes: *FOXP2*

But how do all these genes in our bodies "know" where, when and how much of their product (protein or RNA) to produce? A naïve view that somehow seems to be the default position among many non-biologists is that a gene is active – or "read" – only once, during the development of the organism. This is a reflex of the still widespread metaphor equating the genome to a blueprint or, in a slightly better variant, a recipe that is used to build an organism. And just as you don't really need the recipe to eat the cake after you have cooked it (and only rarely and in exceptional circumstances the blueprint after you have

[1] Of course, assuming that ethnicity plays no role.

built the house), so the logic goes that the genome is not needed any more after the development of the organism has ceased.

However, this is patently false and in fact most genes continue to be expressed throughout life and to respond, sometimes instantly, to changes in the external or internal conditions. As many of the organism's responses to environmental stimuli ultimately depend on the right gene products (hormones, neurotransmitters, etc.), genes must adapt rapidly to changes in the internal and external conditions. For example, the so-called **immediate early genes** are activated immediately in response to various types of changes such as hormones, ionizing radiation or viral infections (Arlt and Schäfer, 2011) and some are essential to the plasticity of our nervous system and its capacity for fast learning and memory formation, even being useful as markers for neural activity (Loebrich and Nedivi, 2009; Pérez-Cadahía et al., 2011; Moorman et al., 2011; Clayton, 2013). Thus, the genome must be seen as an intrinsically dynamic system, always active, always responding to changes in the environment, but also capable of memory.

However, we should not fall into the other extreme, thinking that all genes are always active in every cell of our bodies. In fact, a moment's reflection shows that this cannot be the case given that all our cells share the same genome[2] and yet they are very different in form, structure, composition, lifespan and function, falling into several hundred types. Thus, somehow a neuron must know it is a neuron and not a muscle cell and act accordingly. More than that: there are many subtypes of neurons that use different neurotransmitters, express different receptors, might look strikingly different and certainly behave differently (Wichterle et al., 2013), and they must somehow keep track of which exact subtype they are. The trick is that a certain neuron expresses a different subset of genes than a certain muscle cell and this subset is relatively stable, ensuring that nerves don't suddenly become muscles or the other way around.

This differential gene expression is essential during development, where it makes sure that the right tissues and organs develop in the right place at the right time, and involves a complex ballet of genes being expressed or not depending on the particular location within the embryo, the time and the context (what the other cells are doing), as well as the history of development so far (for a fascinating yet readable account please see Carroll, 2011). One elegant mechanism through which such precise coordination and context-dependency is achieved is by having some genes being "more equal" than others – they *regulate* the expression of other genes. One way to do this is by controlling whether a gene is *transcribed* (is "on") or not (is "off") or, in a more

[2] With the notable exception of very few types that don't have a nucleus, such as red blood cells, and of any **somatic mutations** affecting only a subset of an organism's cells.

quantitative manner, the amount of transcription that happens; a valid metaphor would be a light switch in the first case and a dimmer in the second.

Such **transcription factors**, thus, are genes that regulate the transcription of other genes (their *targets*), that is they control whether messenger RNA (mRNA) is produced and in what quantities (Lee and Young, 2013). There are many types of transcription factors in the human genome (current estimates put them at well over 1000; Vaquerizas et al., 2009), but in general they act by binding to specific patterns of DNA and recruiting other proteins – so-called *cofactors* – to influence the activity of RNA polymerase II (the molecular machine transcribing DNA into RNA). There is certain specificity in what transcription factor binds to which DNA pattern, and the mechanisms involved are extremely complex and fascinating. We are far from having a complete understanding of these mechanisms, but they seem to involve both the direct recognition of short sequences of nucleotides as well as the actual three-dimensional shape of the DNA molecule (Rohs et al., 2010), highlighting the fact that an abstract, sequence-only view of the genome is very restricted and restrictive.

The patterns of activity of transcription factors are complex, as expected given their importance in every aspect of life including development, normal functioning and disease (Lee and Young, 2013). Some of them seem to be expressed quite generally among tissues ("ubiquitous" or "housekeeping" transcription factors) while others seem to be relatively tissue-specific ("specific" transcription factors), and an extremely important property is their combinatoriality both during development and afterwards (Vaquerizas et al., 2009; Carroll, 2011). Through their combinations and interactions, transcription factors manage to ensure both the general processes necessary for life and the tissue- and developmental stage-specific processes required for the development and functioning of complex organisms such as us.

The *FOrkhead boX transcription factors* (or *FOX*) are just one family of transcription factors characterized by the so-called *forkhead box* (or *winged helix*) domain, a sequence of about 100 amino acids involved in binding to DNA (Tuteja and Kaestner, 2007b). There are about 40 individual FOX genes identified in humans so far and they are involved in a large array of tissues and processes including development, cancer, and speech and language (Tuteja and Kaestner, 2007a,b; `http://www.genenames.org/genefamilies/FOX`). These genes are named using a letter to specify the subfamily and a number to denote the individual genes within a subfamily; thus there are to date 19 such subfamilies (*FOXA* to *FOXS*) and within the P subfamily there are four genes (*FOXP1* to *FOXP4*).

FOXP2's close friends, *FOXP1*, *FOXP3* and *FOXP4*, have a mixed set of functions (as is usual for transcription factors given their highly contextual and cooperative binding). *FOXP1* seems to be involved in lung (Li et al., 2012), heart (Chang et al., 2013) and brain development (there are mutations resulting in intellectual disability, autism and language problems; O'Roak et al., 2011; Hamdan et al., 2010; Horn et al., 2010) as well as some cancers (Katoh et al., 2013). *FOXP3* seems to primarily play an important role in the immune system (OMIM 300292) but also apparently in breast cancer (Douglass et al., 2012). *FOXP4* is less well understood, but it seems to interact with *FOXP1* in lung development (Li et al., 2012), could be involved in normal heart formation in mice (Li et al., 2004), the neural tube (Rousso et al., 2012), and seems to play a role in the Purkinje cells (a type of neuron) in the cerebellum, a structure important for the precise coordination of movement (Tam et al., 2011). Interestingly, the FOXP proteins seem to interact, forming *heterodimers* (a complex molecular machine formed by two different components) that regulate the transcription of other genes.

FOXP2 itself has benefited from an enormous amount of attention following its identification in 2001 (Lai et al., 2001) as the gene behind the KE family's (Sections 4.2 and 5.4) language and speech deficit, *Developmental Verbal Dyspraxia* (**DVD**; OMIM 602081), and it is now regarded as a true "molecular window" (Fisher and Scharff, 2009) into the genetic architecture of speech and language. Briefly summarizing the large literature on *FOXP2* (see also OMIM 605317, Fisher and Scharff, 2009, and Enard, 2011), there are at least 17 exons (`http://genome.ucsc.edu/cgi-bin/hgGene?hgg_chrom=chr7&hgg_gene=uc003vgz.3&hgg_start=113726364&hgg_end=114333826&hgg_type=knownGene&db=hg19`, July 2013) producing several isoforms (`http://www.uniprot.org/uniprot/O15409`, July 2013) through alternative splicing (see Section 3.7.2). These isoforms differ not only in length and composition (i.e. how many and which exons are kept in the mature mRNA resulting in the final protein; see also Section 3.7) but may also be expressed in different tissues and at different times. As far as we know, *FOXP2* is involved in brain development and functioning, but also in the development of the lungs and oesophagus (together with *FOXP1*; Shu et al., 2007).

The specific mutation found in the KE family is a single base substitution replacing a G in the normal allele with an A in exon 14, resulting in a protein that has a histidine at position 553 where an arginine should have been (R553H). This single amino acid change alters the forkhead (or winged-helix) domain of the protein, which is essential for the normal DNA binding of the FOXP2 protein.[3] More precisely, this change affects the region of the protein (helix H3) that makes direct contact with the target DNA (Lai et al., 2001;

[3] Concerning the notation of the gene and protein in various species, as a general rule, gene names use italic (*FOXP2*), the protein names the normal font (FOXP2); the mouse homologue (corresponding) gene and protein are *Foxp2* and Foxp2 respectively, and in all other species they are *FoxP2* and FoxP2. There is a single homologue gene in *Drosophila* (fruit fly), *FoxP*, for the whole subfamiliy *FOXP1–FOXP4* (Santos et al., 2011), suggesting that these four members originated by gene duplications and later divergence of function.

Stroud et al., 2006), disrupting the way FOXP2 binds to it (Vernes et al., 2006; Nelson et al., 2013). It is important to highlight again (Section 5.4) that the affected members of the KE family have one normal copy of the *FOXP2* gene and one copy with this mutation, being thus heterozygous. Given the general importance of *FOXP2*, this might explain why the affected members of the KE family have such a relatively circumscribed pathology instead of more generalized syndromes (or even fail to be born altogether): it might be the case that the single copy of normal *FOXP2* produces enough functional protein to allow a mostly normal development and functioning, but that some aspects of brain development cannot proceed normally with such a reduced amount of functional protein. This is termed **haploinsufficiency** and is related to *gene dosage*, the idea that the number of copies of a gene might have a phenotypic effect (see Section 4.3 for a discussion in the context of sex-linked genes); we have encountered a similar phenomenon in Section 6.2 when discussing the effects of certain mtDNA mutations on hearing.

There are currently several known mutations involving *FOXP2* and resulting in speech and language deficits, some affecting large regions of chromosome 7 such as deletions or rearrangements (e.g., Feuk et al., 2006; Shriberg et al., 2006; Lennon et al., 2007; Palka et al., 2012; Rice et al., 2012), but at least another one (MacDermot et al., 2005) is a point mutation from C to T in exon 7, leading to a stop codon and resulting in a truncated (shorter than normal) FOXP2 protein (R328X), which co-segregates with a speech and language pathology in three members of a family (not KE). Figure 6.4 shows the structure of the normal FOXP2 protein and the R553H and R328X mutations.

FOXP2 clearly plays an important role in brain development as shown by its involvement in neural development (Vernes et al., 2011), motor learning and

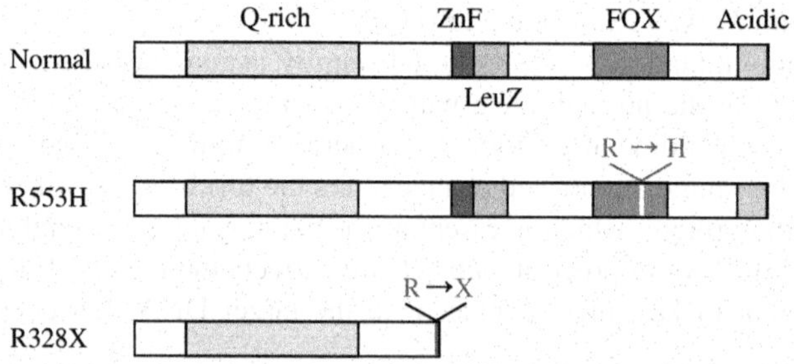

Figure 6.4 The schematic structure of the normal FOXP2 protein and the R553H (KE family; Lai et al., 2001) and R328X (MacDermot et al., 2005) mutations. The interaction of the protein with its target DNA happens through the FOX domain (dark grey). Adapted from Figure 1 in Vernes et al. (2006).

synaptic plasticity (Groszer et al., 2008; French et al., 2012), and the abnormal patterns of fMRI activation during language tasks in carriers of *FOXP2* mutations (Ligeois et al., 2003). Moreover, there are about 300 identified targets of *FOXP2* in the human brain (Spiteri et al., 2007; Vernes et al., 2007). Interestingly, even normal alleles of *FOXP2* present at high frequencies in the population (polymorphisms) seem to have an effect on brain activation without any obvious behavioural effects (Pinel et al., 2012); if replicated, these findings could provide a way to link pathology and normal variation, but more work is needed. *Foxp2* has also been experimentally studied in non-human animals such as mice and birds by manipulating the gene, resulting in very interesting findings: in songbirds *FoxP2* is involved in song learning (Haesler et al., 2007) and in mice the "KE" version of the gene[4] results in impaired motor learning and synaptic plasticity (Groszer et al., 2008), abnormal activity, neural plasticity and temporal coordination in the striatum, a subcortical brain area (French et al., 2012), as well as impaired learning of auditory-motor associations (Kurt et al., 2012). As we will shortly see, the normal human FOXP2 protein is characterized by two amino acid changes when compared to chimpanzees, and it has been suggested that these so-called "human-specific" changes are relevant for understanding the evolution of language; Enard et al. (2009) have produced a "humanized" version of the mouse gene carrying these two "human-specific" amino acids (thus, this is still a mouse gene!) and showed that it results in altered ultrasonic vocalizations and increased dendritic development and synaptic plasticity in the mouse. Work in bats also suggests that it might play a role in echolocation (Li et al., 2007) but more evidence is needed.

The evolutionary history of *FOXP2* is also intriguing and potentially informative for its role in speech and language. Enard et al. (2002) have shown that FOXP2 is one of the most conserved proteins in mammals, being in the top 5% in a direct comparison of about 1800 corresponding genes between humans and mice. In fact, there are very few changes in the protein since our split from the lineage leading to present-day mice (about 70 million years ago), suggesting that FOXP2 is under very strong selective pressures. This means that, whatever it does, it does not tolerate changes in this protein (we will discuss more about evolution and selection in later sections of this book). Thus, there is one amino acid change between mice and chimpanzees but, astonishingly, two more changes happened in our lineage since we split from the chimpanzees about 6 million years ago. These two amino acid changes are located in *FOXP2*'s exon 7 and seemed to be human-specific (i.e., no other

[4] This mouse mutation (R552H) has the same change as the KE family mutation (R553H) despite the slightly different position at which this occurs (552 vs 553); these two positions are equivalent (homologous) between mouse and human and the mouse protein is one amino acid shorter.

animal apparently had them), but Zhang et al. (2002) found one of them in carnivores (these include mammals such as cats, dogs, bears and seals). Enard and colleagues also observed that there seemed to be signatures of natural selection on the human form of *FOXP2* which they dated to about 200,000 years ago, apparently coinciding with the emergence of modern humans, and speculated that the two amino acid changes might be connected to it, possibly by influencing speech and language. However, this nice simple story turned out to be much more complex with the discovery that our cousins the Neandertals have had the same two amino acids in their *FOXP2* (Krause et al., 2007), suggesting that this "modern" form is actually much older, probably preceding our last common ancestor with the Neandertals about 400–600 thousand years ago. Moreover, Ptak et al. (2009) found that most probably the two "human-specific" amino acids were not the reason for the selective pressure on *FOXP2* in the human lineage, and later work (Maricic et al., 2013) suggests that the locus of the selective advantage might have been within intron 8. This locus is a potential binding site for another transcription factor (*POU3F2*) that might regulate the expression of *FOXP2* (yes, *FOXP2* itself is regulated by other transcription factors!) but the jury is still out concerning its exact effects. Thus, far from being a simple story, what we currently understand about the evolutionary history of *FOXP2* supports the idea that this is emphatically not "the" gene "for" language and that the KE family, and the other cases of *FOXP2*-related disorders, are not some sort of throwback to an earlier stage of human evolution (Fisher, 2006; Fisher and Scharff, 2009).

What *FOXP2* represents is nevertheless one of the best entry points into complex gene networks involved in language and speech. It can be argued that, in itself, the phenotypic effects of the deleterious *FOXP2* mutations such as seen in the KE family (Developmental Verbal Dyspraxia) are but a very rare problem, potentially not very relevant to understanding more common types of speech and language problems such as dyslexia or SLI, and even less so to the patterns of normal variation. But, as we've seen, *FOXP2* is interesting because it is a hub in a large network of other genes and, while damaging *FOXP2* might result in catastrophic and rare phenotypes, other genes in this network might have other types of effects on language and speech, being involved in more common types of disorders or even normal variation. In fact, we start to see that this might indeed be the case. Thus, it is important to remember that *FOXP2* exerts all these effects through the genes it regulates – therefore it is of the highest importance to try to identify these targets and how, when and where FOXP2 regulates them, and what "downstream" effects these genes might have. And it is exactly to one such target, the gene known as *CNTNAP2*, that we turn our attention in the next section.

6.8 Neural development and neural impulse conduction: *CNTNAP2*

As we have just seen, *FOXP2* is a gene that regulates other (*downstream*) genes either by reducing their expression (**down-regulation**, *negative* regulation or *repression*) or increasing it (**up-regulation**, *positive* regulation or *activation*), in the extreme turning them off completely or turning them on. This relationship between a transcription factor and its targets can be represented as a *network* of gene regulation, where the nodes represent genes and the links represent which gene regulates which targets, either up (usually represented as arrows, →) or down (usually represented as ⊣). Thus, *FOXP2*'s downstream regulatory network can be visualized as in Figure 6.5. Thus, an increase in *FOXP2*'s activity (resulting in more FOXP2 protein being available) would lead to more *target 1* and *target n* activation (thus more of their gene products being available) and less for *target 2* and *target 3*. However, the exact shape of the relationship can be complicated and dependent on other factors such as the type of tissue and exact location within the organism, the developmental stage and the activity of other molecules, including other genes and signals from the environment. Moreover, given the complexity of the interactions and the number of molecules usually involved in biological systems, such relationships are subject to fluctuations and should be imagined as being probabilistic rather than deterministic. For example, it could be that the concentration of FOXP2 must be over a certain threshold for turning *target 1* on (this can be due to a certain "preference" of FOXP2 for binding to *target 1*'s regulatory sequence) resulting in more or less a step function (as in Figure 6.6 panel A). Or the activation of *target 1* could be approximately linear with the amount of FOXP2 present up to a saturation point where the maximum possible number of FOXP2 proteins is already bound to *target 1*'s regulatory sequence(s) as shown in Figure 6.6 panel B. Or there can be much more complicated patterns due to interactions between FOXP2 and other cofactors.

Various techniques are available for identifying such regulatory networks such as *bioinformatic* techniques that look ***in silico*** (i.e., using computational

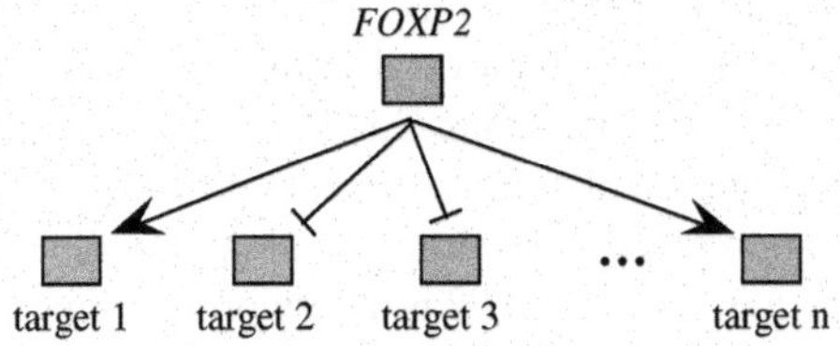

Figure 6.5 An abstract representation of the downstream part of *FOXP2*'s gene regulatory network showing several target genes and both up-regulating (→) and down-regulating (⊣) links.

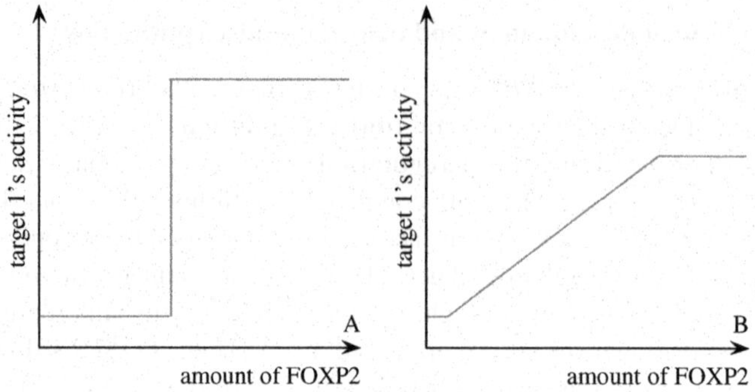

Figure 6.6 An abstract representation of two types of up-regulation: (**A**) step function (switching on and off) of *target 1* by FOXP2, and (**B**) a linear dependency of *target 1*'s activity function on the amount of FOXP2 with an upper bound.

techniques and databases) for probable DNA patterns that might bind the regulatory protein of interest, generating hypotheses that can be further tested using *wet lab* experiments investigating the actual binding of a regulatory protein to locations on the genome or the influence of the amount of regulatory protein in a cell on the quantity of other proteins built by the cell.

The *contactin-associated protein-like 2* gene, or *CNTNAP2*, was found by Vernes et al. (2008) to be down-regulated by FOXP2. In this wet-lab-based experiment, they used neuronal-like **cell lines** (populations of cells grown under controlled conditions and that have specific properties, including genetic makeup) expressing FOXP2 to look for regions in their genome to which FOXP2 actually binds. This was done using antibodies that recognize specifically FOXP2, allowing the extraction of pieces of DNA to which FOXP2 binds (the so-called "chromatin immunoprecipitation") and their subsequent sequencing and identification. Among other genomic regions, it was found that FOXP2 binds to a region in intron 1 of *CNTNAP2* and further experiments showed that it inhibits the production of CASPR2 (its protein product; please note that in this case the protein and the gene have different names).

CASPR2 is a member of the **neurexin** family, usually involved in the nervous system, and seems to participate in several crucial aspects of nervous system development and functioning (Rodenas-Cuadrado et al., 2013), such as neuronal migration and the formation of neural connections, as well as possibly in ensuring the fast transmission of the nerve impulse along myelinated axons. As discussed above (Section 5.3.7), *CNTNAP2* is a very promising gene, being involved in several disorders such as autism, intellectual disability and, especially important for us here, dyslexia, specific language impairment (SLI) and communicative behaviour in the normal population (see Rodenas-Cuadrado et al., 2013, for a recent and comprehensive review).

6.9 Post-transcriptional gene regulation: microRNAs

The *FOXP2*-style of gene regulation, by binding to control regions on the DNA and influencing whether – and how much of – its target genes are transcribed into messenger RNA (mRNA) that is further translated into protein, is not the only one possible. Because it affects the transcription of the DNA message into mRNA, this type of gene regulation is known as *transcriptional*. However, the transcribed RNA must mature into mRNA (including by getting rid of the introns and joining the remaining exons) and it must further be translated into protein, a complex process where regulation can take place. This type of gene regulation is known as *post-transcriptional* and can happen in several ways, one of them extremely interesting and involving very short single-stranded molecules of RNA.

These *microRNAs* (or **miRNAs**) are about 22 nucleotides in length, are encoded by specific genes in the genome,[5] and result from a complex process involving cutting the ~ 80-nucleotide-long precursor[6] RNA molecules (Yates et al., 2013; Wilson and Doudna, 2013). Their regulatory mechanism is based on base-pairing with complementary patterns of nucleotides in the target mRNA and, when such a pairing occurs, it can prevent the mRNA from being translated into protein or it can determine its decay. The base-pairing is imperfect and can happen to a relatively large set of similar sequences. Therefore, microRNAs can target hundreds or even thousands of mRNAs, regulating the expression of vast numbers of genes.

6.9.1 microRNAs and FOXP2

In mice, *Foxp2* is regulated by at least two miRNAs: *miR-9*[7] and *miR-132* (Clovis et al., 2012). Interestingly, *miR-9* is encoded by three genes: *MIR9-1* on chromosome 1q22, *MIR9-2* on 5q14.3 and *MIR9-3* on 15q26.1, all resulting in the same mature miRNA sequence (see OMIM 611186–611188), while *miR-132* is encoded by the gene *MIR132* on chromosome 17p13.3 (OMIM 610016). In turn, Foxp2 regulates *miR-124a* (*MIR124-1* on 8p23.1; OMIM 609327), *miR-137* (MIR137 on 1p21.3; OMIM 614304) and *miR-9* (Vernes et al., 2011). Therefore, there could be complex regulation cascades and feedback cycles whereby *FOXP2* regulates certain miRNAs which in turn regulate other targets (including *FOXP2* itself), and some of these other targets could be also directly regulated by *FOXP2*. *miR-132* seems to be involved in neural development, and currently miRNAs are a very active area of research as

[5] The end product of these genes is not represented by one (or more) protein(s) but instead by these RNAs.

[6] There is quite a bit of variation around these average numbers.

[7] Usually the names of miRNAs are composed of "miR-" followed by a number in sequence.

they are potentially involved in various aspects of neural development such as neurogenesis and neural proliferation, neural differentiation, the growth of dendrites and axons and synaptic plasticity (Vernes et al., 2011).

By identifying these miRNAs, both downstream but also upstream, *FOXP2*'s regulatory network was further expanded, delivering on its characterization as a "molecular window" into the genetics of speech and language (Fisher and Scharff, 2009). It becomes clearer that *FOXP2* is a hub in a complex network of genes controlling multiple aspects of development, normal functioning and disease, being far from "language-specific". This entails, as we will discuss more in the following chapter, that the evolutionary story behind modern language and speech cannot be reduced to a simplistic "*FOXP2* mutated and gave us modern cognition, speech and language in one package", but that it involves a complex evolutionary process affecting not only *FOXP2* but also many other genes and, crucially, regulatory elements.

6.9.2 *miR-96 and hearing loss*

For example, *miR-96*, a miRNA encoded by a gene[8] on chromosome 7 has been recently implicated in genetic hearing loss. More precisely, Mencía et al. (2009) found that mutations in a very conserved region of this gene (one involving a G to A transition at position 13 and another a C to A change at position 14) co-segregate with hearing loss in two Spanish families (one mutation in each family) and do not occur in healthy controls. This pathology is an autosomal dominant non-syndromic progressive hearing loss (*DFNA50*; OMIM 613074; Mencía et al., 2009) and the suggestion is that this miRNA (which is expressed in a tissue-specific manner) regulates the expression of genes essential for the correct functioning of the hair cells. In support of *miR-96's* role in hearing are studies using animal models (Lewis et al., 2009) which show that homozygous mice (carrying both copies of the mutation) are deaf while heterozygous mice show progressive hearing loss similar to that seen in humans.

However, miRNAs are not a common cause of hearing loss even if they seem to play essential roles in the development and functioning of the inner ear (Hildebrand et al., 2010; Patel and Hu, 2012), and, as expected for such powerful post-transcriptional regulators, *miR-96* seems to play roles in other processes such as cancer (Haflidadóttir et al., 2013). It will be very interesting to uncover *miR-96's* targets as this will further our understanding of the normal development of hearing as well as provide new candidate genes for various forms of hearing loss.

[8] Interestingly, the gene for miR-96 and the genes for two other miRNAs, miR-182 and miR-183, are transcribed as a single unit that is further processed to produce these three individual molecules.

7 The way forward: exome and genome sequencing

This very short chapter introduces whole exome and genome sequencing, which will probably represent a large section of future studies. However, at present their potential has not yet been used for uncovering the genetic architecture of language and speech. We also touch here on a very important issue, namely the nature of the genetic architecture of complex traits such as language and speech: are these controlled by a few genes of large effect or many genes of small effect? And do we have a full account of the quantitative genetic estimates of heritability in terms of genetic loci?

7.1 Exome and genome sequencing

How about the new and hot developments in **high-throughput sequencing** (including the "next-generation" and "third-generation" methods) that promise to make sequencing thousands of whole genomes feasible in terms of time, infrastructure required and costs? At the time of writing there is only one published pioneering study using such methodologies to investigate the genetic architecture of language and speech (Worthey et al., 2013), but it is to be expected that in the near future they will provide a sizeable proportion of unexpected and exciting new discoveries (see also Deriziotis and Fisher, 2013).

In a nutshell, it is feasible to sequence somebody's whole **genome** (or just the protein-coding part of it covering only the exomes – the *whole* ***exome***) and, in principle, we could look at any characteristics of the person's genome/exome and her/his phenotype of interest. However, this soon runs against a possibly unexpected problem, namely that individual genomes harbour a lot of variation, some of it shared by a sizeable proportion of the population (representing the so-called **polymorphisms**, usually taken to be those variants appearing in more than 1% or 5% of the population, SNPs being one example), but also some appearing only in a very few other individuals (such as other family members), being thus rare, and yet others specific to the individual (*de novo* mutations). Moreover, currently there are many false negatives and positives and the results look different when the same data is analysed using different methods and pipelines (e.g., O'Rawe et al., 2013, report a less than 60% agreement for single nucleotide variants and less than 30% for insertions and deletions between several pipelines). Which of these variants are responsible for the phenotype of interest and how can we find out?

One idea is to assume that certain severe phenotypes (pathologies such as autism or intellectual disability) that appear sporadically are due to new, private mutations appearing in the affected individuals and not in the other family members (O'Roak et al., 2011; Vissers et al., 2010). Given that these phenotypes usually result in a very low fitness (i.e., affect the individual in quite negative ways), natural selection actively removes the causative mutations from the population but they are constantly reintroduced through such individuals (we will discuss more about these evolutionary forces in Chapter 8). Thus, a sample of such sporadic cases and their parents is sequenced and the *de novo* variants identified, but usually their list is longer than just the variants actually involved. However, using various filters (such as assuming that the variant must change the protein produced by the affected gene or that the gene is involved in processes relevant to the phenotype of interest such as neurocognitive development and functioning), the list of candidate variants can be reduced to a manageable number that can be further tested (after checking that they are not false positives) using animal models or cultured cells, for example.

Another idea is to sequence several individuals with the same phenotype and hope that the same rare variant is involved in a subset of them. Thus, one would hope that low-frequency variants will pop up in several affected individuals but not (or with a much lower frequency) in unaffected controls, suggesting that this variant is involved in the phenotype. Yet another idea is to combine linkage and sequencing and look for families segregating the phenotype of interest and looking for rare variants (even family-specific ones) that are shared by the affected family members but not by the non-affected ones or the non-affected general population. It is important to highlight that all these strategies aim to find rare variants that are co-transmitted with the disease but the methods used are slightly different depending on the particular setting. In the end, they will produce lists of candidate rare variants that must be further investigated using other methods (bioinformatics, involvement in relevant pathologies, wet lab-based, etc.), a painstaking process whose ultimate success is very dependent on the procedure used to generate and filter the candidate list.

Despite these difficulties, whole genome (and, until this becomes cheap and reliable enough, exome) sequencing[1] clearly represents the way forward in the investigation of the genetic architecture of speech and language. The data

[1] Besides the (still) higher costs, whole genome sequencing identifies vastly more variants than whole exome, and it is currently unclear how to deal with this deluge of data (most of which is noise for the purposes of a particular study).

produced by it can then be used in linkage and association studies just as today we use genotype data at a limited number of polymorphic markers, but we can also use methods designed to look for rare variants; the power of these new data-generation methods is that we will have access to vast amounts of variation of all sorts and we can afterwards choose which types to focus on.

7.2 The "missing heritability" and the genetic architecture of complex traits

Classic quantitative genetic approaches (Section 2.4) allow us to obtain estimates of the heritability of a given phenotype using designs such as twin and adoption studies. Moreover, multivariate techniques also allow the estimation of the genetic correlation between phenotypes, resulting in the proposal that there might be "generalist" genes involved in a large set of traits, but also "specialist" genes whose influence is more limited. It is therefore expected that there are actual molecular mechanisms behind these estimates, resulting in the prediction that such genetic loci will be revealed through linkage, association and whole exome/genome sequencing studies, and that the amount of variation explained by these loci should account for the estimated heritability. Therefore, some years ago, it came as a surprise for some that the then known loci involved in well-understood, well-behaved and highly heritable traits such as height seemed to only explain a tiny fraction of the estimated heritability. The term **missing heritability** was coined to highlight this paradox (Manolio et al., 2009) and much work has been dedicated to solving it, not only because geneticists hate not understanding what has been called the "dark matter of the genome"[2] (just like physicists do with their own version of the stuff), but more importantly because it might provide a better understanding of the genetic architecture of complex traits (Gratten et al., 2014; Gibson, 2012), language and speech included.

As it happens, height is a very good model to study for several reasons (Durand and Rappold, 2013; Visscher et al., 2010b). First, height is (almost) normally distributed in humans, with the bulk of the population around the mean and with only a small proportion at the extremes (both very short and very tall). Second, it is highly heritable, with modern and very reliable estimates at $h^2 \approx 0.80$ (Yang et al., 2010; Allen et al., 2010), that is (Section 2.4.5), about 80% of the inter-individual variation in height is due to variation in the genotype among those individuals. Finally, it is easy and cheap to obtain reliable measures, it is routinely included in many medical studies and

[2] Please note that there's at least another "dark matter of the genome" namely the non-protein-coding majority of our DNA.

databases as well as other sources such as conscription records, and even self-report is highly reliable (Visscher et al., 2010b), meaning that we can access many hundreds of thousands of participants with high-quality phenotypic data (unfortunately, something we are quite far from when it comes to language and speech). These characteristics have made height an important case study in genetics and evolutionary theory since at least the foundational works of Francis Galton (at the end of the nineteenth century) and Ronald Fisher (the beginning of the twentieth; Fisher, 1918).

Thus, the first two characteristics suggest that a large number of genes are involved (i.e., it is a quantitative trait) and that the genetic influence is quite important. Of course (Section 2.4.5) the environment is very important as well as shown by the obvious effects that nutrition and health care have on adult height. In fact, there is a so-called *secular trend* (Cole, 2003; Danubio and Sanna, 2008) representing the increase in average height across several generations in the recent past (from about the middle of the nineteeth century), especially visible in Europe and North America, but also starting to become manifest in other regions as well (such as China). In some cases this increase was spectacular (Cole, 2003), with the Dutch being famously the tallest nation in the world. This secular trend is mostly attributed to environmental changes, especially to better nutrition and health care during the early years of life.

There are several known genes where mutations have a large impact on height (Durand and Rappold, 2013), either resulting in very short stature (or dwarfism; currently about 130 genes) or in extremely tall individuals (gigantism; currently about 15 genes are known), with some of these genes being involved in bone growth or growth hormone signalling (Durand and Rappold, 2013). These mutations tend to be rare in the population, have high penetrance, large effects and tend to be responsible for the extremes of the distribution.

However, the bulk of the normal variation in height across individuals in the population cannot be explained by such mutations of large effect. In fact, recent genome-wide association studies (Section 5.3) involving thousands of individuals (for example, 183,727 in the GIANT consortium meta-analysis; Allen et al., 2010) have identified more than 180 loci associated with variation in height. Notably, each such locus explains only a tiny fraction of the observed variation in height; for example, the effect sizes of such loci are tiny, explaining up to 4 mm of variation in height (Visscher et al., 2010b). When combining together all these 180 loci, they explain only about 10% of the observed variation in height (Allen et al., 2010), thus accounting for only one eighth ($\frac{1}{8}$) of the estimated heritability (Visscher et al., 2010b).

So, where is the "missing heritability"? Various proposals exist (Manolio et al., 2009; Visscher et al., 2010b; Gibson, 2012). First, the current approaches are not well suited to discovering gene–gene and gene–environment interactions and we suspect that these play a major role at least for some phenotypes;

maybe the missing heritability is hidden by our incapacity to detect these interactions? Second, the current association studies do not consider rare variants (remember that a requirement for a polymorphism to be considered a SNP is to have a minor allele frequency – or MAF – of at least 1%), so maybe the missing heritability is due to these? Third, as discussed in section 5.3, we are rarely lucky enough to include the actual causative locus in an association study but instead use proxies, that is SNPs in linkage disequilibrium with the untested causative loci; however, LD is in most cases less than perfect and maybe some of the heritability is simply lost due to these imperfect correlations. Fourth, accepting that a SNP is associated with the trait of interest requires that the association passes the draconian requirement imposed by multiple testing correction (e.g. a p-value of less than 5×10^{-8}) and we simply do not have the statistical power to detect all SNPs with tiny effect sizes with the current sample sizes; maybe there are thousands of such loci, each explaining too little of the variation in height to be individually significant (and thus to be included in lists of accepted associations) but collectively explaining a sizeable fraction. Finally, maybe our estimates of heritability, mostly derived from various family designs (Section 2.4.4), are inflated and there's really nothing to be explained.

The first proposal, highlighting gene–gene and gene–environment interactions, is not relevant when discussing narrow-sense heritability, which estimates the proportion of phenotypic variation due to additive genetic effects (Section 2.4.2), because, by definition, it does not include any non-additive effects such as gene–gene and gene–environment interactions (Visscher et al., 2010b). As it happens, most heritability estimates published are of narrow-sense heritability. This does not exclude the importance of these effects but, for height at least, given that its narrow-sense heritability is about 80% it would mean that these effects would explain far less than the remaining 20% (as the environment shapes variation in height as well).

The second requires that we move beyond the current association study methodology and sequence the participants in order to also detect and consider any rare (and even *de novo*, i.e., not inherited from the parents but representing a new mutation arising in a particular individual) variants. With the spread of whole genome and whole exome sequencing (Section 7.1) this will become more and more feasible and fascinating findings will no doubt emerge. However, it is unclear what exactly to expect (Gibson, 2012) and most probably their impact differs between phenotypes. For example, rare variants (especially variation in copy numbers or CNVs) seem to be involved in several neurological and psychiatric phenotypes such as autism, intellectual disability or schizophrenia possibly by contributing to a genetic load resulting in an increased liability which, given environmental or other genetic factors, can result in the onset of disease (Gratten et al., 2014; O'Roak et al., 2011; Vissers et al., 2010). For height (and maybe other forms of normal variation), rare variants (with MAF < 1%) seem to also explain some of the variation (Yang et al., 2010; Visscher et al., 2010a).

How about three and four, incomplete LD between the tag SNPs and the causative loci, and the statistical impossibility of detecting most associations with small effect sizes with current samples and multiple testing correction requirements? A recent investigation of associations for height (Yang et al., 2010; Visscher et al., 2010a) showed that when considering incomplete LD and taking into account the influence of all genotyped SNPs (in this study 294,831 SNPs in 3925 participants) on the phenotype, then 45% of the variance in height can be accounted for. However, the remaining half of the unaccounted variation is probably hiding in infrequent variants that are not well covered by current methods.

Finally, how can we be sure that we actually have a problem to explain: what if our estimates of heritability are wrong? There are strong criticisms of the classic quantitative methods for estimating heritability, especially of the twin designs (e.g., Charney, 2012) but, while these clearly have a point, heritability estimates are usually quite robust across methods and designs (e.g., Plomin and Simpson, 2013). Moreover, the recent introduction of Genome-wide Complex Trait Analysis (or GCTA; Yang et al., 2011, Section 2.4.4), which allows the estimation of heritability from genetic (SNP) data in large samples of unrelated participants, has largely confirmed the heritability of complex traits such as intelligence (Plomin and Simpson, 2013; Davies et al., 2011; Plomin et al., 2013), but usually with lower estimates probably due to variants that were not covered by the genotyping platforms.

Interestingly, there is some overlap between the genes with mutations of large effect resulting in extremes of height and the lists generated by association studies of normal variation in the population (Durand and Rappold, 2013) but of course the effect sizes differ drastically between the two. Nevertheless it shows that the same loci can have both dramatic, large effect size, as well as small effects, depending on the particular variant. This is paralleled by findings that the same genes can be involved in both speech and language pathologies as well as the normal variation (e.g., *CNTNAP2*, *FOXP2*, or *ROBO1*).

In conclusion, using height as a model trait (but also supported by recent advances in understanding neurological and psychiatric disorders; Gratten et al., 2014), we can expect that language and speech have a complex genetic architecture involving many loci (very probably many more than height). Some very rare variants at these loci might have large effects producing speech and language pathologies such as DVD in the KE family, but most of the observed variation will be due to alleles of relatively small effect, both common and rare in the population. Understanding both the devastating rare mutations (such as *FOXP2* in the KE family) and the very small ones involved in normal variation (such as *CNTNAP2*, *ROBO1*) will provide entry points to unravelling the genetic architecture subtending language and speech. It is impossible a priori to decide that the study of a gene discovered through the major effects of some of its variants on rare diseases is irrelevant for understanding normal variation (and vice versa); we need to be open-minded and grab any opportunity nature gives us at cracking this extremely tough problem.

8 Population and evolutionary genetics

In order to properly appreciate the genetic bases of language and speech it is necessary to not only uncover the molecular and neuro-cognitive mechanisms involved, but to also shed light on the evolutionary pressures that have shaped them. This chapter introduces fundamental concepts of population and evolutionary genetics such as genetic drift, population structure and the various types of selection, and how we can infer their action from genetic data. These will provide the background for discussing the evolution of humans, the present-day genetic structure of our species and some cases where natural selection is (surprisingly for some) still shaping us.

What we have discussed so far about genes and their effects made relatively little reference to **evolution** and instead focused on the methods used to discover genes, the **proximate** molecular mechanisms through which they affect the observable phenotype, and their complex inter-relationships. However, as Dobzhansky (1973) famously argued more than 40 years go, "nothing in biology makes sense except in the light of evolution" and the genetic bases of speech and language are no exception. Quite the opposite, we need to understand the *why* questions behind the *how*'s, such as: why are there so few genes in humans? Why are the interactions between genes so complex? And why are there differences in interactional complexity among genes?

The next sections will introduce the notions of evolutionary biology needed in order to understand questions relevant for the genetic bases of language and speech but will also briefly sketch, where feasible, the larger context into which these notions, methods and results fit. Therefore, it is not intended to be a general introduction to these topics as there are many excellent textbooks such as Halliburton (2004) and Jobling et al. (2013), nor does it cover all possible aspects or go into great depth, but instead tries to sample a selection of relevant topics, with an accent on recent developments that might help shed light on language evolution and change.

8.1 Foundations of population genetics: loci, alleles, individuals and populations

Until now, we have focused on the molecular aspects of our genome, on how it influences the phenotype and on some of the methods used to find and study such influences. However, individuals are embedded within **populations** and it is to the populations and their dynamics through space and time that we now turn our attention. From a very abstract point of view, populations are composed of similar *individuals*, each individual having a **genome** that can be usefully reduced to a set of **loci**, each copy of the locus being occupied by one of a set of possible **alleles**.

For example, we might focus on a single autosomal locus that has two alleles, denoted *A* and *a*; therefore, each individual's genotype at this locus can be either *AA*, *aa* (both known as **homozygous**) or *Aa* (**heterozygous** – please note that it is usually assumed that the genotypes *Aa* and *aA* are equivalent). Thus, a population with nine individuals might be composed of two homozygous *aa*, five heterozygous *aA*, and two homozygous *AA* (see Figure 8.1). In this case, the proportion (or *frequency*) of *aa* individuals in the population is $p_{aa} = \frac{2}{9} \approx 0.22 = 22\%$, that of *aA* individuals is $p_{aA} = \frac{5}{9} \approx 0.55 = 55\%$, and that of *AA* individuals is $p_{AA} = \frac{2}{9} \approx 0.22 = 22\%$. Please note that the sum of the *genotype frequencies* of all genotypes present in the population (with possible but absent genotypes having a frequency of 0%) equals 100% (here, $p_{aa} + p_{aA} + p_{AA} = \frac{2}{9} + \frac{5}{9} + \frac{2}{9} = \frac{9}{9} = 1 = 100\%$). Another way of looking at this is to "dissolve" the individuals and look at the frequency of the alleles in the resulting *gene pool*: thus there are nine *a* alleles and nine *A* alleles, with *allele*

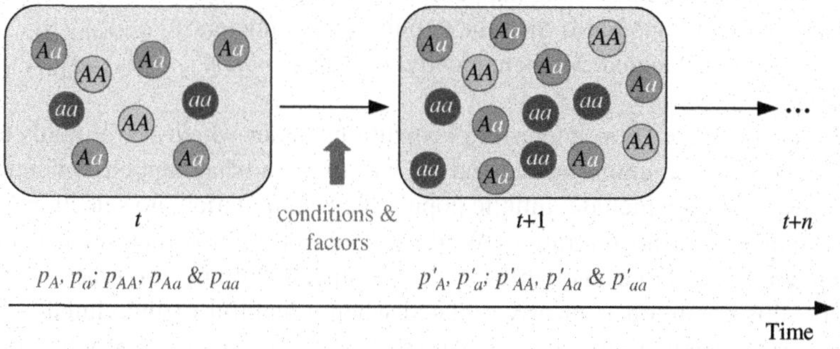

Figure 8.1 An idealized population composed of individuals each having a specific genotype at a given autosomal locus with two alleles, *a* and *A* (thus, a biallelic locus). Shown are the population-level allele and genotype frequencies and the evolution of the population from timestep *t* to the next timestep *t*+1.

frequencies of $p_a = \frac{9}{18} = 0.5 = 50\%$ and $p_A = \frac{9}{18} = 0.5 = 50\%$. It is easy to see that the allele frequencies can be obtained from the genotype frequencies:

$$\begin{cases} p_a = p_{aa} + \frac{p_{aA}}{2} \\ p_A = \frac{p_{aA}}{2} + p_{AA} \end{cases} \tag{8.1}$$

but the reverse is trickier. For our example, $p_a = \frac{2}{9} + \frac{1}{2} \cdot \frac{5}{9} = \frac{2 \cdot 2 + 5}{2 \cdot 9} = \frac{9}{18} = 0.5$, but nine a and nine A alleles can be arranged into several different sets of nine genotypes, such as all nine heterozygous aA, or our original 2:5:2 aa:aA:AA arrangement.[1] For such biallelic loci (i.e., loci with only two possible alleles), a usual shorthand notation is to denote p_a by p and p_A by q, and given that a and A are the only possible alleles, it follows that $p + q = 1$, or equivalently that $p = 1 - q$.

One of the main goals of this chapter will be to understand the *evolution* (i.e., change) across time (and space) of these population parameters, namely what happens to the allele and genotype frequencies (here, p, q, p_{aa}, p_{aA} and p_{AA}) when certain factors and conditions apply to a population, resulting in possibly new values in the next generation (here, p', q', p'_{aa}, p'_{aA} and p'_{AA}). When do these parameters change and in what ways? Can we predict these changes? And, more importantly for us, can we use the observed values of such parameters to say something about the past history of the population?

8.2 A useful baseline: the Hardy-Weinberg equilibrium

The so-called **Hardy-Weinberg equilibrium** (or **HWE**, named after its discoverers at the beginning of the last century) represents a very useful baseline against which to understand the dynamics of populations. Briefly put, it states that when nothing interesting happens, well, then nothing interesting happens: if a set of conditions holds then the population parameters p, q, p_{aa}, p_{aA} and p_{AA} remain unchanged. Mathematically, $p' = p, q' = q$, $p'_{aa} = p_{aa}$, $p'_{aA} = p_{aA}$ and $p'_{AA} = p_{AA}$.

However, these "non-interesting" conditions under which HWE holds turn out to be far from uninteresting and we will explore what happens when they are violated. They include: infinite populations, no mutation, no selection, random mating and no migration. While it could be argued that some natural populations are large enough as to be effectively infinite for practical purposes, this condition is clearly violated when a handful of individuals leave their source population to colonize a new habitat (as has frequently happened during human history when new islands and even whole continents have been settled by relatively small groups) or when various constraints prohibit large

[1] The notation 2:5:2 aa:aA:AA is a shorthand for two objects of type aa, five of type aA and two of type AA.

populations (like living in a marginal environment such as a desert). As we saw in the previous chapters, mutation is ubiquitous and a fundamental force in driving biological evolution; likewise, while there might be loci evolving below the radar of selective forces, since Darwin we know that selection is the "blind watchmaker" that explains adaptation in the living world. Random mating assumes that everybody in the population can have children with everybody else with equal probability, but we all know that not to be true due to simple geographical constraints or to more subtle, but equally powerful, cultural phenomena. Finally, while some adherents of right-wing ideologies would like to dream of a world without immigrants and populated only by "pure natives" (however arbitrarily defined), we will see that migration is a widespread phenomenon and an essential force shaping genetic diversity.

Within HWE's assumptions, it can be shown that the following derivations hold. Given the genotype frequencies in the population p_{aa}, p_{aA}, and p_{AA}, these represent the **diploid** individuals that produce **haploid gametes** (or sex cells; spermatozoa for males and ova for females; Section 3.5) of only two types: *a* and *A*: an *aa* individual can produce only *a* gametes, an *AA* only *A* gametes, while *aA* produces *a* and *A* in equal proportions. Thus, the frequencies of the gametes are the same as the allele frequencies, p for *a* and q for *A* (equation 8.1). Now, assuming random mating, any gamete is equally likely to combine with any other gamete, meaning that there will be $p \cdot p = p^2$ offspring with genotype *aa*, $q \cdot q = q^2$ offspring with genotype *AA*, and $p \cdot q + q \cdot p = 2pq$ offspring with genotype *aA*. Thus, in the next generation there will be $p'_{aa} = p^2$ *aa* individuals, $p'_{aA} = 2pq$ *aA*, and $p'_{AA} = q^2$ *AA*, and (by 8.1) $p' = p^2 + \frac{2pq}{2} = p^2 + p(1-p) = p(p+1-p) = p$ and $q' = q^2 + \frac{2pq}{2} = q^2 + (1-q)q = q(q+1-q) = q$. It can be seen that the allele frequencies remain the same ($p' = p$ and $q' = q$) even after one generation of HWE, but the genotype frequencies might change in this first generation as it is not necessary that initially $p_{aa} = p^2$, $p_{aA} = 2pq$ and $p_{AA} = q^2$, but they will stabilize afterwards at these values.

The HWE is used as a null hypothesis against which observations from real populations are tested. Let us assume that we went out there and gathered both allele and genotype frequencies at a biallelic locus of interest, say a SNP with two possible alleles *T* and *G*, in 1000 people, obtaining the following frequencies: $p = p_T = 0.3 = 30\%$, $q = p_G = 0.7 = 70\%$, $p_{TT} = 0.1 = 10\%$, $p_{TG} = 0.4 = 40\%$ and $p_{GG} = 0.5 = 50\%$. First, the sanity checks: $p + q = 0.3 + 0.7 = 1.0$ and $p_{TT} + p_{TG} + p_{GG} = 0.1 + 0.4 + 0.5 = 1.0$, as it should be. Second, the HWE: we would expect theoretically to have $p^2 = 0.3 \cdot 0.3 = 0.09 = 9\%$ TT individuals, $2pq = 2 \cdot 0.3 \cdot 0.7 = 0.42 = 42\%$ TG individuals, and $q^2 = 0.7 \cdot 0.7 = 0.49 = 49\%$ GG individuals. These expected frequencies turn out to be very close to the actually observed frequencies (10%, 40% and 50% respectively), suggesting that the population is indeed under

HWE. We can statistically test this using a χ^2 test or Fisher's exact test (see Section 4.2 for details); for our example, we have to compare the observed genotype frequencies (10%, 40% and 50%) in the 1000 individuals with the expected genotype frequencies given the observed allele frequencies if HWE were to hold (9%, 42% and 49%). This results in a small and insignificant $\chi_2^2 = 2.3$, $p = 0.32$, failing to reject the hypothesis that the population from which these individuals came evolves under HWE. Alternatively, let us say that we genotype at the same locus another 1000 individuals coming from a different population and we find the same allele frequencies $p = p_T = 0.3 = 30\%$ and $q = p_G = 0.7 = 70\%$, but different genotype frequencies $p_{TT} = 0.2 = 20\%$, $p_{TG} = 0.2 = 20\%$ and $p_{GG} = 0.6 = 60\%$. Applying a χ^2 test now gives an enormous $\chi_2^2 = 274.34$ and an extremely significant $p < 2.2 \cdot 10^{-16}$, clearly rejecting HWE for this second population. In practice HWE can be tested using a large variety of software such as `PLINK` (`http://pngu.mgh.harvard.edu/~purcell/plink/`) or the `HardyWeinberg` package of the statistical environment `R` (`http://www.r-project.org/`).

A population can deviate from HWE for a multitude of reasons, from the most trivial (genotyping or data manipulation errors) to the most interesting (selection, population structure, etc.), which we will explore in the following sections. Therefore, finding a deviation from HWE does not in itself constitute an indication of which such reason(s) might be at work, more specific tests and caution in jumping to interpretations being required.

8.3 Genetic drift: the power of chance

Many things can go wrong in a small population, but the most relevant for us here is that *random fluctuations* become important, so important in fact that they are the main factor behind the dynamics of very small populations. To understand why, let us remember that a main condition behind HWE is that everybody leaves the same number of offspring in the next generation (otherwise the allele and genotype frequencies would be different between generations). This is ensured by the equality of fitness between individuals (all are equally good at surviving and reproducing) and the very large (infinite) size of the population. If the population is small, even if everybody is as good at surviving and reproducing, inevitably some will just happen – not because they are better in any sense – to leave more offspring than others.

In the extreme, imagine a population with two individuals with genotype *aA*, producing the next generation also limited to two individuals. The possible types of offspring that these two can produce are *aa*, *aA* and *AA* with a ratio of 1:2:1 as shown in the table below displaying the gametes produced by each parent and their possible combinations (see also Section 4.1):

$aA \times aA$	a	A
a	aa	aA
A	aA	AA

Thus, if we are trying to keep the same allele frequencies in the next generation, we need either two *aA* individuals or one *aa* and one *AA*. By using this table, the probability of randomly picking two *aA* offspring is $50\% \cdot 50\% = 25\%$, and another 25% to pick an *aa* and an *AA*; thus, there's a $100\% - (25\% + 25\%) = 50\%$ chance that the next generation will have different allele frequencies from the parental one (and 75% that it will have different genotypes). If we keep following this population for several generations, it becomes clear that the probability of keeping both alleles soon dwindles to almost zero, resulting in one of them becoming *lost* and the other *fixated*.

A graphical illustration is given in Figure 8.2, which shows the evolution of the allele and genotype frequencies across 500 generations in populations of

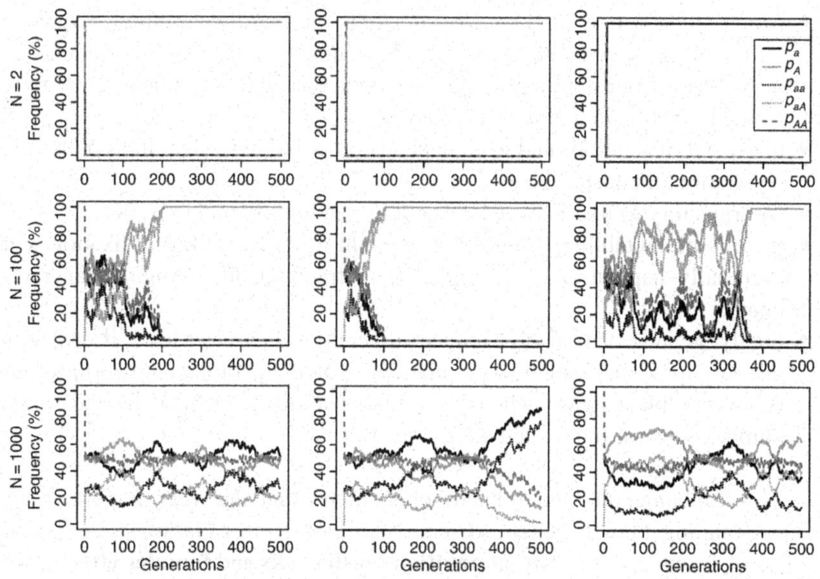

Figure 8.2 Simulations of genetic drift in a population with N individuals for a diploid biallelic locus. Initially (generation 0) the population is composed only of heterozygous individuals *aA* and evolves under the assumptions of HWE (except for finite population size). Each row represents three independent runs (the columns) for a given population size (top to bottom $N = 2$, $N = 100$ and $N = 1000$). Shown are the allele frequencies p_a and p_A, and the genotype frequencies p_{aa}, p_{aA} and p_{AA} (see also the legend in top rightmost plot).

various sizes (2, 100 and 1000, respectively), showing three independent runs for each population size (the R code is in Appendix A.3). It can be seen that while all populations start the same, composed only of heterozygous individuals *aA* (thus initially, in generation 0, $p_a = p_A = 50\%$, $p_{aA} = 100\%$ and $p_{aa} = p_{AA} = 0\%$), the allele and genotype frequencies start drifting away at different speeds. In the smallest possible populations ($N = 2$), one of the alleles immediately goes to **fixation** (frequency 100%; *A* two times, *a* once in this simulation) while the other is lost (frequency 0%), even before generation 10. In the larger $N = 100$, fixation (and loss) also happens but much more slowly (by about generations 200, 100 and 400 in each of the runs), while for the largest ($N = 1000$), fixation (and loss) still had not happened by the end of the simulation (500 generations).

There are several important points to note about genetic drift:

- the fixation of one allele and the loss of the complementary allele are unavoidable in finite populations;
- which allele gets fixated (or lost) depends on the allele's initial frequency but is also dictated by chance (i.e. more frequent alleles have a lower chance of being lost but nothing is guaranteed except when the allele is the only game in town with a 100% frequency);
- the speed of fixation (and loss) depends on population size: the smaller the population, the faster this is;
- the trajectory to fixation is dominated by random fluctuations, but
- there is predictability behind this apparent drunken walk (i.e. fixation will eventually happen and will affect more frequent alleles more than rarer ones);
- finally, genetic drift always reduces diversity (alleles get lost); if drift was the only force dictating how populations evolve, the biological world would be a very bleak place where everybody would be genetically the same, homozygous for the fixated allele.

Two fundamental (and related) phenomena that are very important for understanding human genetic diversity are sometimes classified as special cases of strong genetic drift: **population bottlenecks** and **founder effects** (see Figure 8.3). When a population's size crashes due to, for example, epidemics (think of the Black Death in medieval Europe with an estimated 40% of people dying or the 1918 flu pandemic – aka the "Spanish influenza" – which killed 3–5% of the world's population), wars (not only modern wars, but also, importantly, traditional warfare as well; Diamond, 2012) and many other reasons, the survivors will represent a small sample of the parent population's genetic diversity. Such population bottlenecks can thus result in major loss

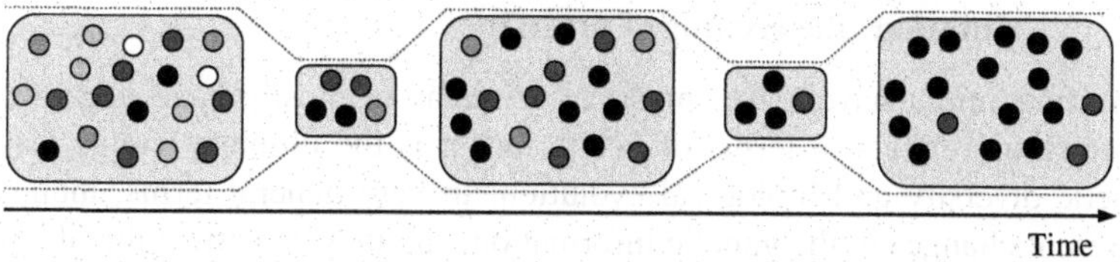

Figure 8.3 Population bottlenecks and (repeated) founder effects result in a loss of genetic diversity through random sampling: in the first bottleneck the "white" and "light grey" variants are lost, while through the second bottleneck only the "black" and "dark grey" variants survive.

of genetic diversity and leave their signature in the genetic structure of the survivors, with important consequences, one being that population geneticists can recover such episodes.

Likewise, when a small daughter population splits off and colonizes a new environment (such as a new island, continent or the land beyond those mountains or desert), the colonists will inevitably represent a small sample of the original population's genetic diversity. Thus, the populations that result from them will carry the signature of this initial founder effect. When daughter populations continue to split off and colonize new territories, settle, grow and then split off again, the result of such a repeated (or *serial*) founder effect is a cline of decreasing genetic diversity radiating from the place of origin (where the most diversity is found) and the further away one moves the less genetic diversity there is. An important example of a serial founder effect is given by the strong decrease in human genetic and phenotypic diversity the further away one moves from Africa, probably resulting from the process through which our ancestors left Africa many tens of thousands of years ago and colonized the whole world by repeatedly splitting off, expanding and settling into new territory (Manica et al., 2007; Betti et al., 2009).

Intriguingly, Atkinson (2011) has recently proposed that a similar cline of decreasing diversity with increasing distance from Africa can be seen in the phonological complexity of the world's languages, and that this cline results from a similar process whereby daughter languages lose phonological contrasts present in the parent language. However, this result and explanation have been heavily criticized on multiple grounds (such as the quality of the data, the definition of phonological complexity, the methodology used to search for such clines) and the consensus now is that while the proposal was bold and represented a major impetus for methodological and conceptual advance in linguistic typology, it is most probably flawed (e.g., Cysouw et al., 2012; Sproat, 2011; Jaeger et al., 2011).

8.4 Mutation: the creator of diversity

Mutation might have a bad press, as it seems to mostly have negative (or **deleterious**) effects in terms of health, but it is the ultimate engine driving genetic diversity and biological evolution. In very general terms, mutations represent changes to the genetic information, and they can affect the offspring when they occur in the **germ line** (the **gametes** or sex cells, the spermatozoa and ova in humans), which is the most commonly known type and the focus of this section, or they can affect cell lineages within an organism (**somatic mutations**) without being transmitted to the organism's offspring, and that can have various effects, the best known being cancer.

There are many types of mutations affecting the germ cell and thus potentially transmitted to the offspring. The simplest type is represented by a change in a single DNA letter (nucleotide), known as a **point mutation**, such as the replacement of a G with an A in exon 14 of *FOXP2* in the KE family (see Section 6.7). When there are enough copies of this new variant (or allele) in a population such that the *Minor Allele Frequency* (or **MAF**) goes over a certain threshold such as 1% or 5%, the mutation becomes known as a *Single Nucleotide Polymorphism* (or **SNP**).

Alternatively, a single nucleotide can be deleted or a new one inserted, potentially resulting in a **frameshift** mutation if they happen to change the three-letter message in a gene that is then translated and transcribed into an almost completely garbled protein (as happens in several diseases such as in some cases of the Tay-Sachs disease).

Other types of mutation involve more extensive changes to the genetic information, the biggest ones being represented by whole chromosomes being *deleted* (these mutations are usually not compatible with life, i.e., the organism dies before or shortly after birth, but persons with a single X chromosome do survive but suffer from Turner syndrome) or *duplicated* (there are multiple copies of a given chromosome, such as in the classic Down Syndrome – or trisomy 21 – where there are three copies of chromosome 21, or the XYY syndrome where a male has two copies of the Y chromosome instead of the normal single copy), the so-called *aneuploidies*. Other types of mutation such as chromosomal *rearrangements* might involve the deletion or duplication of (relatively large) parts of a chromosome, the change in the orientation of part of a chromosome (**inversion**) or the exchange of genetic material between chromosomes (or **translocations**) such that part of one chromosome ends up on another and vice versa. While these changes might result in various pathologies (especially if they happen to affect an important region such as a gene or a regulatory element), they might also result in relatively normal phenotypes, but can be associated with low fertility or even infertility.

Finally, we have to realize that *Copy Number Variants* (or **CNVs**) might play a very important role in both normal variation and pathologies. CNVs represent changes in the number of copies of a certain DNA region; thus, some people might have a single copy of the region while others might have more than two or none. When the region contains genes or regulatory elements it might affect the phenotype such as apparently is the case for autism, or the capacity to efficiently digest carbohydrates through higher levels of the amylase enzyme in the saliva due to some of us having more copies of the salivary amylase gene *AMY1* (Section 9.2).

However, from an abstract point of view, what mutation does is create a new allele which might have certain phenotypic effects (or not). This new allele, also known as the **derived allele** in opposition to the older, original **ancestral allele**, is initially very rare in the population and its fate is governed by a multitude of factors. As a simple example, imagine a population of fixed size N and a single locus with only two possible alleles a and A with frequencies p and $q = 1 - p$. Thus, there are only two possible types of mutation: a change from the a allele to the A allele ($a \to A$) and the reverse ($A \to a$); assuming that these processes happen with fixed rates (or probabilities) u and v respectively (i.e., any a can become an A with probability u in every generation, etc.), then we have the model in Figure 8.4. In each generation there is a $1 - u$ probability that an a will not change (and $1 - v$ for A).

Assuming all the HWE conditions except for the presence of mutation, then there will be an equilibrium in which the number of a alleles mutating into A will equal that of A alleles mutating into a, and it can be shown that the equilibrium frequencies of a, p^*, and A, $q^* = 1 - p^*$, depend only on the mutation rates: $p^* = \frac{v}{u+v}$. If the population is finite, then genetic drift will "try" to get rid of one of the two alleles, with the less frequent one being at a disadvantage. However, in the long run, drift will fail to completely eliminate one of the alleles (and fixate the other) as mutation will continuously reintroduce it, restarting the whole process of allele frequency fluctuations. Eventually (how often depends on the mutation rates and population size), it will happen that the newly introduced mutation rapidly progresses to fixation itself, but most of the time it will hover at low frequencies.

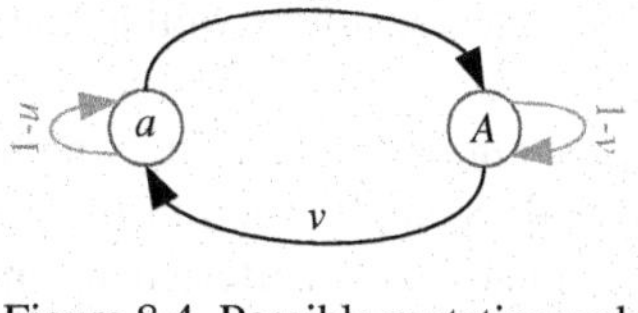

Figure 8.4 Possible mutations when only two alleles, a and A, are allowed.

Of course, when there are multiple possible alleles the process is even more complex but the important take-home message is that in the absence of mutation the world would be a boring place, with a jammed evolutionary process stuck due to lack of new variants to test.

8.5 Selection: when differences do matter

Since Charles Darwin we have all known that selection is the main shaper of life, and that it is based on a few ingredients such as the presence of *variation*, its *inheritance* across generations and the unequal **fitness** it entails, which roughly translates into how many copies are left in the next generation. Of course, life – with its genetic basis for inheritance – is the prime example of selection at work, but it can mould many other systems that fit these conditions, as we will discuss shortly.

Selection in the biological world can be classified in many ways, depending on several criteria, but we will approach this issue here from a pragmatic point of view. First, we can distinguish **artificial selection** from **natural selection** in that the first one is directed by human agency and can be said to be *teleological* (goal-directed, purposeful). Breeders are interested in selecting various features in animals and plants such as milk production in cows, yield in crops, tameness in foxes, or fancy looks in dogs. Breeding programmes work by allowing certain individuals with the desired characteristics to preferentially mate and reproduce based on the intuitive idea that it is their offspring that will most probably feature the same characteristics of interest. By consistently applying this rule across many generations and capitalizing on the chance fluctuations that make some offspring have more of that characteristic as well as by deciding to preferentially mate certain parents whose characteristics will positively combine in the offspring, breeding programmes have successfully managed to result in amazing agricultural productivities and unbelievably shaped dogs and cats, showing that this sort of *directional* selection certainly works. As is well known, this intensive breeding often produces side effects such as the increased risk of cancer in certain dog breeds (Dobson, 2013) or hip dysplasia in large ones. These side effects could be due either to the small populations from which the breeds came (strong founder effects), in which the sought-after characters happen to co-exist with harmful ones (which then piggy-back or *hitchhike* on the selected feature), or to *pleiotropic* effects whereby the same genes produce both the selected-for as well as the negative characters.

In opposition to artificial selection is *natural* selection, which can be argued to include **sexual selection** as well. Sexual selection happens when the selective pressures refer to securing mates, is manifested in anatomical and behavioural features that increase sexual attractiveness to the other sex or improve the ability to fight off rivals (classic examples being the peacock's tail, women's breasts and deer's antlers), and might result in characteristics apparently deleterious for the individual (for example, the peacock's tail impairs the individual's capacity to escape predators).

8.5.1 Positive, stabilizing and disruptive selection

Another way of thinking about selection is in terms of its effects on the population across generations (Hurst, 2009). **Positive selection** (or "*Darwinian*" or *directional*) is probably what most of us think of when selection is mentioned: a consistent advantage for individuals leaning towards one of the extremes of the trait values. For example, giraffes with longer necks have a consistent advantage in that they can consume food (leaves) outside the reach of shorter individuals, having thus, on average, a higher fitness. Therefore, the following generations will feature more of the descendants of the taller giraffes, which will themselves tend to be taller, moving the average tallness of these generations further from the mean of the previous ones (Figure 8.5). Thus, the effect of this type of selection is to move the population away from the previous mean, producing consistent "improvements". The reason for these consistent fitness differences can be rooted in sexual selection (taller males might be preferred as mates) or natural selection (taller males might be better able at securing food).

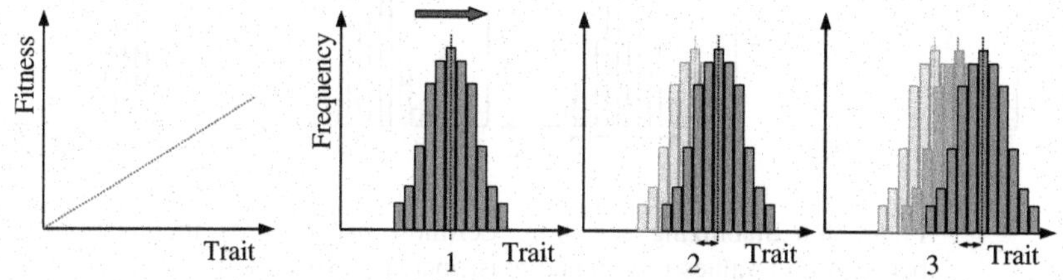

Figure 8.5 Positive selection. Leftmost panel shows the relationship between trait values (horizontal axis, increasing from left to right) and fitness (vertical axis, increasing bottom to top); in this case bigger values of the trait (say, height) have a higher fitness, therefore selection favours taller individuals. Please note that the relationship between trait and fitness can be non-linear. Right panel shows the distribution of the trait values in the population (each bar shows the frequency of a value, taller bars representing higher frequencies) across three successive generations (1, 2 and 3). The top arrow shows the direction of selection and the dotted lines show the population means. The graph for generation 2 also shows the shadow of the distribution in generation 1, while that for generation 3 also shows the shadows of generations 1 and 2, for easier comparison. It can be seen that the mean trait value consistently moves rightwards.

Stabilizing selection (also known as *purifying* or *negative*) acts against deleterious variation and ensures that the population "stays the same". This happens for example when the extremes of a trait are disfavoured compared to the mean values, as might be the case with height or body mass. In this case, the individuals further away from the mean value (the bulk of the distribution; see Figure 8.6) have a lower fitness, being actively removed by selection such that the future generations are closer and closer to this "desirable" mean, resulting in a stabilized distribution of the trait across generations, keeping the status quo. As above, this stabilization against extremes can be due to sexual selection (extremely short individuals might be less favoured as mates) or to natural selection (extremely tall individuals might be more prone to health issues or too heavy to run away from predators).

Both stabilizing and positive selection in a sense promote homogeneity by disfavouring variants away from the mean (stabilizing selection) or from the preferred extreme (positive selection): everybody must be the same. But a third type, **disruptive selection** (or *balancing*), actively promotes diversity. This can happen in various ways. So-called **frequency-dependent selection** occurs when the fitness of an individual depends on the frequency at which other individuals with the same (or similar) trait exist in the population. In

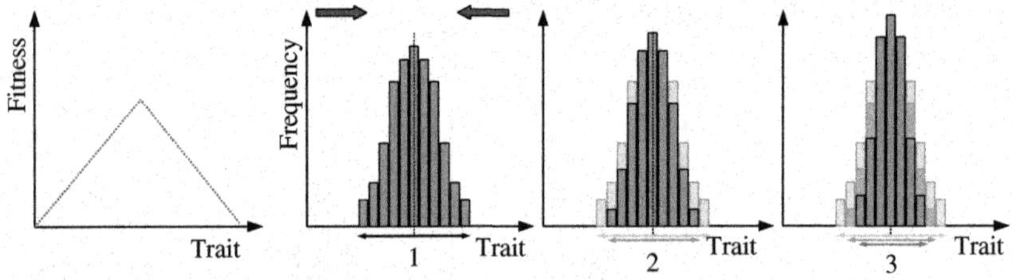

Figure 8.6 Stabilizing selection. Leftmost panel shows the relationship between trait values (horizontal axis, increasing from left to right) and fitness (vertical axis, increasing bottom to top); in this case average values of the trait (say, height) have a higher fitness than the extremes, therefore selection favours individuals of average height. Please note that the relationship between trait and fitness can be more complex. Right panel shows the distribution of the trait values in the population (each bar shows the frequency of a value, taller bars representing higher frequencies) across three successive generations (1, 2 and 3). The top arrows show the direction of selection and the dotted line the population mean. The graph for generation 2 also shows the shadow of the distribution in generation 1, while that for generation 3 also shows the shadows of generations 1 and 2, for easier comparison. It can be seen that the mean trait value remains the same across generations, but the spread of the distribution consistently shrinks as extremes are disfavoured, producing narrower and narrower distributions as more individuals have traits closer to the favoured intermediate value.

positive frequency-dependent selection the fitness increases with increasing frequency of the individuals with the trait, while in the negative type the fitness decreases the more frequent these individuals are. Frequency-dependent selection is involved in the evolution of *mimicry*, a complex evolutionary process (Ruxton et al., 2004), one example being the imitation of the body shape and colouring of a stinging wasp by an edible fly so that the would-be predators are fooled into not eating it. This strategy works well as long as the frequency of the mimetic flies is relatively low, but as soon as they become too frequent birds might learn that most of the time this body shape and colouring are actually fine to eat except in the rare cases when a real wasp is attacked. This produces an increased fitness for the mimetic fly when rare relative to wasps (not eaten) but a decreased fitness when too frequent, resulting in a fluctuating population of mimetic flies. Similar fluctuating dynamics is produced by prey-predator interactions and parasite-host relationships: too much prey allows the predator population to expand, leading to fewer prey, leading to predator population crashes in a cyclical manner.

Another important type of stabilizing selection is represented by **heterozygote advantage** or *overdominance* where the alleles at the same locus interact in a non-linear way. A classic example is represented by **sickle cell anaemia** (Rees et al., 2010): simplifying, a single nucleotide change in the *β-globin* gene on chromosome 11 results in the so-called *HbS* allele (as opposed to the normal *HbA* allele). Homozygous individuals carrying two copies of this allele, *HbS/HbS*, will develop a blood disease (characterized by a specific abnormal, sickle-like shape of the red blood cells) that lowers their life expectancy, especially without modern medical help. However, heterozygous *HbA/HbS* individuals have a mostly normal phenotype, but in regions where *malaria* (a serious infectious disease caused by parasites of the genus *Plasmodium* and spread by mosquito bites) is endemic, *HbA/HbS* heterozygotes have less severe symptoms of the malaria disease than normal *HbA/HbA* homozygotes. Thus, as shown in Figure 8.7, the fitness of the heterozygote is higher than that of both homozygotes under malaria pressure, ensuring that despite its deleterious effects in homozygous form the *HbS* allele will persist in the population.[2]

8.5.2 *Hiking the fitness landscape*

A much-used metaphor for visualizing how selection works is to think of the fitness of all possible genomes as a "hilly landscape" on which populations of organisms walk, trying to get to the highest "peaks". As an example, think of two genetic loci, one controlling height and one weight, each with many

[2] As a side note, cases such as these are a very powerful reminder that evolution is not about the well-being or the "good" of the individual.

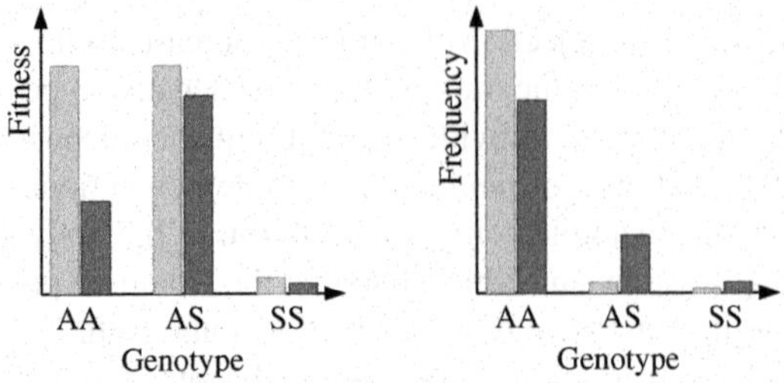

Figure 8.7 Heterozygote advantage such as in sickle-cell anaemia. Leftmost panel shows the fitness (vertical axis, increasing bottom to top) of each genotype (horizontal axis), where A stands for the normal *HbA* allele and S for the sickle-cell disease allele *HbS*, in environments without malaria (light bars) and with malaria (dark bars). Right panel shows the frequencies of the genotypes in a population without (light bars) and with (dark bars) malaria (each bar shows the frequency of a genotype, taller bars representing higher frequencies). It can be seen that when malaria is present, despite the low fitness of the homozygous *HbS/HbS*, the allele is still frequent due to the higher fitness of the heterozygote *HbA/HbS*, but in populations without malaria the allele is virtually absent.

possible alleles (so many in fact that we can abstract them as being essentially continuous), and each genotype combining one height and one weight allele has a unique fitness represented by a number between 0 (lowest possible fitness, namely incompatible with life) and 1 (maximum possible fitness). With these, we can represent (Figure 8.8) the relationship between genotype and fitness as a 3D landscape where the higher the peak the higher the fitness of that set of genomes and deep valleys separate such peaks of high fitness. Individuals (genotypes) are points on such a landscape and whole populations are clouds of such points.

The **fitness landscape** (or adaptive landscape – the distinction need not concern us here) was introduced in the early 1930s by Sewall Wright and is usually taken as a good metaphor for how evolution works: populations "climb" towards fitness "peaks" (genomic regions of maximal adaptation) surrounded by fitness "valleys" (genomes that are selected against), and populations might get "trapped" in "local maxima" (hills higher than their surroundings) even if an even higher "global maximum" exists, as there is no way to descend into the valleys cross-cutting the landscape.

However, no matter how useful this metaphor might be, especially for the non-mathematically minded, it has been forcefully argued that this is more misleading than helpful. Gavrilets (1997, 2003) showed using computer models that these 3D representations have properties that do not hold for the realistic cases where many more loci with many alleles interact, resulting in

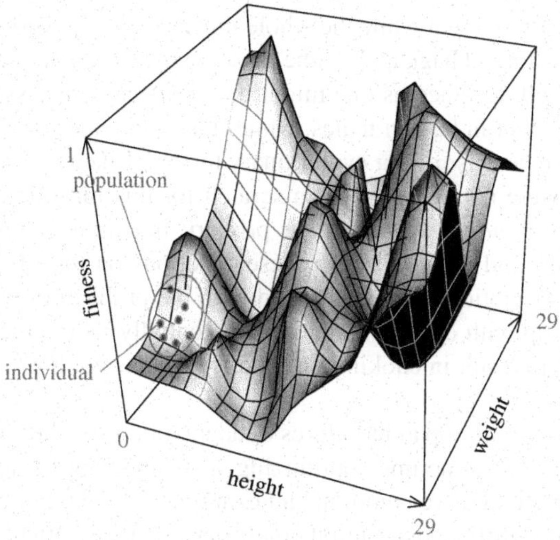

Figure 8.8 An idealized fitness landscape for two loci with many alleles, say controlling height and weight (the x and y axes, each with 30 alleles numbered from 0 to 29). The fitness is represented on the vertical (z) axis increasing upwards towards 1. Individuals are points and populations clouds of points. Adaptation pushes the population upwards towards peaks of (local) fitness maxima while drift and mutation can also explore other directions.

highly multi-dimensional spaces. In such spaces the dreaded fitness valleys might not even exist as there are narrow ridges connecting multiple peaks of high fitness and populations can explore vast regions of the genomic space more easily. Kaplan (2008) argues that if the assumptions behind these computer simulations hold, some issues in evolutionary biology might be artefacts of metaphors such as the fitness (adaptive) landscape and advises against their use in thinking about evolution.

However, there are deeper issues with such metaphors. It is fundamentally wrong to assume that there is a singe somehow predetermined absolute fitness value for a genotype, given that an individual's fitness depends crucially on its development and environment. In the real world, an organism's fitness results from complex interactions with other organisms and is intrinsically context-sensitive and dynamic.

8.5.3 *When selection seemingly fails*

Evolution is not a rational designer that starts from scratch for every new problem it has to solve but instead is a tinkerer modifying and cobbling together what is already present while the whole thing must keep working – not an easy job! This "historical baggage" is one reason why sometimes organisms seem not to be as well designed as one might have expected, a classic example being the sperm duct in human males, which takes an unnecessary detour through the abdomen because of our evolutionary history. Likewise, the organs we use for speech were not designed from scratch for this function but were rather **exapted** (or *pre-adapted*) from their primary functions in respiration and food processing (and they still keep these functions on top of modulating sound). Probably a rational designer faced with the problem of producing sound would come up with different solutions, potentially more efficient and certainly less likely to result in choking on food, but this is not how evolution works.

Another reason for some apparent failures of adaptation is the fact that traits are not independent and sometimes optimizing for something breaks something else. These **trade-offs** are common and examples abound. For example, the dangers of human birth, under natural conditions relatively often resulting in the death of the mother, the child or both, would be much alleviated by either reducing the newborn's brain size or enlarging the mother's birth canal. The first solution alas would have negative effects on the cognitive capacities and developmental trajectory of the child, resulting in heavy fitness penalties for a member of our species,[3] while the latter is counterbalanced by the mechanical inefficiency of walking with very wide hips. Concerning speech, the vocal tract is a trade-off between multiple pressures, trying to reach the best compromise between breathing, eating, speaking, and probably singing. Thus, while some features might seem sub-optimal when analysed individually and separated from the rest of the organism and its environment, it usually turns out that they represent one of the best compromises possible when considering this whole system.

However, some traits seem not to have been selected at all but simply exist as a *by-product* of other traits. Mammalian blood is red simply because haemoglobin (the iron-rich molecule responsible for carrying oxygen around the body) has this colour; in fact, other animals such as lobsters have blue blood due to a different oxygen-carrying molecule, haemocyanin, based on copper. Of course, such by-products can become functional (e.g., the importance of redness as a signal) or can be exapted and fine-tuned for other functions. For example, it can be argued that the capacity of the tongue to modulate the shape of the vocal tract and thus of its acoustic output was initially a by-product of its high mobility for food-processing later exapted for speech.

[3] Which does not mean that selection has not already tinkered with this as much as it probably could: human children are born prematurely and require a prolonged period of development.

Finally, true failures of selection might result from the conflicting forces of genetic drift and mutation, especially powerful in small populations. As we will detail in Section 8.5.5, sometimes selection is simply unable to do its job, resulting in populations with arguably sub-optimal features.

Thus, while a very powerful agent and the only explanation for the apparent design prevalent in the biological world, selection is far from being a perfect optimizer always producing perfect designs. But its apparent failures usually turn out to be, in fact, solutions to extremely complex problems that we simply fail to fully appreciate and may even turn out to be the starting point of new evolutionary developments, just as feathers (most probably used by early dinosaurs for display and thermal insulation) opened the sky to the extremely successful birds.

8.5.4 What is selected: de-focusing the individual

Despite the intuitive appeal of the idea that selection acts on individuals, this is clearly wrong. As we saw above when discussing heterozygote advantage, those homozygous individuals with sickle-cell anaemia are paying the price for the increased fitness of the heterozygotes in regions with malaria. Thus, what seem to be selected here are the two alleles, *HbA* and *HbS*, at the expense of the individuals they happen to temporarily "inhabit". This *gene-centred view* of evolution was strongly promoted by Richard Dawkins (1989) among others and elegantly captures a number of otherwise puzzling phenomena ranging from parent–offspring conflicts to sperm competition and parasitism.

However, as we saw previously, genes are notoriously hard to define and their effects on the phenotype extremely complex and interactive, calling into question even the foundation of this gene-centred view: if we don't really know what a gene is then how can they be the locus of evolution? This is brought into focus by phenomena that extend beyond the individual. Individuals, sometimes whole communities, alter their environment in such a way that this becomes an **extended phenotype** (Dawkins, 1996) which in turn affects the fitness of the individuals themselves. Thus, a beaver's dam is certainly constructed by beavers but it in turn profoundly alters the life of the beavers by creating a relatively sheltered and controlled environment, being thus part of the animal's phenotype.

However, things are more complex than that: building and maintaining a dam is an active process, similar to the development and maintenance of a body. But, as opposed to the individual body, the dam will be inherited by the next generations, altering their developmental and living environment in profound ways. Thus, organisms do not simply inhabit a given environmental niche but they actively engage in **niche construction** (Odling-Smee et al., 2003), altering their own environment. Humans are probably the most powerful niche constructors currently alive as we use our language, culture and technology to massively reshape our environment across many generations, but there are other notable species involved in such efforts, such as the earthworms (creating the soil) and the leaf-cutter ants (involved in farming their own fungal food).

Thus, not only are the boundaries between individual phenotype, the community's effects on the environment and the environment itself becoming quite blurred, but it even becomes unclear what channels the information transmitted across generations takes. It could be argued that genetics still plays a special role in faithfully transmitting it, but the constructed, maintained and continuously modified niches also transmit very important information. Arguably a beaver without its dam would still be a beaver (even if not a very happy one) but a leaf-cutter ant colony would die out without its gardens, and the human population would crash without its cultural and technological niche. Moreover, the recent realization that epigenetic changes affecting gene expression without altering the genetic information play a very important role in cross-generational adaptive responses further erodes the primacy of genetic information.

There are multiple attempts at addressing these (and other) issues within a more comprehensive evolutionary theory, but it is currently unclear how these theoretical developments will turn out, with *Developmental Systems Theory* (Oyama et al., 2003) being probably one of the most promising frameworks. These developments again highlight the inadequacy of the simplistic fitness landscape, evolution-as-optimizer and gene-centred evolution metaphors, as there is not a single measure of success, selection acts simultaneously at multiple levels (Okasha, 2008), development is essential for evolution (Carroll, 2011), and the landscape (whatever it really represents) is not static but dynamic, co-constructed and fundamentally involved in trans-generational inheritance.

8.5.5 The (nearly) neutral theory of evolution

Some alleles are **neutral**, that is they are "invisible" to selection in the sense that they make no difference to the fitness. Thus, their evolution is entirely governed by mutation (which introduces them into the population) and drift (which either eliminates them or drives them to fixation given enough time), and it can be shown that such a neutral allele has a probability of becoming fixated in the population equal to its mutation rate. In other words, if the mutation rate is approximately constant then neutral mutations will accumulate in the population at a speed dictated by the mutation rate, resulting in what has been called a **molecular clock**. If we compare two populations and count the number of fixed neutral mutations differing between them, then, by knowing the mutation rate or by using *calibration points* of known age (e.g., from the fossil record), we can estimate the time since these populations split from each other. This technique has been applied to, for example, estimating the split of the lineage leading to us from that leading to the chimpanzees (Jensen-Seaman and Hooper-Boyd, 2001) and to dating the age of the last common ancestor of all existing mitochondrial DNAs in modern humans (Cann et al., 1987).

This **neutral theory** of evolution was quite controversial when introduced by Kimura (1968) (and independently by King and Jukes, 1969) but is now regarded as one of the essential pillars of evolutionary theory and the null hypothesis of molecular evolution against which more complex proposals (such as various types of selection) must be tested.

However, it can be argued that there is no such thing as a really neutral mutation (but some do come close enough) and that different alleles always result in ever so slight fitness differences, but under what conditions is selection able to "see" them? The **nearly neutral theory** of evolution (Ohta, 1973, 1992) shows that, perhaps counter-intuitively, when an allele has a small effect on fitness, it will be under selection in large populations but neutral in small enough ones. This threshold population size depends on the fitness effect of the allele and, thus, implies that even alleles with quite a strong effect on fitness might escape selection in small enough populations and might be able to rapidly progress, against all expectations, to fixation.

This explains why in populations that have been small for a very long time, such as on isolated islands, mountains or other environments that cannot support large human populations, or that have cultural practices promoting strong reproductive inequality, there might be very high frequencies of genetic diseases otherwise rare in other populations (a well-known example is Finland, where the so-called "Finnish disease heritage" covers 30–40 otherwise rare diseases; Norio, 2003a,b,c). It has even been used recently to explain otherwise puzzling but essential features of genome architecture such as the existence and prevalence of introns and other non-coding DNA in the genomes of complex organisms, features that, once there, allow (through exaptation) complex evolutionary processes, such as gene duplication and regulatory change, to happen (Lynch, 2007; Koonin, 2012).

8.5.6 Detecting selection from genetic data

While there are some cases where we can actually see selection working in real time, such as when bacteria adapt to new antibiotics (Rosenthal and Elowitz, 2011), we usually must infer its presence, type, strength, direction and sources from a variety of data, including fossils, apparent matches to the environment and genetic information, the latter being of special concern for us here. However, the detection of selection from genetic data is far from simple, and is currently of enormous interest and undergoing very fast development (for reviews see, for example, Quintana-Murci and Clark, 2013; Sabeti et al., 2006; Kelley and Swanson, 2008). Thus, we will overview here the main basic ideas and techniques and we will use the search for selection on *FOXP2* as a very instructive and relevant example.

Protein-coding regions

One of the most-used classes of methods is based on the idea that not all mutations are equal when it comes to selection. As we saw in Section 3.7 above, the message in the protein-coding **exons** (given by their sequence of nucleotides) is translated into a sequence of amino acids using the genetic code which defines the relationship between one triplet (three nucleotides) and one amino acid. Because there are more possible triplets ($4^3 = 64$) than amino acids (20), there can be more than one triplet mapping to one amino acid, a property called *degeneracy* (or redundancy). Thus, for example, all four codons GGU, GGC, GGA and GGG map to glycine (Gly), while GAU and GAC map to aspartic acid (Asp) and GAA and GAG map to glutamic acid (Glu), and so on (see Table 3.2 for the whole code). This degeneracy means that some mutations (such as GG<u>U</u> → GG<u>A</u>) would not affect the resulting protein (still with Gly at the relevant position), being thus **synonymous mutations** (or substitutions), but others do change the resulting protein (e.g., G<u>G</u>U → G<u>A</u>U, changing Gly to Asp), being **non-synonymous mutations**. While synonymous substitutions are arguably invisible to natural selection[4] as they result in the same protein, the non-synonymous ones code for different proteins that might have properties dissimilar enough for selection to act upon them.[5]

Thus, to check whether a protein-coding region is evolving under selection, we could compare the number of synonymous substitutions (usually denoted K_s or d_S) to the number of non-synonymous ones (K_a or d_N): if the region is not under selection (neutral) then the ratio K_a/K_s should be about 1 (the two types of change happen at similar rates), if it is under positive selection then K_a/K_s should be more than 1 (more protein-changing non-synonymous substitutions accumulate as they prove advantageous), and finally if it is under negative selection K_a/K_s should be less than 1 (protein-changing non-synonymous substitutions are mostly bad). A related method is the McDonald-Kreitman test (McDonald and Kreitman, 1991). In practice, one picks one or more protein-coding regions (genes) in two related species (say

4 However, they might not be fully equivalent from the point of view of the translation/transcription or other processes affecting the DNA or mRNA, resulting in relatively weak selective pressures (Sharp et al., 1995; Hershberg and Petrov, 2008) that we will nevertheless ignore here.

5 Given that there are amino acids with similar properties and their effect depends on many factors such as the particular region of the protein, it could be possible for the two mutations to have results similar enough so that they are effectively neutral; see also the influence of population size on neutrality in Section 8.5.5.

humans and chimpanzees), estimates the synonymous and non-synonymous substitutions and computes their (averaged) ratio, estimating as well the odds that the deviance from neutrality is real (statistically significant) using specialized software such as MEGA (`http://www.megasoftware.net/mega4/index.html`) or PAML (`http://abacus.gene.ucl.ac.uk/software/paml.html`). However, despite its elegance and power, this method can be applied only to protein-coding regions (exons) and works only if enough substitutions of both types have accumulated, leaving out the very interesting regulatory regions of the genome where probably the most important chapters of human evolution have taken place.

Inter-population differences

Populations could differ in their genetic structure for a number of reasons, one being that selection has shaped them differently. These differences in selective pressures could originate from the environment or the cultural practices of the populations (or a combination thereof), such as different pathogens to which the immune system must adapt, or different food sources to which the digestive system has to adapt, foods that may be products of cultural practices such as agriculture. There are many ways to estimate genetic differences between populations, one of the most popular being the *Fixation Index* (or $\mathbf{F_{ST}}$; Holsinger and Weir, 2009). F_{ST} can take values between 0 (the populations are interbreeding freely, there being no genetic differences between them) and 1 (the populations are maximally differentiated), and on average across many loci human populations show an F_{ST} of about 0.12 (but more on this later).

Thus, higher than expected genetic distances for a locus might be an indication that selection is at work. For example, there are large differences in the frequency of the *HbS* sickle-cell anaemia mutation between populations living in regions with malaria and those living in regions without it, suggesting that this environmental factor might be the source of selective pressure at this locus. Another fascinating example (which we will discuss in detail later) concerns the large differences in the frequency of an allele allowing adults to digest fresh milk between, on one hand, populations practising farming resulting in the availability of milk, and, on the other, populations that do not, suggesting that this ultimately cultural difference generates selective pressures on the digestive system.

However, it is very difficult to be sure that large inter-population genetic differences are a result of selection and not of different demographic histories involving mostly the accumulation of different neutral mutations. As an example, assume we observe a biallelic locus with alleles a and A in two populations, allele a having a frequency $p_1 = 0.7$ in the first population and $p_2 = 0.3$ in the second, which could be interpreted as due to selection for a in population 1 but against it in population 2. However, it could as well be that these two populations descended from the same ancestral population 0 with allele *a* frequency of $p_0 = 0.5$ but a strong founder effect (or bottleneck) has resulted in more carriers of *a* than expected making it into population 1, while more carriers of *A* are found in 2. Thus, strong genetic differences between populations are to be taken as suggestive of selection but more is required before a conclusion can be drawn.

Selective sweeps

Initially, when an advantageous mutation is introduced in the population, through either mutation or introgression/admixture (Section 8.6), it is necessarily at very low frequencies. As generations go by, positive selection pushes it to higher frequencies as its carriers tend to leave more children than non-carriers. Let us assume first that this mutation is located on a non-recombining piece of DNA such as mitochondrial DNA (mtDNA) or the non-recombining region of the Y chromosome (NRY); then the whole molecule of DNA on which it is found will increase in frequency in unison with the mutation as the mutation cannot be "cut" from the rest of the molecule. Thus, the whole mtDNA (or NRY) molecule hitch-hikes on the back of the selective pressure generated by the mutation.

However, things are more interesting if the mutation happens to be located on an autosome affected by recombination (Figure 8.9 panel A). In this case, the DNA regions close to the mutation are in linkage disequilibrium (LD) with it, tending to be transmitted together with it across generations (in a sense, mtDNA and NRY are extreme cases of such LD blocks). The strength of LD between the mutation and another location on the DNA molecule decreases with the distance from the mutation, because the chance of a recombination occurring between them and cutting their link increases with distance (see Sections 3.6 and 5.1). Thus, the effect is that regions closer to the mutation have a stronger tendency to hitch-hike (or be swept) to higher frequencies than regions further away. If selection runs its course to completion then the mutation will be present in effectively all individuals in the population, reaching fixation (this is a complete **selective sweep**), but it can also falter midway due to changes in the source of the selective pressure (e.g., the responsible pathogen dies out), the genetic environment (another mutation does a better job, making this one redundant) or a trade-off is reached, resulting in an incomplete or partial sweep.

Such sweeps leave specific signatures on the pattern of intra-population inter-individual genetic diversity around the mutation's location, signatures that can be detected by appropriate methods (Sabeti et al., 2006; Kelley and Swanson, 2008). Because the selective sweep leads to everybody in the population having the advantageous allele and its linked hitch-hiking neighbours, it results in very low genetic diversity around the selected variant. However, mutation (and recombination) slowly recovers the lost diversity by constantly

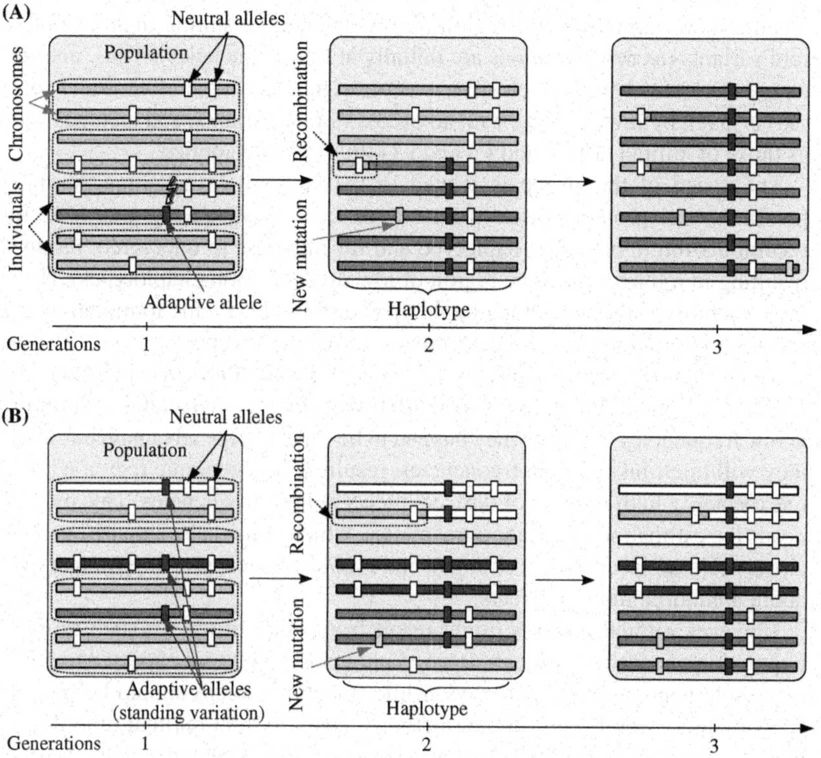

Figure 8.9 The effects of a selective sweep at an autosomal locus on genetic diversity. Shown are three generations of chromosomes (horizontal bars) in a population, abstracted away from the individuals they are in (dotted rectangles in generation 1), each chromosome carrying alleles at various loci. Panel **A**: hard sweep. In generation 1, mutation (ϟ) introduces an advantageous new allele (dark grey) on a chromosome (also darkened for convenience). This allele's frequency is pushed up by selection in generation 2, carrying with it the linked regions of the dark chromosome (hitch-hiking regions); this linked region is called a haplotype. However, recombination can break this haplotype, and mutation can inject new variants. If selection stops now, we have a partial selective sweep, but if it runs to completion (generation 3) we have a complete sweep. Panel **B**: soft sweep from standing variation. Multiple alleles already in the population become selectively advantageous and spread to fixation carrying with them their genomic neighbours (coloured chromosomes), resulting in a more complex and more diverse genetic structure.

creating new variants in this region of low diversity, resulting in an excess of rare variants (as new mutations are initially at low frequencies). Thus, one signature of relatively recent selective sweeps (at most a couple of thousand years old) is given by low diversity with an excess of rare variants, and methods such as those of Tajima (1989) and Fu and Li (1993) can be applied.

The speed of the sweep is related to the strength of selection: stronger pressures will promote faster increases in frequency, leaving less time for recombination to break the haplotype and for mutation to inject new variants, resulting in a longer genomic region of low diversity (long haplotypes). However, such high-frequency long haplotypes rapidly decay due to mutation and recombination, limiting their detectability to recent sweeps.

Mutation produces new variants, so-called **derived alleles** (as opposed to the original variants called the **ancestral alleles**). These derived alleles are initially at low frequencies but when they happen to be linked to the advantageous allele they will hitch-hike to high frequencies, resulting in a genomic region with an unexpectedly high number of derived alleles at high frequencies. This method needs the estimation of the ancestral alleles, which could use comparisons with closely related species (e.g., chimps), and is able to detect sweeps younger than about a hundred thousand years.

However, similar patterns might result from other processes such as rapid population expansion, but such demographic phenomena are expected to affect the whole genome in similar ways, while selective sweeps should be specific to a few loci. Moreover, different methods have different optimal time-depths but new developments attempt to provide more sensitive and reliable tests of selection for specific genes as well as scans of the whole genome for regions under selection (see for example Chen et al., 2010; Sabeti et al., 2002; Voight et al., 2006).

What we described above are sometimes called hard sweeps and their main feature is that selection pushes towards fixation a variant that is initially at very low frequencies (usually a *de novo* mutation in a single individual). In contrast, soft sweeps (Pritchard et al., 2010; Novembre and Han, 2012) happen when some change in the environment generates selective pressure on one or more variants already present in the population (so-called **standing variation**), resulting in their increase in frequency across generations together with their respective (multiple) hitch-hiking genomic regions. Therefore, the resulting picture is not as clear-cut as that produced by a hard sweep (Figure 8.9 panel B) and current methods are not as good at detecting such soft sweeps, even if they may be quite frequent in humans (Novembre and Han, 2012).

Therefore, it is fair to say that it is very difficult not only to detect selective sweeps, but it is even harder to date them, to estimate their strength and direction, to locate them to a specific region on the genome and, in particular, to offer meaningful interpretations that do not stretch the available data and the reliability of the methods. We will exemplify these difficulties by turning now to *FOXP2* and its evolutionary history.

8.5.7 *The evolutionary tale of FOXP2*

FOXP2 is a fascinating gene, a true window (but hopefully not the only one) into the molecular underpinnings of speech and language (Fisher and Scharff, 2009). Thus, it is understandable that soon after its discovery (Lai et al., 2001) there was a lot of speculation on its evolutionary history and its impact on the emergence of language in our lineage. Enard et al. (2002) compared this gene across several species including humans, chimpanzees, gorillas and mice and concluded that the protein produced is one of the most conserved, being in the top 5% most stable. For example, during the whole evolutionary time separating mice and the last common ancestor of humans and chimpanzees (totalling about 130 million years of separate evolution), there was a single amino acid change. Intriguingly, since our split from the chimpanzees just about 6 million years ago, humans have accumulated two amino acid changes in exon 7 (these changes have been called "human specific"), one of these independently appearing in carnivores (Zhang et al., 2002). The most interesting finding, however, was that in modern humans there is an excess of rare alleles (using Tajima's D), more derived alleles at high frequency in a region of about 14 kb neighbouring exon 7, prompting Enard and colleagues to conclude that there was a selective sweep on *FOXP2* and that "the best candidates for the selected sites are the two amino acid substitutions specific to humans in exon 7" (p. 871). Next, they estimated the age of this complete selective sweep in humans assuming demographic growth and concluded that this fixation occurred in the last 200,000 years, "concomitant with or subsequent to the emergence of anatomically modern humans [and] compatible with a model in which the expansion of modern humans was driven by the appearance of a more-proficient spoken language" (p. 871).

However, the story of *FOXP2* being "the" gene was too nice (and simple) to be true; only five years later, Krause et al. (2007) extracted nuclear DNA from Neandertal fossils and found not only that the two "human-specific" amino acids are present, but that the whole modern human haplotype is shared with the Neandertals. This strongly suggested that the previous age estimate (younger than 200,000 years) must be wrong, and that the haplotype must be much older than that, predating the last common ancestor of modern humans and Neandertals about half a million years ago.[6]

[6] The alternative, that Neandertals might have got it from modern humans through admixture (Coop et al., 2008), was also considered but is probably very unlikely.

So, well, the dating might have been wrong but what about the signal of selection itself? Further analyses by Ptak et al. (2009) and more sequencing in a set of individuals from Nigeria (this choice of population was due to an interesting pattern observed in one Nigerian individual originally analysed by Enard et al., 2002, and re-analysed in this paper) revealed that the pattern of LD around *FOXP2*'s exon 7 is most probably incompatible with a selective sweep caused by the two "human-specific" amino acids (p. 2183). Instead, they propose a two-step model in which an old selective sweep fixated the two amino acids before the split between modern humans and Neandertals, while a second (possibly still ongoing) sweep affects only modern humans and is driven by an unknown locus near exon 7.

Very recently, Maricic et al. (2013) claimed to have finally identified the locus of this second selective sweep. More precisely, they compared *FOXP2* sequences in Neandertals and Denisovans (a sister group of the Neandertals; Meyer et al., 2012) with 50 present-day humans, looking for substitutions that are (almost) fixed in moderns but absent (or variable) in these remote cousins. One of the most promising was a locus in intron 8, seemingly a potential binding site for a transcription factor (named *POU3F2*) that is proposed to affect the expression of *FOXP2* in neurons, and where most modern humans carry the derived allele and the Neandertals and Denisovans the ancestral one (p. 6). Again, a nice solution to this puzzle but it turns out that the ancestral allele (the one present in Neandertals and Denisovans) is also present in modern living Africans at a frequency of about 10%, raising the question of what effects it might have in them, especially in homozygous individuals (about 1% of the population). It remains to be seen if these 1% of modern humans have any interesting phenotypes but it might turn out that this is yet another blind alley.

An intriguing twist to this story (see also Section 8.6.4) is presented by the finding (Vernot and Akey, 2014) that, when compared to the whole genome, a large region encompassing *FOXP2* is strongly depleted in Neandertal DNA, apparently suggesting that the specific Neandertal sequence in this region did not fare well in a modern human genome. It remains to be seen what exactly was the cause of this incompatibility, but even if it turns out to be related to *FOXP2* it might not necessarily point to language and speech given the multiple regulatory roles played by this gene.

This story about the putative selective sweep(s) affecting *FOXP2* is a cautionary tale clearly showing that things are very complex and conclusions should almost always be tentative. The evolutionary history of *FOXP2* is but a small (even if currently very important) part of the larger debates concerning the evolution of humans, of language and speech, a debate bitter at times and where multiple hypotheses are almost continuously proposed and tested (within the limits of the available data). In this debate, my current view, based on the available data and theoretical proposals from palaeoanthropology, archaeology, genetics and language sciences, is (controversially) that recognizably modern speech and language are ancient, predating the last common ancestor of modern humans and Neandertals about half a million years ago (Dediu and Levinson, 2013).

8.6 Population structure: love is not blind

Yet another way to violate the Hardy-Weinberg conditions is by interfering with **panmixia** (or random mating), the assumption that within a population everybody has the same chance of mating with everybody else. While arguably a nice ideal to hold, it is immediately contradicted by reality. There are several ways in which panmixia can be violated, but we will briefly discuss here inbreeding, (dis-)assortative mating and population structure.

8.6.1 Inbreeding: mating among genetic relatives

Inbreeding refers to mating preferentially between genetically closely related partners, such as first cousins (Charles Darwin himself married his first cousin Emma Wedgwood in the context of a high rate of consanguineous marriages between the two families; Berra et al., 2010) or members of the same village or clan (generally termed **endogamy**). Inbreeding can be measured (Wright, 1922) using the *inbreeding coefficient* (F), which is based on the concept of *identity by descent* (IBD), representing the probability that the two alleles at a locus within an individual are inherited from the same allele in an ancestor from a previous generation. For example, all alleles in a pair of monozygotic (identical) twins are IBD (thus the probability of IBD is 100%) as they all descend from the same fertilized egg, while for a pair of dizygotic twins and a pair of non-twin siblings the IBD probability is on average 50% (but in particular cases this varies due to the stochastic nature of the allocation of alleles to gametes – Mendel's Law of Segregation or his first law, see Section 3.5 – between about 40% and 60%; Visscher et al., 2006; Gagnon et al., 2005). A related concept is the *coancestry* (or *kinship*) coefficient between two individuals, which represents the probability that one of A's alleles and one of B's (where A and B are individuals) are IBD.

Inbreeding increases the chances that the two alleles at a locus are identical, resulting in a higher probability of homozygosity. Several deleterious alleles are recessive, meaning that they affect the phenotype only in homozygous individuals, and therefore inbreeding increases the chances that such deleterious alleles will be expressed, increasing the frequency of the resulting genetic diseases. Classic examples include the Finnish disease heritage (Section 8.5.5), the increased frequency of intellectual disability and microcephaly in certain highly endogamous communities in Pakistan (and their daughter immigrant communities in Britain) and the high prevalence of various forms of recessive hearing loss in endogamous communities in Turkey and Indonesia as well as in the deaf community in the USA (we will discuss these in detail in Section 9.3). *Inbreeding depression* is a general decrease in fitness experienced by inbred individuals which results from a large number of slightly deleterious recessive alleles (or a few severe ones) in homozygous state, cumulatively affecting the phenotype.

The issues concerning inbreeding in humans are complex, and range from the legal attitudes against incestuous relationships to the practices of endo- or exogamy and their cultural and environmental causes. A number of traditional societies practise *linguistic* ***exogamy*** (Jackson, 1983; Fleming, 2010) whereby the partners must not share the same language and this practice could be a factor explaining high rates of linguistic diversity.

8.6.2 (Dis-)assortative mating: choosing partners (un)like yourself

While inbreeding is based on mating with genetic relatives (thus partners are genetically more similar than expected by chance), *mate choice* can be non-random with respect to the phenotype. Thus, partners can be more similar than expected in height, intelligence, skin colour, wealth, religion or deafness status (what is called **assortative mating**), or they can be more dissimilar than expected in other traits. Such **disassortative mating** seems to be much rarer (Jiang et al., 2013), but an interesting case, still controversial, could be represented by human partners apparently tending to have different alleles at the major histocompatibility complex (MHC, an important component of the immune system) driven by disassortative mating for body odour and supposedly increasing the effectiveness of the offspring's immune system.

Assortative mating is very important as it alters the allele and genotype frequencies from those expected under Hardy-Weinberg equilibrium, and can promote homozygosity and increase the incidence of certain traits such as genetic hearing loss (as will be discussed later). It has been proposed that assortative mating can also create correlations between the loci involved in expressing the trait and those involved in expressing the choice of that trait.

8.6.3 Structured populations: choosing the partners you can actually choose

A major source of violations of panmixia is represented by population structure: not every opposite-sex reproductive-age currently living individual is actually eligible as (s)he might live across the big river, on a different island too far away or over impassable mountains. Thus, geography plays an important role in creating population structure, with closer individuals being in general more genetically similar as a consequence. Populations separated by greater distances tend to have more differentiated genetic structures. Moreover, humans have other means to create powerful barriers to free mating, namely culture: it is well known that it is generally more difficult to mate across religions, ethnicities or social strata, with extreme examples being the recently abolished South African apartheid, or the Indian caste system (old and pervasive enough to have left discernible traces in the genetic structure of the Indian subcontinent; Moorjani et al., 2013).

The degree and duration of genetic isolation between two populations are important factors (together with population size, bottlenecks and founder effects and possible selective pressures) in determining their genetic divergence, in extremis resulting in **speciation** (*allopatric* if the two populations inhabit geographically separate territories and *sympatric* if this differentiation happens within the same geographical region due to for example specialization for different ecological niches) whereby the two populations have diverged so much that viable offspring cannot result from their mating any more. Genetic differentiation between populations is usually measured using F_{ST} and related measures (see Holsinger and Weir, 2009, for details of estimation methods and their interpretation), where smaller F_{ST} values point to similar allele frequencies in the populations while higher values characterize populations with different distributions of alleles.

However, sometimes a trickle of genes might continue to connect the two populations (**gene flow**), fighting the opposing forces of genetic differentiation, and its success depends on their relative strength versus the amount and pattern of mate exchange between populations. This gene flow can show various amounts of *asymmetry*, such as when more people emigrate from the country of origin into a newly established colony than the reverse, and *sex-biasedness* (sex-biased migration). Sex-biased migration can be inferred by comparing patterns of diversity at autosomal (bi-parentally inherited), mtDNA (maternal transmission only) and NRY (paternal transmission only) loci (Wilkins and Marlowe, 2006; Langergraber et al., 2007), and can result from different *post-marital residence patterns* (Jordan et al., 2009; Fortunato and Jordan, 2010): *matrilocality* will promote preferential male migration, *patrilocality* female migration, while *neolocality* would result in sex-unbiased migration.

When the two populations have not been in contact for a while but the exchange of migrants is re-established, we talk of **admixture** and the resulting *admixed population* has a *mixed ancestry*. When the two populations are considered to be separate species, then their mating is known as **hybridization** and the resulting individuals are known as *hybrids*. Hybrids can be infertile (e.g., mules) or fertile (e.g., between wolves and dogs) and the latter play an important role in **introgression**, the phenomenon whereby genes from one (source) species enter another (destination) species through repeated *backcrossing* of hybrids into the destination species. While hybridization and introgression play no role in modern humans, they might have been important in our evolutionary past when several different species (or varieties) of humans co-existed and possibly exchanged genetic material, as has recently been shown for Neandertals, Denisovans and their contemporaneous modern humans (Green et al., 2010; Meyer et al., 2012).

8.6.4 The genetic structure of modern humans

So, how much genetic diversity is there in modern humans, how is it patterned and why? These are very important questions not only for illuminating our ancient and more recent history, or for investigating the genetic bases of human traits such as speech and language, but also for understanding the correlation (or lack thereof) between linguistic and genetic diversities. Here we will address these fascinating and complex issues only briefly but the interested reader can consult the excellent and comprehensive introduction and review provided by Jobling et al. (2013).

When compared to other mammals we are a relatively homogeneous species (Barbujani and Colonna, 2010), but we are no clones; the average nucleotide diversity between two living humans is under half a percent but given the size of the human genome (about 3 billion nucleotides) that amounts to several million differences. However, not all these differences might have a phenotypic effect at all, and for those that do raw numbers are not very informative as they disregard the complexity of gene expression that we surveyed in the previous chapters.

To put things into their proper evolutionary perspective, we share with our closest living relatives, the chimpanzees, from which we split some 6 million years ago, about 98% of our genetic information; about 85% we share with mice (last common ancestor about 70 million years ago; Douzery et al., 2003), and 50% with the very remotely related fruit fly (*Drosophila*), with which we shared a common ancestor some time 550–580 million years ago. There are many **orthologous genes** (*orthologs*), namely genes that derive from the same ancestral gene, and some of them have been conserved since the deepest evolutionary times when complex animals first emerged. One very well-known example is presented by the *Hox* genes, which play essential roles in animal development as "master genes" controlling and coordinating the complex sequence of events that must unfold for the right developmental outcomes. These genes control the expression of other genes (just as *FOXP2* does) but on a grand scale, controlling the patterning of the body and the growth of the various appendages. Another fascinating case of extreme conservatism is represented by the *PAX6* gene, which is a master switch in eye development from fruit flies (where the orthologous gene is *eyeless*) to humans; for example, the mouse *Pax6* can still induce eye formation in *Drosophila* (Gehring and Ikeo,

1999). For a very good introduction to these fascinating topics of evolutionary developmental biology (or Evo-Devo) please see Carroll (2011).

This very small amount of genetic diversity within our species (we are the least diverse of all primates; Barbujani and Colonna, 2010) is due to our peculiar evolutionary history. To simplify a very complex (and still unclear) story (see, for example, Klein, 2009, for its grand lines, and Dediu and Levinson, 2013, for a brief summary focusing on the more recent events), our own genus, ***Homo***, appeared most probably about 1.8 million years ago in Africa from some earlier *Australopithecus* lineage, and its earliest forms, known as ***Homo erectus***, expanded outside Africa to reach present-day Georgia almost immediately (Lordkipanidze et al., 2013), and as far as western Europe, Java and China. Later, about 400–600 thousand years ago, **Neandertals** (Neanderthals being an earlier spelling) split from the lineage that would lead to modern humans and would inhabit western Eurasia continuously until about 30,000 years ago. Denisovans represent a sister group to the Neandertals and are very poorly known currently.

In the meantime, African humans continued to evolve largely separated from their non-African cousins and, at various times and places, biological and cultural innovations emerged (and died out to re-emerge again) that would later come together to form the so-called "human package", including a characteristic *gracile* morphology, a large brain, art, personal ornamentation, complex tool kits, and large-scale trade, among other qualities. It has been proposed that this represented a true "modern human revolution" whereby possibly a single genetic mutation (or a small number thereof) changed something in our ancestors that gave them language, art and complex culture (e.g., Mithen, 1996, or Chomsky, 2010) in one go. However, this myth was soon dispelled as more data and better dating, especially from Africa, became available (McBrearty and Brooks, 2000), on one hand, and by our growing understanding of the genetic bases and evolution of complex traits such as speech and language, on the other.[7] These African *modern humans* started a fast colonization of the whole world by first emerging from Africa some time around 50,000–70,000 years ago, reaching as far as Australia and New Guinea about 50,000 years ago, only later, about 12,000 years ago (or a little earlier), reaching the Americas, and Oceania only about 1000 years ago (Jobling et al., 2013).

The main process driving this colonization of the world was represented by a *serial **founder effect***, whereby populations split off, migrate and then split off again, in a process of repeated demographic growth and fission. As discussed earlier (Section 8.3), the daughter populations experience a founder effect whereby their genetic diversity is a subset of the mother population's diversity,

[7] But old habits die hard – see for example Chomsky (2010) very recent claims; see Dediu and Levinson (2013) for a critique.

repeated again and again and again. This should result in a gradual decrease of genetic diversity the further away from the origin of migration one is, with the genetic diversity outside the origin being a subset of the diversity at the origin – thus, the most genetically diverse populations should be those near the origin of migration and the least diverse those that experienced the most founder effects, in general those the furthest away (see Figure 8.10). In fact, this is exactly what we see: African populations have the highest genetic diversity (Barbujani and Colonna, 2010; Jakobsson et al., 2008), with India the second with its genetic diversity a subset of the African one (Majumder, 2010), and one of the least genetically diverse being the Americas (and within the Americas, there is a gradient from north to south; Wang et al., 2007). Moreover, this decrease in genetic (and phenotypic) diversity is almost linear with an estimate of travelled distance from Africa (Manica et al., 2007; Betti et al., 2009).

One might naïvely expect most of the genetic diversity within our species to be distributed between large-scale groups (or what are sometimes called "races"), a view especially driven by the "obvious" perceived large phenotypic differences between such groups compared to the differences between members of the same group. However, this is emphatically not the case and on average about 85% of genetic diversity is distributed among members of the same population, with an extra $\approx$7% between populations within the same continent and only about 8% being explained by membership of continental-size groups (Barbujani and Colonna, 2010; Jobling et al., 2013). Moreover, these differences are distributed as continuous *gradients* (or *clines*) among populations instead of clear-cut abrupt changes or discrete boundaries marking an unquestionable demarcation between different groups. Many of these differences are due to the same "ubiquitous" alleles being present at different frequencies in most populations and very rarely to "private" alleles being present in just a few.

Of course, this distribution of genetic diversity rules out any view of humanity as composed of discrete "races" characterized by well-defined genetically based traits, but it does not mean either that there is no structure to our species. In fact, using large numbers of so-called **Ancestry Informative Markers** (or AIMs), that is, loci with known large allele frequency differences between populations, allows the relative identification of the population of origin of an individual. For example, using just a few hundred AIMs is enough for placing individuals within continent-sized populations with acceptable accuracy (Nassir et al., 2009; Paschou et al., 2010) and presumably as more data becomes available the more precise this will be. A striking illustration of the human genetic structure and its strong relationship with geography is given by the very precise reconstruction of the map of Europe by Novembre et al. (2008) using only genetic information. More precisely, they genotyped about 3000

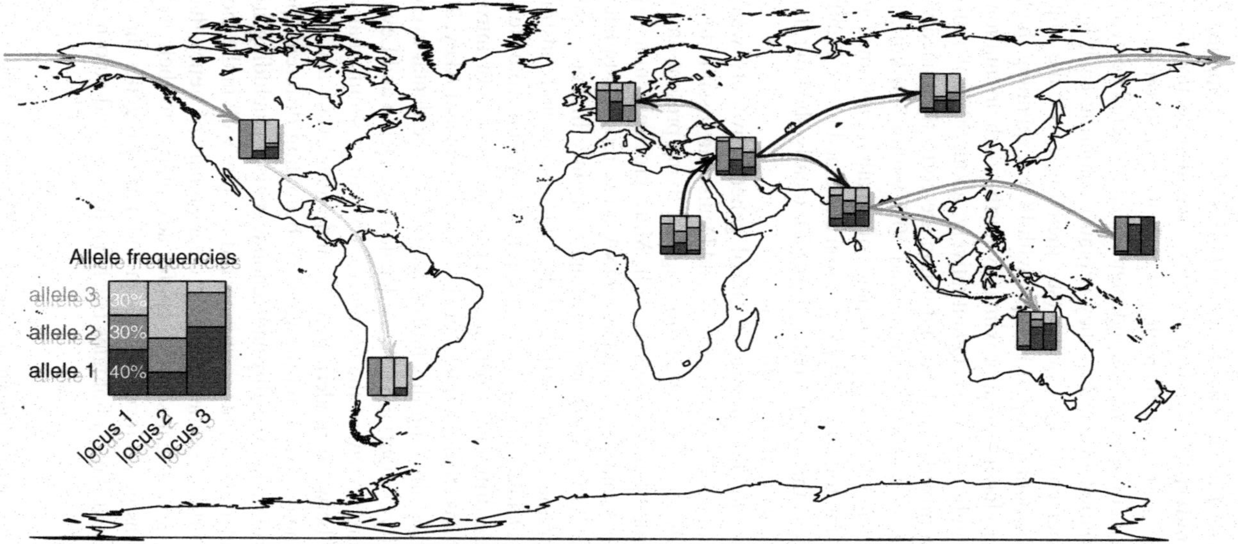

Figure 8.10 Serial founder effect during modern humans' expansion out of Africa (please note that the actual migration was much more complex and this is just for illustration purposes). Shown are three loci each with three possible alleles (shown with different shades of grey) and their frequencies for a given population are represented by the heights of the corresponding bars. It can be seen that Africa harbours most genetic diversity, and the more founder effects (represented by arrows) there are, the less diversity there is due to allelic loss and fixation. Arrow colour represents the amount of diversity left.

individuals from across Europe for about 500,000 SNPs and applied **Principal Component Analysis** (PCA; Jolliffe, 2002; see also Section 5.3.3) to the resulting genotype data. They then plotted each individual on the PC1 × PC2 axes (the principal components are independent and ordered decreasingly by the amount of variation in the data they explain) colouring each individual according to his/her country of origin (see their Figure 1); the image is strikingly similar to the actual map of Europe, with PC1 (explaining the highest amount of variation) having an approximately north–south orientation, and PC2 being approximately east–west. Quantitatively, about 90% of individuals who do not have a mixed grandparental origin are correctly placed within 700 km of their actual geographical origin (and 50% within 310 km). However, despite this amazing level of geographic information contained in the Europeans' genomes, the total amount of variation explained by PC1 and PC2 together amounts to less than half a percent (0.5%), clearly showing that the vast majority of the variation is local in scale.

An interesting twist to this picture of global genetic diversity was provided by the recent sequencing of Neandertal and Denisovan fossil individuals (Green et al., 2010; Meyer et al., 2012) showing that living present-day modern humans carry some of their genes as well. More precisely, non-African humans have between 1 and 4% Neandertal DNA (very recent estimates put this at an average of 1.15% for Europeans and a slightly higher 1.38% in East Asians; Sankararaman et al., 2014) while Melanesians and Australians carry a further 3–4% Denisovan DNA. However, things are far from clear and there seem to be large differences between loci in the amount of introgression into modern humans (Sankararaman et al., 2014; Vernot and Akey, 2014), and the actual scenarios allowing this introgression (rare accidental events or long-term low-intensity contact) are heavily debated (see Dediu and Levinson, 2013, for an up-to-date review of the evidence). Overall, it seems that while there was admixture between modern humans and Neandertals and Denisovans, these human lineages were on the verge of becoming different species with evidence of hybrid male infertility and genomic regions (including around *FOXP2*) where Neandertal variants seemingly did not fare well in the modern human genome. Nevertheless, there are also clear signals of genes that we acquired from them, genes that seem to confer advantages to their modern human carriers (such as genes involved in skin and hair) and genes that seem to be involved in present-day pathologies (such as diabetes and auto-immune disorders[8]).

[8] However, care should be taken when jumping to conclusions about the phenotypic effects of these genes in the pre-modern environment in which they were acquired by our ancestors from their evolutionary cousins; genes involved in some modern-day pathologies had very strong benefits before the advent of plentiful food, lack of physical exercise, widespread hygiene and advanced medical care (see also Sections 9.1 and 9.2).

8.7 Recent and ongoing evolution in humans

Are we, the mightiest biological creatures ever to walk the Earth, capable of colonizing almost every niche the planet has to offer, of literally moving mountains, changing the planet's climate and sending robots to Mars, still shaped by biological evolution, or have we somehow managed to escape its laws? One popular opinion, recently exemplified by Sir David Attenborough's interview for *Radio Times* (`http://www.radiotimes.com/news/2013-09-09/david-attenborough-i-dont-ever-want-to-stop-work`, 9 September 2013), has it that thanks to our cultural, scientific and technological advances we (at least in the Western world) have freed ourselves from biological evolution, which has been replaced by cultural evolution. A controversial proposal in this vein is Crabtree's (2013a,b) view that human intelligence has been steadily declining since about 2000–6000 years ago, a view that nevertheless is most probably wrong. On the other hand, there are arguments that despite our cultural "insulation", biological evolution still shapes us; that, in fact, it is even this culture that creates new pressures on our biology!

8.7.1 Skin colour

Despite the difficulties of detecting and interpreting the signs of the evolutionary processes in our genomes, recent approaches leave no doubt: there are clear cases of very recent and even ongoing selective pressures affecting us. One of them is skin-deep and yet probably the most used characteristic in classifying people into "races": skin (and hair and eye) colour. The genetics of these traits is unexpectedly complex (Rees, 2003; Sturm, 2009), and the distribution of skin colour among human populations has a striking geographic pattern, with darker shades being present at lower latitudes. Simplifying, this can be understood as the result of a trade-off between protection against the harmful effects of ultraviolet light (such as certain skin cancers and possibly folate deficiency; Borradale and Kimlin, 2012), accomplished through a darker skin, and allowing enough ultraviolet light to penetrate the skin and produce its positive effects (it is required for vitamin D synthesis, the absence of which results in rickets and may affect the immune system as well; van der Mei et al., 2003; Dorr et al., 2013). Thus, in lower latitudes there is enough UV light so that protection against it is more important (resulting in darker-skinned populations), while at higher latitudes there is much less UV light available and less protection is required (resulting in lighter skins).

Genome scans and candidate gene studies have identified signals of selection on genes that affect skin pigmentation in Europeans and Asians (such as *SLC24A5*, *SLC42A2*, *MC1R*, *MATP*, *OCA2*, *TYR*, *KITLG* and *TYRP1*; Lao et al., 2007; Norton et al., 2007; Sturm, 2009; Voight et al., 2006). Moreover, it seems the pattern of selection was complex, with some genes affected in both Europeans and Asians, while others show signs in one population only, suggesting that convergent evolution for light skin is affecting different genes (Norton et al., 2007; McEvoy et al., 2006). Importantly, this selective pressure must have acted after our dispersal out of Africa on the ancestral phenotype of dark skin, hair and eye colour, and the identification of convergent evolution in Europeans and Asians suggests that this selection is recent, probably acting during the last couple of tens of thousands of years.

Interestingly, at least some Neandertals (Cerqueira et al., 2012) also seem to have had red hair and light skin (Lalueza-Fox et al., 2007) but the particular mutation apparently affecting the gene *MC1R* is different from the known mutations in modern humans in this gene resulting in light skin, again suggesting convergent evolution for the light-skin trait at high latitudes.

8.7.2 Hair, sweat and ear wax

It might seem laughable but there seems to be a pretty robust signal of recent selection for ear wax type in humans! Ear wax (Ishikawa et al., 2013) comes in two main forms, dry and wet (or sticky), and they form a Mendelian phenotype with wet being dominant. The frequency of these two types shows marked differences between human populations, with the dry form being frequent among East Asians (>80%) but infrequent among others (<5% in Europeans), forming roughly a north–south and east–west gradient (Yoshiura et al., 2006). It was recently found that a single SNP (with A and G alleles) in the *ABCC11* gene is causative of these differences, with the AA genotype determining the dry form, while AG and GG the wet form (Yoshiura et al., 2006). The *ABC* (ATP-binding cassette) family of genes is generally involved in transporting various molecules across the cell membrane, and *ABCC11* in particular is involved in the functioning of the *apocrine glands*. These glands are present in the ear canal (where their excretion helps produce the ear wax), the axillary regions (where they contribute to the characteristic sweat smell) and the breast (milk production).

The AA genotype of the "ear wax SNP" results in a non-functional form of the transporter protein (this form is in fact quickly degraded by the **endoplasmic reticulum** before it can do its job; Toyoda et al., 2009), which fails to secrete the oils required for wet earwax, but also the odoriferous substances involved in the axillary sweat smell, and the production of colostrum. Moreover, there seems to be a link between this SNP and breast cancer in Japanese (but not Europeans; Ishikawa et al., 2013), and the effectiveness of certain cancer chemotherapeutic drugs (Toyoda and Ishikawa, 2010), but more work is required.

Recently, genome scans (Kimura et al., 2007) and more specific candidate gene studies (Ohashi et al., 2011) have found that there was selective pressure on the AA genotype in East Asians, resulting in a selective sweep. The selection coefficient was estimated at a relatively weak 0.01[9] and the age of the mutation approximately 2000 (1000–4000) generations ago (assuming a generation time of 15 years that would be $\approx 30,000$ years ago), well after the migration out of Africa. The selective pressure itself, however, is currently unclear and is probably not related to ear wax per se, but rather to sweating in cold climates (Ohashi et al., 2011). Interestingly, the dry/wet ear wax genotype has been used to trace the history of Japan (Super Science High School Consortium, 2009; Sato et al., 2009), and to make inferences about a ≈ 4000-years-old individual from Greenland whose DNA was extracted from hair preserved in permafrost (he had dry ear wax and his genotype was interpreted as pointing to a recent wave of migration from Siberia; Rasmussen et al., 2010).

Finally, an allele of the *EDAR* gene, 370A, seems to have been under positive selection in East Asians some time before 10,000 years ago, and could be involved in phenotypes such as hair thickness, sweat glands and tooth shape, but it is currently unclear what exactly the selective pressure might have been (Bryk et al., 2008; Kimura et al., 2007).

8.7.3 Living at high altitude

Populations living at very high altitudes above 3500–4000 meters experience very low oxygen concentrations compared to those living at sea level (about 60% lower; Yi et al., 2010; Peng et al., 2011). This chronic *hypoxia* has negative effects on non-adapted individuals, such as low capacity for effort and low birth weight (which is associated with high infant mortality rates and reduced health and cognitive development). However, populations adapted to high-altitude conditions, such as those living on the Tibetan plateau, in the Andes and the Ethiopian highlands, show genetic traits that allow them to tolerate such low concentrations of oxygen. Selection scans have identified several genes such as *EPAS1*, a transcription factor known to be involved in hypoxia (Yi et al., 2010), *EGLN1*, which regulates *EPAS1* (Peng et al., 2011), *PRKAA1* and *NOS2A*, a gene involved in the synthesis of the ubiquitous signalling molecule NO (nitric oxide; Bigham et al., 2010), and *BHLHE41*, also involved in hypoxia (Huerta-Sánchez et al., 2013).

However, probably the most interesting aspect is represented by the pattern of selected genes found in the populations studied to date: Himalaya and Tibet (Yi et al., 2010; Peng et al., 2011; Bigham et al., 2010; Hanaoka et al., 2012),

[9] The *selection coefficient* s represents the relative fitness ratio between two phenotypes/genotypes, and varies between 0 (no selection, neutral evolution) and 1 (the selected-against variant is completely lethal, having thus a fitness of 0). Here $s = 0.01$ means a 100:99 fitness ratio between the selected (AA) and the other (AG and GG) genotypes.

the Andes (Bigham et al., 2010) and Ethiopians (Huerta-Sánchez et al., 2013; Alkorta-Aranburu et al., 2012). Given that these adaptations are fairly recent and certainly happened after the out-of-Africa expansion (in the Andes they must postdate the colonization of the Americas about 12,000 years ago), it is not a surprise that different genes are involved in each of these populations, pointing to convergent evolution. However, it is interesting to note *EGLN1* is involved in adaptations in both the Andes and Tibet (Bigham et al., 2010), while *BHLHE41* in Ethiopians belongs to the same hypoxia pathway (Huerta-Sánchez et al., 2013).

8.7.4 Is culture shielding or selecting us?

It is clear from the examples briefly discussed above that we, at least until relatively recently in evolutionary terms, have been under quite strong selective pressures that shaped our genomes in such a way that the signature of their influence is clearly detectable with current techniques. These signatures are sometimes hard to interpret (for example, "ear wax"), but sometimes reveal a pretty clear environmental pressure (e.g., skin colour and altitude-related hypoxia), and show the differential influence of selection on different populations, sometimes resulting in convergent evolution. However, this highlights the importance of being able to detect selective sweeps in sub-populations of a larger population. For example, at the level of our whole species, the selective sweeps from hypoxia in Tibetan, Andean and Ethiopian populations would not be detectable as hard sweeps as they do not affect the same variant, and can be seen instead as a soft sweep composed of several local hard sweeps (Novembre and Han, 2012).

In the next sections we will look at more recent times and even more fascinating examples where culture plays a major role in shaping our genomes.

9 Interactions between genetic and cultural evolution

Here we discover that a new force, cultural evolution, is at work when it comes to humans and we discuss some cases where biological and cultural evolution interact, resulting in outcomes that are very hard to explain otherwise. Besides examples related to pathogens and nutrition, we will focus here on the fascinating cases of de novo emergence of sign languages. These cases clearly show that language is not shielded from the influence of our genetic background and might provide a useful model for understanding the emergence and evolution of language in our lineage. We end by investigating the idea that weak biases with a genetic component might be able to affect the process of cultural evolution of language and result in universal tendencies but also in patterns of linguistic diversity.

We saw in the last section (8.7) of Chapter 8 that humans certainly were not free from strong selective pressures acting mostly on local populations after they left Africa to conquer the whole world. Thus, it seems that, despite widespread ideas claiming otherwise, our culture did not completely shield us from the rigours of biological evolution, at least until relatively recently. Here we will continue this list of examples, but we will focus on selective pressures that were generated by our own culture, and that fed back on our genomes, shaping them. Sometimes, this moulding of our genomes through selection has favoured the cultural practices that generated those selective pressures in the first place, or even allowed the development of new cultural practices, creating a full *co-evolutionary spiral* between biological and cultural evolution. For an up-to-date and comprehensive overview of **cultural evolution** see Richerson and Christiansen (2013); gene–culture evolution is nicely introduced in Richerson and Boyd (2008) and Feldman and Laland (1996), while more recent developments also covering niche construction (discussed in Odling-Smee et al., 2003) are described in Laland et al. (2010), Richerson et al. (2010), and the special issue (number 366) "Human niche construction" of the *Philosophical Transactions of the Royal Society B* (Biological Sciences) from 27 March 2011.

The more we understand our evolutionary history the more we realize how important and powerful this process has been in shaping us, and how early it started in our lineage (Richerson et al., 2010). For example, even the earliest stone tools (the Oldowan) made around 2.5 million years ago undoubtedly changed the selective environment of our ancestors by opening up new sources of food and new modes of defence and aggression, while the invention of cooking, at least 250,000 years ago but most probably much earlier, likewise seems to have profoundly affected our genomes by increasing the quality of our diet, which, in turn, allowed the reduction in chewing and digestion and the increase in brain size that characterize our evolution (Wrangham, 2009; Carmody and Wrangham, 2009).

But what about more recent times, when arguably culture had advanced well enough to protect us from the elements? How about after agriculture was discovered (or invented) and allowed an unprecedented demographic growth and increase in social and cultural complexity, culminating in the emergence of chiefdoms, kingdoms, empires, nations and finally the globalized world we currently live in? On one hand this certainly relaxed some very strong pressures (as shown by the massive demographic explosion that accompanied it), but on the other it amplified others and even generated new ones. From one perspective, culture is but a greatly exaggerated form of niche construction and we know that niches, whatever they are, bring selective regimes with them (Kendal et al., 2011; Laland et al., 2010).

We will now turn to some of these cases where cultural evolution (including language and speech) and biological evolution interact in complex ways to produce novel and unexpected phenomena. We will look at selection on our immune systems and how farming and its consequences have created – and continue to create – enormous pressures to adapt to new deadly diseases. We will then see that agriculture has also shaped our digestive system through the kind of foods available and their processing, focusing on the celebrated case of lactase persistence. Finally, we will turn to language and speech and look at the evolution of the vocal tract (shaped in large part by the pressure to articulate) and the fascinating case of the emergent village sign languages. A particularly interesting case is represented by the influence our biology has on cultural evolution, and we will discuss some proposed cases of *genetic biasing* in speech and language.

9.1 Fighting pathogens

The complex arms race between pathogens and their (usually unwilling) hosts is one of the most powerful drivers of biological evolution, resulting in wonderful adaptations on both sides in a co-evolutionary dynamics playing out continuously since the dawn of life. However, the adoption of farming (a very recent phenomenon by evolutionary standards, having taken place only in the last 12,000 years; Diamond and Bellwood, 2003; Diamond, 1998; Mithen, 2003) has had several important consequences for our fight against disease.

First, this development resulted in more food being available in a more predictable way, allowing a massive demographic explosion. Importantly, these people did not disperse uniformly across the available landscape in small mobile bands (as the majority of hunter-gatherers do) but instead settled in sedentary communities with high population densities (and, for most of our history, with pretty bad sanitation), ranging in size from small villages to fortified and highly structured city-states and the present-day mega-cities, all inter-connected by trade, marriage, war and, nowadays, tourism. This allowed the emergence and spread of highly infectious diseases resulting in epidemics of massive proportions (such as the Black Death in the fourteenth century and the 1918 "Spanish flu") but also the relatively benign seasonal flu.

Second, living in close contact with domestic animals facilitated the jump of pathogens to this new host, the humans, resulting in deadly diseases such as anthrax and (possibly) tuberculosis, and more recently the outbreaks of swine and especially avian flu.

Third, agriculture, especially before the "green revolution" of the 1940s–1960s, was severely affected by weather patterns and pest outbreaks, resulting quite often in widespread famines. Such famines would not only directly cause loss of life, but would also result in a drastic reduction in the survivors' capacity to fight off infectious diseases, increasing the probability of deadly epidemics.

All these taken together suggest that the onset and spread of agriculture must have resulted in new and very strong selective pressures on our immune system (Diamond, 1998; Mithen, 2003). Indeed, genome scans clearly support this inference (e.g., Fumagalli et al., 2011; Novembre and Han, 2012) by identifying pathogens (viruses, bacteria but also parasites) as one of the major sources of selection in humans. Therefore, a cultural development (agriculture) generated novel selective pressures that have shaped our genomes.

Interestingly, another set of cultural developments resulting in the modern obsession with hygiene might have inadvertently, by eliminating most of the naturally occurring pathogens and parasites, resulted in the "misfiring" of our immune system. The *hygiene hypothesis* (Strachan, 1989; Bufford and Gern, 2005; Okada et al., 2010) suggests that this can produce *auto-immune diseases* such as hay fever, asthma, multiple sclerosis, diabetes (type 1) and even some forms of cancer that, given their impact on health, might in turn generate new selective pressures. Another explanation for the signal of selection on genes related to auto-immune (inflammatory) diseases (Raj et al., 2013) might be that these are by-products of genes selected for resistance to pathogens, but it is clear that the two hypotheses are not mutually incompatible. Yet another very recent development concerns the over-use of antibiotics, which generates enormous selective pressures on bacteria, resulting in the emergence of multi-drug-resistant strains (Davies and Davies, 2010). If the current failure to develop new classes of antibiotics and/or the pathogens' speed of adaptation to new drugs persists, then we might see a return to the pre-antibiotic era and a re-ignition of strong selective pressures on our immune systems.

In conclusion, the widespread signal of recent selection on genes involved in the immune system witnesses the effectiveness and complexity of the co-evolutionary dynamics between our culture and our genes. New cultural developments, such as agriculture and modern medicine, affect our relationship with pathogenic organisms, drastically altering the selective landscape to which our genome must answer.

9.2 Eating well

As we saw in the previous section, cooking seems to have played a significant role in our early evolution by providing a higher-quality diet while reducing the need for chewing and digestion. Another set of cultural innovations related to diet happened with the transition to agriculture and involved the switch from typical hunter-gatherer diets (except in arid environments) to a starch-rich diet. The digestion of starch starts in the mouth with enzymes secreted in the saliva (*salivary amlyase*) and continues in the small intestine. It was recently found (Perry et al., 2007) that populations with a diet rich in starch tend to have more copies of the *AMY1* gene responsible for the secretion of salivary amylase, and that this increased copy number results in more enzyme being produced.

Several other genes involved in diet seem to show signals of selection (Laland et al., 2010), an intriguing case being represented by the *alcohol dehydrogenase* (*ADH*) gene cluster in East Asians (Li et al., 2008; Peng et al., 2010). Here, the derived allele *ADH1B*47His* is present at very high frequencies and there are signs of recent selection either on this allele or on related regulatory regions (Li et al., 2008; Peng et al., 2010), but the resulting phenotype includes fast alcohol metabolism and the production of high quantities of aldehyde, a toxic intermediate product with nasty side effects. Therefore, carriers of these alleles are protected against alcoholism and it has been argued that the pattern of selection and inferred age squares well with the emergence of rice-based agriculture (and alcohol production) in East Asia, suggesting that the selective pressure behind their spread is represented by the high cost of culturally driven alcohol availability (Peng et al., 2010; Laland et al., 2010).

Another example related to food consumption concerns the unequal distribution of *metabolic disorders* (such as type 2 diabetes and obesity) and *salt retention disorders* (such as hypertension) across the world. Recent studies (Hancock et al., 2008, 2011) seem to suggest that there is a correlation between genes involved in metabolic disorders and cold climate. These are consistent with the older "*thrifty genotype*" proposal (Neel, 1962) which suggests that episodes of food scarcity (either due to natural causes such as climatic variability or, importantly for us, due to cultural factors such as the long voyages undertaken by Polynesians, where starvation was also compounded by intense cold stress; McGarvey, 1991; Myles et al., 2011) generated very strong pressure for genotypes capable of quickly storing energy in times of plenty, but which in today's environment of a reliable food supply result in disease.

But the best-studied and most impressive case is represented by the presence at high frequency, in some populations, of adults who are still capable of digesting fresh milk without ill effects. Being mammals, human babies can digest milk, but this ability is usually lost by adults and the ingestion of fresh milk results in various symptoms including flatulence, diarrhoea and abdominal pain (Gerbault et al., 2011). This is due to the drastic reduction in the synthesis of the enzyme responsible for digesting the main carbohydrate in milk (lactose), an enzyme called *lactase*. When digested, lactose is an important source of energy, but undigested it reaches the colon, where it is fermented by bacteria, resulting in the ill effects mentioned above. The adults who retain the capacity to digest fresh milk are **lactose tolerant** or *lactase persistent* (the opposite, "normal" phenotype is called *lactose intolerance*) and their frequency varies dramatically, being generally low, but in some populations it reaches close to 90% (for a map see Figure 1 panel a on p. 865 in Gerbault et al., 2011).

Lactose persistence involves several genetic variants that affect the *lactase* gene (*LCT*) on chromosome 2q21 but, interestingly, they seem to do so by affecting its regulation. For example, the most common in Europeans is a SNP known as *–13910*C/T*, with the T allele being lactose-tolerant; it occurs 14 kb upstream of the *LCT* gene actually in an intron of a different gene (*MCM6*), and it has been shown that this variant does indeed affect *LCT*'s promoter region. Other variants are also located in the same intron as *–13910*C/T* and seem to affect the expression of *LCT* through the same mechanism. However, these variants have arisen independently in different populations (Tishkoff et al., 2006; Gerbault et al., 2011) in the last 10,000 years and have been subjected to some of the strongest and clearest recent selective sweeps detectable in the human genome (Tishkoff et al., 2006; Gerbault et al., 2011; Laland et al., 2010; Itan et al., 2009). This suggests that very strong convergent evolution during our recent history has favoured adults who were lactose-tolerant in *some* populations.

Interestingly, the populations with the highest frequency of lactose tolerance also have a history of animal husbandry, prompting the proposal that it was the availability of fresh milk in these populations that generated the selective pressures. The advantages of fresh milk consumption include its high energetic value, calcium and vitamin D content (especially important for agriculturalists at high latitudes), but also its water content, and probably lactose-tolerant adults enjoyed these advantages to a high level during episodes of famine and/or drought, when the lack of other sources of food and water exposed lactose-intolerant adults to intense stresses (Gerbault et al., 2011). Nevertheless, it is clear that the introduction of animal husbandry – a cultural innovation – changed the ecological niche of these populations and resulted in strong selection for lactose persistence. In turn, the resulting increase in lactose-tolerant adults resulted in the persistence of this cultural practice, closing the feedback loop between the cultural and genetic evolutionary systems.

Of course, the story is much more complex and fascinating (see for example Gerbault et al., 2011) and includes the fact that not all adults incapable of digesting lactose suffer intense side effects, and that populations with high frequencies of lactose-intolerant adults developed other cultural means of using milk by fermenting it into yogurts and cheeses (thus culturally shielding themselves from this selective pressure) and, more recently, by developing lactase food supplements that help digest lactose.

9.3 Creating new languages

Most of us have invented "new" languages as children but, alas, those do not really count as independent cases of *de novo* emergence that could inform us about fundamental questions on the origin and evolution of language and the interaction between genetic, environmental and cultural factors in these processes. Luckily, there are such natural experiments available, where language is created anew with little or no influence from pre-existing languages. These new languages are not spoken but use sign as their medium and usually emerge in relatively closed communities (villages) in which there is a high incidence of deafness persisting for several generations, and where the deaf members are well integrated in the community. Some of these *village sign languages* have been intensely studied both from a genetic point of view, seeking to identify the mutation(s) responsible, and from a linguistic point of view, as they might offer an invaluable glimpse into the early stages of language evolution. Here we will focus on two examples only, **Kata Kolok** on the island of Bali in Indonesia, and **Al-Sayyid Bedouin Sign Language** (or ABSL) in the Negev desert of Israel. Other interesting cases are represented by the famous emergence of the **Nicaraguan Sign Language** (or NSL) and the equally fascinating processes accompanying the establishment of deaf institutions and the development of the *American Sign Language* in the USA.

We have already encountered Kata Kolok in Section 4.2 as an example of a phenotype caused by recessive mutation, and we studied the actual gene involved (the unconventional myosin *MYO15A*) and how its mutation produces recessive deafness by interfering with the protein's function as a cellular transporter and its interaction with two other proteins, whirlin and EPS8, in the stereocilia of the hair cells in the ear. While congenital deafness in the general population is somewhere between 1% and 7‰ [1], about 2.2% of the villagers in Bengkala are profoundly deaf, with ≈17% of hearing people being heterozygous carriers of the mutation (Winata et al., 1995). The deaf members of the community are relatively well integrated in the society and their marriage rates and numbers of children are comparable to those of the hearing members, even if there is a high level of assortative mating (i.e, deaf people usually inter-marry). It is important to note that hearing members of the community are also second-language signers in Kata Kolok. The sign language is relatively old at about five generations (the mutation has probably existed for about 100–300 years) and shows interesting linguistic features (for more details and discussions see De Vos, forthcoming, 2011, 2012, and Levinson and Dediu, 2013).

The congenital recessive deafness in *ABSL* is due to a mutation at the *DFNB1* locus, which maps to two tightly linked genes, *GJB2* (*gap junction beta-2* or *connexin 26*; OMIM 121011) and *GJB6* (*gap junction beta-6* or *connexin 30*; OMIM 604418). These genes encode proteins that are components of the so-called *gap junctions* – tiny channels that connect neighbouring cells and allow the exchange of molecules and information between the cells. Mutations in these genes are the most frequent cause of non-syndromic recessive deafness. The exact mechanism through which mutations in *GJB2* and *GJB6* result in deafness is not clear but could involve deficits in the regulation of potassium in the inner ear. About 3.4% of the Al-Sayyid villagers are affected by recessive deafness due to a mutation at this locus, a mutation that appeared about four generations ago (Scott et al., 1995). The community is highly inbred and polygynous, with the deaf members well integrated and many of the hearing members also using ABSL as a second language. ABSL is extremely interesting from a linguistic point of view (Meir et al., 2010) as it seems to be at an early stage of cultural evolution, for example still lacking full phonology (Sandler et al., 2011), which allows the study of the emergence of linguistic structure through use and transmission (Sandler et al., 2005).

These two cases (which are, most probably, just the tip of the iceberg as many more village sign languages are known to exist; Zeshan and de Vos, 2012) seem to highlight several very important points (Levinson and Dediu, 2013; Gialluisi et al., 2013). First, given the right biological and social environment, a completely new language may emerge as the primary means of communication for a whole community, largely independent of neighbouring languages or the other language(s) that some of its speakers might know and

[1] Aggregate estimates in newborns in the USA; `http://www.asha.org/public/hearing/Prevalence-and-Incidence-of-Hearing-Loss-in-Children`, November 2013.

use. Second, the acceptance and social integration of the deaf members of the community, and the use of the emergent sign language as an effective medium of communication, relaxes the otherwise strong selective pressures against deafness and allows the increase in the frequency of the causative alleles. This is a particularly powerful process where inbreeding and/or assortative mating by deafness are also involved. In turn, this increase in allele frequency leads to an increase in the proportion of deaf members in the community, resulting in a positive feedback loop, closing the co-evolutionary cycle.[2] Third, this clearly shows that language does not emerge fully formed in one step and that there is no sort of qualitative complexity threshold that a communication system must cross to qualify as "language" (like a sort of phase transition) but that, instead, the process is gradual and driven by use and transmission, even in what concerns fundamental properties of language such as the duality of patterning. The number of generations and the size and structure of the effective pool of individuals using the language (including its L_2 users) seem to be fundamental parameters explaining its complexification.

This special case of co-evolution between the genetic bases of recessive hearing loss and language in the signed modality clearly shows that the general concept of co-evolution between the genetic and cultural systems is effective and has explanatory power, and strongly suggests that other cases might exist as well. However, the story is far from simple and earlier models (e.g., Aoki and Feldman, 1991, and Feldman and Aoki, 1992) must be extended to take into account the complex genetic, cultural and linguistic realities (Gialluisi et al., 2013) revealed by painstaking fieldwork (Zeshan and de Vos, 2012).

A different twist to this fascinating co-evolutionary story is given by the case of the deaf in large, open societies. A well-studied example is represented by the USA before and after the introduction of the institutions for the deaf and the systematic use of the American Sign Language (ASL), about two centuries ago (Nance et al., 2000; Arnos et al., 2008). This process reduced the negative selection against deafness due to social exclusion and reduced mating opportunities, not only by allowing the deaf members of society to become better integrated, to access better jobs and attain better social status, but also by allowing them to meet other individuals with similar experiences, promoting *assortative mating* by deafness status. This, in turn, favoured the increase in frequency of deafness (approximately five-fold in the last century), also coupled with what Arnos et al. (2008) call *linguistic homogamy*. This is the tendency in a deaf by hearing couple for the hearing partner also to sign as a result of him/her being the hearing offspring of a deaf family, being therefore a heterozygous carrier. An interesting and unexpected effect

[2] There are notable differences between the effects of assortative mating and random inbreeding but we are glossing over them here.

(Nance and Kearsey, 2004) is that the mutation that happens to be initially more frequent in the population (even slightly so) at the onset of the relaxation in selective pressure is disproportionately pushed to high frequencies, becoming the (numerically) dominant cause of recessive hearing loss. Thus, even in large and open societies there is co-evolution between the mutations resulting in hearing loss and sign language, mediated through social integration, institutions for the deaf, and assortative mating and linguistic homogamy.

However, on one hand, it can be argued that no matter how fascinating these emerging sign languages might be, they are not true parallels to the emergence of language in our species, as the people generating them are modern humans with a fully modern genome and obviously growing up in an environment where language already exists, even if in a modality that is inaccessible to them. On the other hand, the genetic effect on language is so drastic and "brutal" that it forces a complete change of modality from the universally preferred acoustic channel to the visual, potentially raising the question whether more subtle genetic effects would be capable of actually gently pushing or pulling language in certain directions.

While the answer to the first question is probably more theoretical in nature and depends on one's view of how language might have originated and evolved in our lineage (I personally believe that emerging sign languages represent valid but limited parallels to how language evolved; see also Dediu and Levinson, 2013, and Levinson and Dediu, 2013), the second question is amenable to empirical investigation.

9.4 Genetically biased cultural evolution

The process of language change is a complex cultural phenomenon studied by disciplines such as *historical linguistics* (Campbell, 2004; Campbell and Poser, 2008) and *sociolinguistics* (Wardhaugh, 2010), and until recently it was generally assumed that the biology and neuro-cognition of the language users – that's us – are essentially identical across languages and populations. Thus, the only interesting way in which these factors could be considered in discussions of language change was as constraints and facilitators explaining *language universals*. However, the existence, nature and number of such universals are hotly contested (see for example Christiansen and Chater, 2008, and especially Evans and Levinson, 2009, and the subsequent comments, the latter also collected in a special issue of the journal *Lingua*, Volume 120, Issue 12, 2010) and it is unclear what the actual relationship between the assumed biological and cognitive factors and the linguistic proposals is.

However, if anything, an evolutionary view of our species clearly makes the assumption of a boringly uniform biological and neuro-cognitive infrastructure for language untenable; there is variation between individuals and populations in almost every aspect of our genomes and phenotypes. As we have seen in this book so far, there is inter-individual variation in almost any measure related to language and speech performance and for some of this variation there are good reasons to suspect a genetic component, as shown by non-negligible heritability estimates and, in some cases, actual genes involved in their development and maintenance. Thus, the really interesting question then becomes, "do these differences actually affect language change?"

I will begin with an intuitive example which, until further research, must be taken as hypothetical (Dediu, 2007; Levinson and Dediu, 2013). My native language is Romanian, a Romance language (a member of the Indo-European language family and closely related to Italian, Spanish and French, having derived less than 2000 years ago from vulgar Latin) which uses the alveolar trill /r/ (as do Spanish and Scottish English), a sound that I am unable to produce despite sustained effort, some speech therapy, and in the absence of obvious articulatory, perceptual or cognitive impairments. This in itself is not surprising as in every language that uses /r/ this is among the last sounds to be successfully acquired and there is always a small proportion of the speakers who fail and systematically replace it with some other sound (in my case, a postalveolar approximant [ɹ̠]). But it is clear that the small percentage of Romanian speakers incapable of articulating /r/ are unable to generate enough pressure on the language to drop this sound from its inventory and replace it with something else, be it an approximant such as /ɹ/ or /ɻ/, a tap or a flap /ɾ/, or even a voiced uvular fricative /ʁ/. On the other hand, a quick thought experiment: a whole generation of Romanian babies born such that none can get their tongue tip to vibrate would result in Romanian instantly losing /r/ and replacing it with something else. Thus, the interesting question becomes: how many of these non-/r/ speakers must there exist, what sort of communicative networks must they be part of and in what roles and for how long, for such a sound change to happen? Sociolinguistics suggests that if such speakers hold high prestige then just a few of them could be enough to prompt a sound change, but more modelling work is required to understand the exact interplay between all these factors and to allow us to make informed predictions.

Now, assume there is some genetic basis for this difficulty in (or, in the absence of adequate speech therapy, even the impossibility of) acquiring /r/. As far as we currently know, if there is such a genetic basis it is probably not simple, involving more than one locus and with incomplete penetrance. One interesting aspect is that in languages without the alveolar trill /r/ the alleles responsible would be invisible, and they would be exposed only in those languages with /r/ in their inventory. The other interesting aspect is that now we can frame the discussion above as an inquiry into the characteristics of a genetic system capable of influencing the way languages change, an example of what I call the *genetic biasing of language*.

9.4.1 *Genetic biases in language*

The main idea (which is not entirely new, being foreshadowed by, for example, Darlington, 1947, and especially Brosnahan, 1961) is that language is a cultural system evolving in time (Richerson and Christiansen, 2013; Croft, 2000, 2008; Beckner et al., 2009) and the direction of this evolutionary process is influenced by several factors including forces outside linguistics itself such as the social structure and historical and demographic events. One such source of influences is represented by the anatomy, physiology and neuro-cognition of the language users, which might ultimately have a genetic basis. As in the example above, it could be about a difficulty or incapacity in articulating certain sounds, or it could be much more subtle, as in a preference toward or against certain linguistic structures or phonological distinctions. D. Robert Ladd and I have recently proposed (Dediu and Ladd, 2007) that such a preference towards or against using voice pitch to encode linguistic meaning (what is called **linguistic tone**) might play a role in explaining the distribution of tone languages (such as Chinese and Yoruba) by biasing the preferred direction of language change (Ladd et al., 2008). We have specifically proposed that two genes involved in brain growth and development, *ASPM* and *MCHP1*, influence an individual-level preference for or against using voice pitch to convey linguistic meaning, and that this individual bias would be very weak (virtually undetectable without conducting specific investigations) given that every normal child can acquire the language(s) of the community they are raised in, as strikingly shown by, for example, adopted children. Thus, individuals can override these biases under the influence of the culture of their community and end up as users of the prevalent linguistic systems even if this means that they have to fight against their own biases.

However, we proposed that when several such biased individuals are part of a speech community, their biases might become visible and influence the trajectory of language change, in effect becoming amplified by the process of repeated cultural transmission of language (for a review see Dediu, 2011a). This process of bias amplification and the influence of biases on the process of language transmission has been convincingly shown in computer models (Dediu, 2008, 2009; Kirby et al., 2007) and experiments with human participants (Kirby et al., 2008; Smith and Wonnacott, 2010) and can become manifest either when language is transmitted across generations of biased learners (Ladd et al., 2008; Dediu, 2011a) or when language is repeatedly used within communities of such biased agents (Levinson and Dediu, 2013).

While some biases affecting language change might be cognitive in nature (Dediu and Ladd, 2007; Ladd et al., 2008), it is highly probable that others affect the production and perception of speech, being enacted through the anatomy and physiology of our vocal tracts and hearing. This line of research

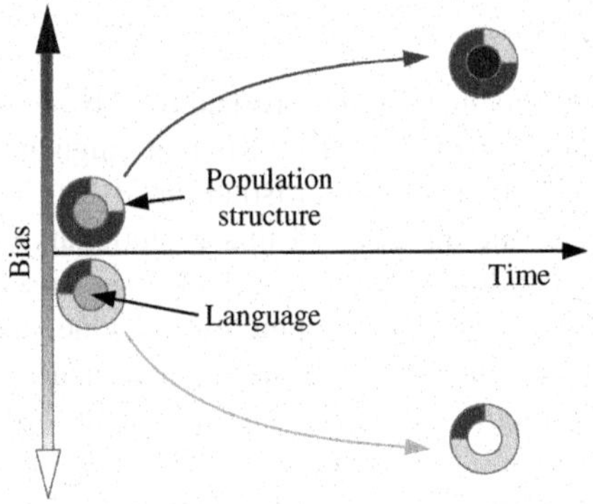

Figure 9.1 The influence of genetic biases on language change and typological patterns. An abstract bias is represented by the vertical axis and two populations with different structures (the top has 75% of members biased towards the darker values and 25% towards the lighter value, while the bottom one has the opposite structure) but initially with the same language (inner circle colour). In time, the languages of the two populations diverge and come to reflect the bias structure of their speakers (the black and white colours used for the languages do not imply that this type of bias inexorably pushes languages towards binary oppositions but is simply a visual device).

has not been much investigated to date, but it is a very promising line of inquiry, as it allows the measurement and quantification of the existing variation as well as its possible influences on phonetic and phonological diversity.

These biases influence language change in the sense that, depending on the pattern of distribution of biases within the population of language users, some directions of language change are more probable than others. For example, a population with a high frequency of non-/r/ articulators would increase the probability of replacing /r/ with other phoneme(s) relative to the probability of maintaining it (or not acquiring /r/ in the first place), while a population with a high frequency of the alleles of *ASPM* and *MCPH1* that bias against using voice pitch to convey linguistic meaning would have a greater chance of losing tone contrasts and replacing them with segmental distinctions (if it has them) or not acquiring them in the first place. Thus, if everybody shares the same biases, language change will tend to favour certain linguistic features and disfavour others, ultimately resulting in *universal tendencies*. For example, all humans seem to share a strong bias against centre-embedding (Karlsson, 2007), probably rooted in general cognitive constraints, which results in a universal tendency for languages not to use more than very few levels of centre-embedding.

However, if populations differ in the distribution of biases in the sense that some are more affected by them than others, then we would expect the trajectory of language change to diverge between the populations, ultimately resulting in *typological* (or structural) differences between the languages spoken by the two populations. Thus, if some populations contain significantly more individuals who are biased against using voice pitch to convey linguistic meaning than others, then we would expect that they would lose tone contrasts faster (or resist the introduction of tone contrasts more) than the others (see Figure 9.1). Dediu and Ladd (2007) propose exactly this scenario to explain (in part) why some areas have many languages using tone while in others there are hardly any tone languages; the fact that these languages do not all belong to the same linguistic family strongly suggests that shared inheritance from a common ancestor is an unlikely explanation, an important confounder for such claims.

In fact, it is extremely important to rule out as many confounders as possible in such *correlational studies* (Roberts and Winters, 2013) lest one discover a "chopstick gene" (Hamer and Sirota, 2000): given how many cultural traits there are and how many variants in the human genome, and that history and demography tend to shape them in similar ways, it is highly probable that a blind search for significant correlations will find heaps of false positives far outnumbering the real cases of causal relationships between genes and cultural (linguistic) phenomena.

10 Conclusions, topics not covered, future directions

This book is intended to give a relatively up-to-date introduction and overview tailored to the needs of speech and language scientists who venture into genetics and do not have the time to delve into more comprehensive general introductions such as Brown and Brown (2012), Snustad and Simmons (2010) and Jobling et al. (2013). It tries to cover selected topics using wherever possible examples from speech and language and tries to give a level of understanding that should be enough for working in cross-disciplinary teams and understanding modern genetic and genomic research. The book tries to refer both to up-to-date reviews and to primary research reports, which should allow the reader to deepen his or her understanding on selected topics or to understand alternative or emerging interpretations and views.

The main aim of the book, if one must be singled out at all costs, is to allow language and speech scientists to appreciate the complexity and beauty of their topic of interest's genetic infrastructure and understand the constraints and opportunities such an understanding has on their day-to-day work, be it designing psycholinguistic experiments into the acquisition of non-native phonological contrasts, studying sound change or the typology of negation.

However, given the target audience, the length of the book and (not least) my own competence, I had to be highly selective in the topics covered, leaving aside a number of extremely fascinating issues. Among them, the use of concepts and methods adapted from evolutionary biology to study language change and evolution such as *phylogenetics* and *phylogeography* (for introduction, discussion and up-to-date research see for example Dunn et al., 2008; Atkinson and Gray, 2005; Gray et al., 2010; Bouckaert et al., 2012; Dediu, 2011b). Likewise, I must only very lightly mention here the use of genetic and linguistic data to reconstruct population history, whereby usually uniparental markers (mitochondrial DNA and the non-recombining region of the Y chromosome) are combined with historical linguistics (for a good overview see Jobling et al., 2013, and the reviews in the special issue "Global genetic history of Homo sapiens" of *Current Biology* 2010 volume 20 issue 4). Strictly on the genetic side I ignored many developments in genomics, proteomics and other -omics as well as most exciting advances in epigenetics and gene regulation, and

I only mentioned in passing the fascinating techniques and findings deriving from investigations using cell cultures and animal models to probe – probably unexpectedly for some – the genetic architecture of speech and language (see for example White et al., 2006; Gabel et al., 2010; Berwick et al., 2012). Finally, I glossed over the current revolution in evolutionary biology brought about mainly by advances in genomics and developmental biology, including such oft-used (and oft-misrepresented) topics as Evo-Devo (for a good and readable introduction see Carroll, 2011), developmental/phenotypic plasticity (West-Eberhard, 2003), the role of neutral processes in evolution (Lynch, 2007; Koonin, 2012) and the importance of horizontal/lateral genetic transfer for evolutionary thinking (Dagan and Martin, 2006; Doolittle and Bapteste, 2007).

The fields associated with genetics in general, and their applications to speech and language, are evolving at a maddening rate (and this is not a tired cliché but the reality one has to confront when trying to keep up-to-date with this literature) and I hazard to predict that in 10 years' time most probably we will understand the regulatory network containing *FOXP2*, would have found some other such "molecular windows" through exome and whole genome sequencing coupled with linkage in special families, that we will have conducted some really large-scale association studies in the general population on language production, processing and perception, and that the investigation of the genetic biases affecting language and speech would have resulted in a better understanding of speech production and perception and their impact on the evolution of phonological systems. Having said that, I expect that this book will still be useful five years from now (and parts of it maybe even in 10 years), as its main proposals and examples are relatively well established and will survive (and will, in fact, be strengthened by) future discoveries.

10.1 Resources for further study

The literature cited throughout the book tries to strike a balance between reference works (usually books), review articles (selected for their comprehensive coverage of a topic and, wherever possible, their accessibility), primary research reports (sometimes quite technical and hard to crack but important for a full understanding of the topic) and online resources (with more frequent updates and giving access to vast databases and tools). Here I will try to list only a small selection of resources that the interested reader might pursue to delve into specific directions but the literature cited in the book's sections should be consulted as well.

Introductions to genetics and genomics: Lewin and Krebs (2011) gives a very comprehensive coverage of genetics, but I find Brown and Brown

(2012) and especially Snustad and Simmons (2010) better for newcomers. Wikipedia (http://www.wikipedia.org) has a comprehensive "Category on Genetics" (http://en.wikipedia.org/wiki/Category:Genetics) and the articles "Genetics" (http://en.wikipedia.org/wiki/Genetics), "Genomics" (http://en.wikipedia.org/wiki/Genomics) and "Introduction to genetics" (http://en.wikipedia.org/wiki/Introduction_to_genetics) are not only good overviews but also good entry points down this deep "rabbit hole"; articles are usually well-written, up-to-date and decently referenced. The National Library of Medicine hosts a "Genetics Home Reference" web portal (http://ghr.nlm.nih.gov/) that provides pointers to various conditions, genes, etc. but also a whole freely available handbook on genetics (http://ghr.nlm.nih.gov/handbook).

Gene discovery: Linkage and association are covered in depth in the various articles cited in the book, but the interested reader can consult the websites (and manuals) of the various software packages such as PLINK (http://pngu.mgh.harvard.edu/~purcell/plink) and MERLIN (http://www.sph.umich.edu/csg/abecasis/Merlin/index.html). Moreover, online portals such as OMIM (http://www.ncbi.nlm.nih.gov/omim), the National Center for Biotechnology Information (NCBI; http://www.ncbi.nlm.nih.gov/) and the UCSC Genome Browser (http://genome.ucsc.edu/index.html) should be consulted as they contain a wealth of information and tools. For particular genes, portals such as Gene (http://www.ncbi.nlm.nih.gov/gene) and GeneCards (http://www.genecards.org) are very useful, and dbSNP (http://www.ncbi.nlm.nih.gov/SNP) and SNPedia (http://snpedia.com/index.php/SNPedia) are informative when it comes to Single Nucleotide Polymorphisms (SNPs).

Genetics of speech and language: Currently there is no introduction to this topic (except for the very book you are reading now ☺) but Dediu and Graham (2014) is an up-to-date annotated bibliography covering this field. There are a few groups working specifically on these topics and their web presence could provide important information on the current directions and developments in the field, and I mention here only the Language and Genetics Department at the Max Planck Institute for Psycholinguistics in Nijmegen, The Netherlands, led by Simon Fisher (and where I am currently based; http://www.mpi.nl/departments/language-and-genetics).

Population and evolutionary genetics: Evolutionary biology is well covered in various books such as Futuyma (2013) but a quick and well-written account of Evo-Devo is Carroll (2011). Koonin (2012) and Lynch (2007) are fascinating windows into modern developments in evolutionary theory. For population and evolutionary human genetics probably Jobling et al. (2013) is the best available resource. The Tree of Life project (`http://tolweb.org/tree`) maintains a comprehensive and up-to-date database of the living world. Joseph Felsenstein's (2004) classic introduction to phylogenetics is supplemented by a comprehensive collection of evolutionary/phylogenetic software maintained by him at `http://evolution.genetics.washington.edu/phylip/software.html` – one is probably bound to find the right package for most evolutionary problems here.

Cultural evolution: Odling-Smee et al. (2003) introduce *niche construction*, and Dawkins (1996) discusses the *extended phenotype*, both worth reading for anybody interested in cultural evolution. Richerson and Boyd (2008) is a very gentle and informative discussion of cultural evolution by two of its most famous proponents. The recent Richerson and Christiansen (2013) is an edited volume covering many aspects of cultural evolution written by experts in the field and highlighting some of the current directions and issues. Zeshan and de Vos (2012) is a comprehensive resource on village sign languages.

Appendix

The computer code

This appendix contains the computer code used in various places throughout the book.

A.1 Simulating samples from a population in which there is no association between a biallelic genetic marker and a dichotomous phenotype

This shows how I generated the random contingency tables in Section 5.3.1. The R code is given below. Please note that long lines broken to fit the page are marked with “↘” and “↘” at the breakage point. Also, spaces within strings are explicitely marked with “”.

```
# Simulate contingency tables from a given matrix distribution ↘
  ↘ :
# (inspired from Rolf Turner's answer to a question: https:// ↘
  ↘ stat.ethz.ch/pipermail/r-help/2009-February/189538.html)

# The distribution of alleles and cases & controls in our ↘
  ↘ hypothetical example:
m <- matrix( c( 1790, 1900, 210, 106 ), ncol=2, byrow=TRUE );
rownames(m) <- c(''A'',''a''); colnames(m) <- c(''Cases'','' ↘
  ↘ Controls'');
# m is:
#     Cases Controls
# A   1790      1900
# a    210       106

# Compute the frequencies in each cell:
p <- (rowSums(m)%o%colSums(m))/(sum(m)^2);
# p is:
#          Cases      Controls
# A 0.45986936 0.46124897
# a 0.03938177 0.03949991

# Simulate 1000 contingency tables with the same distributions ↘
  ↘ as m (each column is one such table):
X <- rmultinom(1000,sum(m),p);

# For each simulated table, compute the chi-square test and, ↘
  ↘ if significant, print the simulated table and the chi- ↘
  ↘ square test:
```

```
simstat <- apply(X,2,function(x,k){
                # Convert the matrix into a pretty form (for printing):
                ml <- matrix(x,nrow=k); rownames(ml) <- c(''A'',''a''); colnames(ml) <- c(''Cases'',''Controls'');
                # Compute the chi-square test:
                c1 <- chisq.test(ml);
                if( c1$p.value < 0.05 )
                {
                  # If the test was significant al alpha=0.05, print it:
                  print(ml);
                  cat( paste( ''chisq('', c1$parameter, '')='', sprintf(''%.2f'',c1$statistic), '', p='', sprintf(''%.3g'', c1$p.value), ''\n\n'', sep='''' ) );
                }
                # Return the test's p-value:
                c1$p.value;
              },k=nrow(m));
simstat <- unlist(simstat);

# Print the number and proportion of significant contingency tables at alpha=0.05 and 0.01, respectively:
cat( ''False positives for alpha=0.05: '', sum( simstat < 0.05 ), '' out of '', length(simstat), '' = '', 100*sum( simstat < 0.05 ) / length(simstat), ''%\n'' );
cat( ''False positives for alpha=0.01: '', sum( simstat < 0.01 ), '' out of '', length(simstat), '' = '', 100*sum( simstat < 0.01 ) / length(simstat), ''%\n'' );

cat( ''False positives for alpha=0.00005: '', sum( simstat < 0.00005 ), '' out of '', length(simstat), '' = '', 100*sum( simstat < 0.00005 ) / length(simstat), ''%\n'' );
```

A.2 Computing (log) odds ratios and their confidence intervals

This shows the R code used to compute (log) odds ratios from a 2 × 2 contingency table and how to obtian confidence intervals for log odds ratios.

```
# Compute (Log) Odds Ratios and confidence intervals for a 2x2 contingency table:
library(vcd); # use library vcd and more specifically its oddsratio() and confint() functions

# The example contingency table:
m <- matrix( c( 1790, 1900, 210, 106 ), ncol=2, byrow=TRUE ); rownames(m) <- c(''A'',''a''); colnames(m) <- c(''Cases'', ''Controls'');
m; # print it

# The Odds Ratio (OR):
OR <- oddsratio( m, log=FALSE );
```

```
summary(OR);

# The Log Odds Ratio (LOR):
LOR <- oddsratio(m, log=TRUE);
summary(LOR); # Please note that the provided information is ↘
  ↘ much richer, including the z-test and its p-value

# LOR confidence intervals:
confint( LOR, level = 0.95 ); # 95%CI (alpha = 0.05)
confint( LOR, level = 0.99 ); # 99%CI (alpha = 0.01)
confint( LOR, level = (1.0 - 1.0e-8) ); # 99.999999%CI (alpha ↘
  ↘ = 10^-8) - the genomewide alpha level
```

A.3 Simulating genetic drift

This shows the `R` code used to simulate genetic drift in populations of different sizes run for a given number of generations. The code is inefficient but (hopefully) clearer.

```
# Simulate genetic drift in populations of various sizes

# Given population size (n) and number of generations (ngen), ↘
  ↘ simulate and draw the population's evolution under drift
# The code is far from optimal but (hopefully) is easy to ↘
  ↘ follow and understand
run.drift <- function( n=1000, ngen=100, show.ngen=FALSE, show ↘
  ↘ .gen=FALSE, show.legend=FALSE )
{
  # Start with a population composed only of heterozygotes:
  pop <- rep( ''aA'', n );

  # Given a population, produce the next generation:
  next.gen <- function( pop )
  {
    new.pop <- NULL;
    for( i in 1:length(pop) )
    {
      # Pick the parents at random:
      p1 <- sample(1:length(pop),1);
      p2 <- sample((1:length(pop))[-p1],1); # make sure they ↘
        ↘ are different individuals
      # Produce gametes:
      g1 <- c( substr(pop[p1],1,1), substr(pop[p1],2,2) );
      g2 <- c( substr(pop[p2],1,1), substr(pop[p2],2,2) );
      # Randomly pick two gametes and combine them to produce ↘
        ↘ the offspring:
      o <- paste( sample( g1, 1 ), sample( g2, 1 ), sep='''' ) ↘
        ↘ ;
      # Save it in the new population:
      new.pop[i] <- o;
    }
    # Return the new population:
    return (new.pop);
  }
```

```
# The allele and genotype frequencies across generations:
p.aa <- numeric(ngen);
p.aA <- numeric(ngen);
p.AA <- numeric(ngen);
p.a  <- numeric(ngen);
p.A  <- numeric(ngen);

# Generate repeatedly new populations:
for( i in 1:ngen )
{
  # compute the allele and genotype frequencies:
  p.aa[i] <- sum( pop == ''aa'' )/n;
  p.aA[i] <- sum( pop %in% c(''aA'',''Aa'') )/n;
  p.AA[i] <- sum( pop == ''AA'' )/n;

  p.a[i]  <- p.aa[i] + p.aA[i]/2;
  p.A[i]  <- p.AA[i] + p.aA[i]/2;

  # Generate the next generation:
  pop <- next.gen(pop);
}

# Plot the frequencies across generations:
plot(   p.a*100,  type=''l'', lty=''solid'',  col=grey(0.0), ↘
 ↘ lwd=2, xlab='''', ylab='''', ylim=c(0,100) );
points( p.A*100,  type=''l'', lty=''solid'',  col=grey(0.6), ↘
 ↘ lwd=2 );
points( p.aa*100, type=''l'', lty=''dotted'', col=grey(0.0), ↘
 ↘ lwd=1 );
points( p.AA*100, type=''l'', lty=''dotted'', col=grey(0.6), ↘
 ↘ lwd=1 );
points( p.aA*100, type=''l'', lty=''dashed'', col=grey(0.4), ↘
 ↘ lwd=2 );
#abline( h=50, lty=''dotted'', col=grey(0.8), lwd=1 );
if( show.ngen )
{
  mtext( paste(''N='',n,sep=''''), side=2, line=3, cex=1.25 ↘
   ↘ );
  mtext( ''Frequency (%)'', side=2, line=2, cex=0.75 );
}
if( show.gen )
{
  mtext( ''Generations'', side=1, line=2, cex=0.75 );
}
if( show.legend )
{
  legend( ''right'', legend=c( expression(italic(p)[italic(a ↘
   ↘ )]), expression(italic(p)[italic(A)]), expression( ↘
   ↘ italic(p)[italic(aa)]), expression(italic(p)[italic(aA ↘
   ↘ )]), expression(italic(p)[italic(AA)]) ),
                    col=c( grey(0.0), grey(0.6), grey(0.0), ↘
                     ↘ grey(0.6), grey(0.4) ),
                    lty=c( ''solid'', ''solid'', ''dotted'', ↘
                     ↘ ''dotted'', ''dashed'' ),
```

```
                                    lwd=c( 2, 2, 1, 1, 2 ) );
    }
}

# Draw runs for 500 generations of three populations with size 2, 100 and 1000 individuals, each run independently 3 times:
par( mfrow=c(3,3), mar=c(3, 4.5, 0, 0) + 0.1 );
ngen <- 500;
for( i in 1:3 ) run.drift( n=2, ngen=ngen, show.ngen=(i==1), show.legend=(i==3) );
for( i in 1:3 ) run.drift( n=100, ngen=ngen, show.ngen=(i==1) );
for( i in 1:3 ) run.drift( n=1000, ngen=ngen, show.ngen=(i==1), show.gen=TRUE );
```

References

Abecasis, G. R., Cherny, S. S., Cookson, W. O., and Cardon, L. R. 2002. Merlin–rapid analysis of dense genetic maps using sparse gene flow trees. *Nature Genetics*, **30**(1), 97–101.

Alarcon, M., Abrahams, B. S., Stone, J. L., Duvall, J. A., Perederiy, J. V., Bomar, J. M., Sebat, J., et al. 2008. Linkage, association, and gene-expression analyses identify CNTNAP2 as an autism-susceptibility gene. *American Journal of Human Genetics*, **82**(1), 150–159.

Alberts, B. 2009. *Essential Cell Biology*. New York: Garland Science.

Alkorta-Aranburu, G., Beall, C. M., Witonsky, D. B., Gebremedhin, A., Pritchard, J. K., and Di Rienzo, A. 2012. The genetic architecture of adaptations to high altitude in Ethiopia. *PLoS Genetics*, **8**(12), e1003110.

Allen, H. L., Estrada, K., Lettre, G., Berndt, S. I., Weedon, M. N., Rivadeneira, F., Willer, C. J., et al. 2010. Hundreds of variants clustered in genomic loci and biological pathways affect human height. *Nature*, **467**(7317), 832–838.

Allison, D. B. 1997. Transmission-disequilibrium tests for quantitative traits. *American Journal of Human Genetics*, **60**(3), 676–690.

Anscombe, F. J. 1973. Graphs in Statistical Analysis. *The American Statistician*, **27**(1), 17.

Aoki, K., and Feldman, M. W. 1991. Recessive hereditary deafness, assortative mating, and persistence of a sign language. *Theoretical Population Biology*, **39**(3), 358–372.

Arlt, A., and Schäfer, H. 2011. Role of the immediate early response 3 (IER3) gene in cellular stress response, inflammation and tumorigenesis. *European Journal of Cell Biology*, **90**(6-7), 545–552.

Arnos, K. S., Welch, K. O., Tekin, M., Norris, V. W., Blanton, S. H., Pandya, A., and Nance, W. E. 2008. A comparative analysis of the genetic epidemiology of deafness in the United States in two sets of pedigrees collected more than a century apart. *American Journal of Human Genetics*, **83**(2), 200–207.

Astle, W., and Balding, D. J. 2009. Population structure and cryptic relatedness in genetic association studies. *Statistical Science*, **24**(4), 451–471.

Atkinson, Q. D. 2011. Phonemic diversity supports a serial founder effect model of language expansion from Africa. *Science*, **332**(6027), 346–349.

Atkinson, Q. D., and Gray, R. D. 2005. Curious parallels, curious connections – phylogenetic thinking in biology and historical linguistics. *Systematic Biology*, **54**(4), 513–526.

Baayen, R. H. 2008. *Analyzing Linguistic Data: A Practical Introduction to Statistics Using R*. Cambridge: Cambridge University Press.

Ballana, E., Morales, E., Rabionet, R., Montserrat, B., Ventayol, M., Bravo, O., Gasparini, P., et al. 2006. Mitochondrial 12S rRNA gene mutations affect RNA secondary structure and lead to variable penetrance in hearing impairment. *Biochemical and Biophysical Research Communications*, **341**(4), 950–957.

Barbi, M., Binda, S., Caroppo, S., Ambrosetti, U., Corbetta, C., and Sergi, P. 2003. A wider role for congenital cytomegalovirus infection in sensorineural hearing loss. *Pediatric Infectious Disease Journal*, **22**(1), 39–42.

Barbujani, G., and Colonna, V. 2010. Human genome diversity: frequently asked questions. *Trends in Genetics*, **26**(7), 285–295.

Barnett, A. G., Pols, J. C. van der, and Dobson, A. J. 2005. Regression to the mean: what it is and how to deal with it. *International Journal of Epidemiology*, **34**(1), 215–220.

Barry, J. G., Yasin, I., and Bishop, D. V. M. 2007. Heritable risk factors associated with language impairments. *Genes, Brain, and Behavior*, **6**(1), 66–76.

Barton, R. A., and Venditti, C. 2013. Human frontal lobes are not relatively large. *Proceedings of the National Academy of Sciences of the USA*, **110**(22), 9001–9006.

Bates, T. C., Luciano, M., Medland, S. E., Montgomery, G. W., Wright, M. J., and Martin, N. G. 2011. Genetic variance in a component of the language acquisition device: ROBO1 polymorphisms associated with phonological buffer deficits. *Behavior Genetics*, **41**(1), 50–57.

Bateson, P., and Mameli, M. 2007. The innate and the acquired: Useful clusters or a residual distinction from folk biology? *Developmental Psychobiology*, **49**(8), 818–831.

Bateson, W., Saunders, E. R., and Punnett, R. 1905. Experimental studies in the physiology of heredity. *Reports to the Evolution Committee of the Royal Society*, **2**, 155, 80–99.

Baudat, F., Buard, J., Grey, C., Fledel-Alon, A., Ober, C., Przeworski, M., Coop, G., and de Massy, B. 2010. PRDM9 is a major determinant of meiotic recombination hotspots in humans and mice. *Science*, **327**(5967), 836–840.

Baudat, F., Imai, Y., and de Massy, B. 2013. Meiotic recombination in mammals: localization and regulation. *Nature Reviews Genetics*, **14**(11), 794–806.

Beaumont, M. A., and Rannala, B. 2004. The Bayesian revolution in genetics. *Nature Reviews Genetics*, **5**, 251–261.

Beck, J. A., Lloyd, S., Hafezparast, M., Lennon-Pierce, M., Eppig, J. T., Festing, M. F., and Fisher, E. M. 2000. Genealogies of mouse inbred strains. *Nature Genetics*, **24**(1), 23–25.

Beckner, C., Blythe, R., Bybee, J., Christiansen, M. H., Croft, W., Ellis, N. C., Holland, J., et al. 2009. Language is a complex adaptive system: Position paper. *Language Learning*, **59**(s1), 126.

Bellis, M. A, Hughes, K., Hughes, S., and Ashton, J. R. 2005. Measuring paternal discrepancy and its public health consequences. *Journal of Epidemiology & Community Health*, **59**(9), 749–754.

Berg, I. L., Neumann, R., Lam, K.-W. G., Sarbajna, S., Odenthal-Hesse, L., May, C. A., and Jeffreys, A. J. 2010. PRDM9 variation strongly influences recombination hotspot activity and meiotic instability in humans. *Nature Genetics*, **42**(10), 859–863.

Bergen, S. E., Gardner, C. O., and Kendler, K. S. 2007. Age-related changes in heritability of behavioral phenotypes over adolescence and young adulthood: a meta-analysis. *Twin Research and Human Genetics*, **10**(3), 423–433.

Bergh, O., Brsheim, K. Y., Bratbak, G., and Heldal, M. 1989. High abundance of viruses found in aquatic environments. *Nature*, **340**(6233), 467–468.

Berlin, B., and Kay, P. 1969. *Basic Color Terms, Their Universality and Evolution*. Oakland, CA: California University Press.

Berra, T. M., Alvarez, G., and Ceballos, F. C. 2010. Was the Darwin/Wedgwood dynasty adversely affected by consanguinity? *BioScience*, **60**(5), 376–383.

Berry, D. A. 1996. *Statistics – A Bayesian Perspective*. Pacific Grove, CA: Duxbury Press.

Berry, D. P., Skuce, R. A., and Coffey, M. P. 2010. Quantitative and molecular genetics, a love story. *Advances in Animal Biosciences*, **1**(01), 358–358.

Berwick, R. C., Beckers, G. J. L., Okanoya, K., and Bolhuis, J. J. 2012. A bird's eye view of human language evolution. *Frontiers in Evolutionary Neuroscience*, **4**, 5.

Betti, L., Balloux, F., Amos, W., Hanihara, T., and Manica, A. 2009. Distance from Africa, not climate, explains within-population phenotypic diversity in humans. *Proceedings of the Royal Society B*, **276**(1658), 809–814.

Bigham, A., Bauchet, M., Pinto, D., Mao, X., Akey, J. M., Mei, R., Scherer, S. W., et al. 2010. Identifying signatures of natural selection in Tibetan and Andean populations using dense genome scan data. *PLoS Genet*, **6**(9), e1001116.

Bindu, L. H., and Reddy, P. P. 2008. Genetics of aminoglycoside-induced and prelingual non-syndromic mitochondrial hearing impairment: a review. *International Journal of Audiology*, **47**(11), 702–707.

Bishop, D. V. M. 2003. Genetic and environmental risks for specific language impairment in children. *International Journal of Pediatric Otorhinolaryngology*, **6751**, S143–S157.

Bishop, D. V. M., and Hayiou-Thomas, M. E. 2008. Heritability of specific language impairment depends on diagnostic criteria. *Genes, Brain, and Behavior*, **7**(3), 365–372.

Bishop, D. V. M., and Snowling, M. J. 2004. Developmental dyslexia and specific language impairment: same or different? *Psychological Bulletin*, **130**(6), 858–886.

Bland, J. M., and Altman, D. G. 2000. The odds ratio. *British Medical Journal*, **320**(7247), 1468.

Bollenbach, T., Vetsigian, K., and Kishony, R. 2007. Evolution and multilevel optimization of the genetic code. *Genome Research*, **17**(4), 401–404.

Borradale, D. C., and Kimlin, M. G. 2012. Folate degradation due to ultraviolet radiation: possible implications for human health and nutrition. *Nutrition Reviews*, **70**(7), 414–422.

Borsboom, D. 2009. *Measuring the Mind: Conceptual Issues in Contemporary Psychometrics*. 1st edn. Cambridge: Cambridge University Press.

Bouckaert, R., Lemey, P., Dunn, M., Greenhill, S. J., Alekseyenko, A. V., Drummond, A. J., Gray, R. D., et al. 2012. Mapping the origins and expansion of the Indo-European language family. *Science*, **337**(6097), 957–960.

Brar, G. A., and Amon, A. 2008. Emerging roles for centromeres in meiosis I chromosome segregation. *Nature Reviews Genetics*, **9**(12), 899–910.

Bricker, D., and Squires, J. 1989. The effectiveness of parental screening of at-risk infants: The Infant Monitoring Questionnaires. *Topics in Early Childhood Special Education*, **9**(3), 67–85.

Brosnahan, L. F. 1961. *The Sounds of Language: An Inquiry into the Role of Genetic Factors in the Development of Sound Systems.* Cambridge: W. Heffer and Sons.

Brown, T. A., and Brown, T. 2012. *Introduction to Genetics: A Molecular Approach.* New York: Garland Science.

Bryk, J., Hardouin, E., Pugach, I., Hughes, D., Strotmann, R., Stoneking, M., and Myles, S. 2008. Positive selection in East Asians for an EDAR allele that enhances NF-B activation. *PLoS ONE*, **3**(5), e2209.

Bufford, J. D., and Gern, J. E. 2005. The hygiene hypothesis revisited. *Immunology and Allergy Clinics of North America*, **25**(2), 247–262, v–vi.

Calado, R., and Young, N. 2012. Telomeres in disease. *F1000 Medicine Reports*, **4**, 8.

Campbell, L. 2004. *Historical Linguistics: An Introduction.* Edinburgh: Edinburgh University Press.

Campbell, L., and Poser, W. J. 2008. *Language Classification: History and Method.* Cambridge: Cambridge University Press.

Cann, R. L., Stoneking, M., and Wilson, A. C. 1987. Mitochondrial DNA and human evolution. *Nature*, **325**(6099), 31–36.

Cardinale, M., Newton, R., and Nosaka, K. 2011. *Strength and Conditioning: Biological Principles and Practical Applications.* New York: John Wiley & Sons.

Cardon, L. R., and Palmer, L. J. 2003. Population stratification and spurious allelic association. *Lancet*, **361**(9357), 598–604.

Carmody, R. N., and Wrangham, R. W. 2009. Cooking and the human commitment to a high-quality diet. *Cold Spring Harbor Symposia on Quantitative Biology*, **74**(0), 427–434.

Carroll, S. B. 2011. *Endless Forms Most Beautiful: The New Science of Evo Devo and the Making of the Animal Kingdom.* London: Quercus Publishing.

Casson, R. W. 1994. Russett, rose, and raspberry: The development of English secondary color terms. *Journal of Linguistic Anthropology*, **4**(1), 5–22.

Cavalli-Sforza, L. L., Menozzi, P., and Piazza, A. 1994. *The History and Geography of Human Genes* (Abridged paperback edition). Princeton NJ: Princeton University Press.

Cerqueira, C. C. S., Paixo-Crtes, V. R., Zambra, F. M. B., Salzano, F. M., Hnemeier, T., and Bortolini, M.-C. 2012. Predicting homo pigmentation phenotype through genomic data: from neanderthal to James Watson. *American Journal of Human Biology*, **24**(5), 705–709.

Chang, S.-W., Mislankar, M., Misra, C., Huang, N., Dajusta, D. G., Harrison, S. M., McBride, K. L., et al. 2013. Genetic abnormalities in FOXP1 are associated with congenital heart defects. *Human Mutation*, **34**(9), 1226–1230.

Chanock, S. J., Manolio, T., Boehnke, M., Boerwinkle, E., Hunter, D. J., Thomas, G., Hirschhorn, J. N., et al. 2007. Replicating genotype-phenotype associations. *Nature*, **447**(7145), 655–660.

Charney, E. 2012. Behavior genetics and postgenomics. *Behavioral and Brain Sciences*, **35**(05), 331–358.

Chater, N., Reali, F., and Christiansen, M. H. 2009. Restrictions on biological adaptation in language evolution. *Proceedings of the National Academy of the Sciences of the USA*, **106**(4), 1015–1020.

Chen, H., Patterson, N., and Reich, D. 2010. Population differentiation as a test for selective sweeps. *Genome Research*, **20**(3), 393–402.

Chen, M., and Manley, J. L. 2009. Mechanisms of alternative splicing regulation: insights from molecular and genomics approaches. *Nature Reviews Molecular Cell Biology*, **10**(11), 741–754.

Chen, T. J., and Ji, C. Y. 2013. Secular change in stature of urban Chinese children and adolescents, 1985–2010. *Biomedical and Environmental Sciences: BES*, **26**(1), 13–22.

Chomsky, N. 1965. *Aspects of the Theory of Syntax*. Cambridge, MA: MIT Press.

Chomsky, N. 1980. *Rules and Representations*. New York: Columbia University Press.

Chomsky, N. 2010. Some simple evo devo theses: how true might they be for language. *The Evolution of Human Language: Biolinguistic Perspectives*, **62**, 54–62.

Christiansen, M. H., and Chater, N. 2008. Language as shaped by the brain. *Behavioral and Brain Sciences*, **31**(5), 489–508; discussion 509–558.

Cirulli, E. T., and Goldstein, D. B. 2010. Uncovering the roles of rare variants in common disease through whole-genome sequencing. *Nature Reviews Genetics*, **11**(6), 415–425.

Clarke, G. M., Anderson, C. A., Pettersson, F. H., Cardon, L. R., Morris, A. P., and Zondervan, K. T. 2011. Basic statistical analysis in genetic case-control studies. *Nature Protocols*, **6**(2), 121–133.

Clayton, D. F. 2013. The genomics of memory and learning in songbirds. *Annual Review of Genomics and Human Genetics*, **14**(1), 45–65.

Clovis, Y. M., Enard, W., Marinaro, F., Huttner, W. B., and De Pietri Tonelli, D. 2012. Convergent repression of Foxp2 3'UTR by miR-9 and miR-132 in embryonic mouse neocortex: implications for radial migration of neurons. *Development*, **139**(18), 3332–3342.

Cohen, J. 1988. *Statistical Power Analysis for the Behavioral Sciences*, 2nd edn. Abingdon: Routledge.

Cole, T. J. 2003. The secular trend in human physical growth: a biological view. *Economics and Human Biology*, **1**(2), 161–168.

Comrie, B. 1989. *Language Universals and Linguistic Typology: Syntax and Morphology*. Chicago, IL: University of Chicago Press.

Conrad, B., and Antonarakis, S. E. 2007. Gene duplication: a drive for phenotypic diversity and cause of human disease. *Annual Review of Genomics and Human Genetics*, **8**(1), 17–35.

Coolican, H. 2004. *Research Methods and Statistics in Psychology*. 4th edn. London: Hodder & Stoughton.

Coop, G., Bullaughey, K., Luca, F., and Przeworski, M. 2008. The timing of selection at the human FOXP2 gene. *Molecular Biology and Evolution*, **25**(7), 1257–1259.

Cordell, H. J. 2002. Epistasis: what it means, what it doesn't mean, and statistical methods to detect it in humans. *Human Molecular Genetics*, **11**(20), 2463–2468.

Cowie, F. 1999. *What's Within?: Nativism Reconsidered*. New York: Oxford University Press.

Cox, T. F., and Cox, M. A. 1994. *Multidimensional Scaling*. London: Chapman & Hall.

Crabtree, G. R. 2013a. Our fragile intellect. Part I. *Trends in Genetics*, **29**(1), 1–3.

Crabtree, G. R. 2013b. Our fragile intellect. Part II. *Trends in Genetics*, **29**(1), 3–5.

Crawley, M. J. 2005. *Statistics: An Introduction Using R*. Oxford: Wiley-Blackwell.

Croft, W. 2000. *Explaining Language Change: An Evolutionary Approach*. Harlow: Pearson.

Croft, W. 2003. *Typology and Universals*. Cambridge: Cambridge University Press.

Croft, W. 2008. Evolutionary linguistics. *Annual Review of Anthropology*, **37**, 219–234.

Cysouw, M., Dediu, D., and Moran, S. 2012. Comment on phonemic diversity supports a serial founder effect model of language expansion from Africa. *Science*, **335**(6069), 657.

Dagan, T., and Martin, W. 2006. The tree of one percent. *Genome Biology*, **7**(10), 118.

Dale, P. S., Harlaar, N., Haworth, C. M. A., and Plomin, R. 2010. Two by two: a twin study of second-language acquisition. *Psychological Science*, **21**(5), 635–640.

Dalgaard, Peter. 2008. *Introductory Statistics with R*. Berlin Springer.

Danubio, M. E., and Sanna, E. 2008. Secular changes in human biological variables in Western countries: an updated review and synthesis. *Journal of Anthropolological Sciences*, **86**, 91–112.

Darlington, C. D. 1947. The genetic component of language. *Heredity*, **1**, 269–286.

Davies, G., Tenesa, A., Payton, A., Yang, J., Harris, S. E., Liewald, D., Ke, X., et al. 2011. Genome-wide association studies establish that human intelligence is highly heritable and polygenic. *Molecular Psychiatry*, **16**(10), 996–1005.

Davies, J., and Davies, D. 2010. Origins and evolution of antibiotic resistance. *Microbiology and Molecular Biology Reviews*, **74**(3), 417–433.

Davis, L., and Chin, J. W. 2012. Designer proteins: applications of genetic code expansion in cell biology. *Nature Reviews Molecular Cell Biology*, **13**(3), 168–182.

Davis, O. S. P., Haworth, C. M. A., and Plomin, R. 2009. Dramatic increase in heritability of cognitive development from early to middle childhood: an 8-year longitudinal study of 8,700 pairs of twins. *Psychological Science*, **20**(10), 1301–1308.

Dawkins, R. 1989. *The Selfish Gene*. Oxford: Oxford University Press.

Dawkins, R. 1996. *The Blind Watchmaker: Why the Evidence of Evolution Reveals a Universe without Design*. New York: W. W. Norton.

Dean, A. 2011. In the loop: long range chromatin interactions and gene regulation. *Briefings in Functional Genomics*, **10**(1), 3–10.

Dediu, D. 2007. *Non-Spurious Correlations between Genetic and Linguistic Diversities in the Context of Human Evolution*. Ph.D. thesis, University of Edinburgh, UK.

Dediu, D. 2008. The role of genetic biases in shaping language-genes correlations. *Journal of Theoretical Biology*, **254**, 400–407.

Dediu, D. 2009. Genetic biasing through cultural transmission: do simple Bayesian models of language evolution generalize? *Jornal of Theoretical Biology*, **259**(3), 552–561.

Dediu, D. 2011a. Are languages really independent from genes? If not, what would a genetic bias affecting language diversity look like? *Human Biology*, **83**(2), 279–296.

Dediu, D. 2011b. A Bayesian phylogenetic approach to estimating the stability of linguistic features and the genetic biasing of tone. *Proceedings of the Royal Society B*, **278**, 474–479.

Dediu, D. 2013. Genes: interactions with language on three levels: inter-individual variation, historical correlations and genetic biasing. In Binder, P.-M., and Smith, K. (eds), *The Language Phenomenon*. Berlin: Springer, pp. 130–161.

Dediu, D., and Graham, S. 2014. Genetics and language. In Aronoff, M. (ed), *Oxford Bibliographies in Linguistics*. New York: Oxford University Press.

Dediu, D., and Ladd, D. R. 2007. Linguistic tone is related to the population frequency of the adaptive haplogroups of two brain size genes, ASPM and Microcephalin. *Proceedings of the National Academy of Sciences of the USA*, **104**(26), 10944–10949.

Dediu, D., and Levinson, S. C. 2012. Abstract profiles of structural stability point to universal tendencies, family-specific factors, and ancient connections between languages. *PLoS ONE*, **7**(9), e45198.

Dediu, D., and Levinson, S. C. 2013. On the antiquity of language: the reinterpretation of Neandertal linguistic capacities and its consequences. *Frontiers in Language Sciences*, **4**, 397.

Dediu, D., Cysouw, M., Levinson, S. C., Baronchelli, A., Christiansen, M. H., Croft, W., Evans, N., et al. 2013. Cultural evolution of language. In Richerson, P. J., and Christiansen, M. H. (eds), *Cultural Evolution: Society, Technology, Language, and Religion*. Strüngmann Forum Reports, vol. 12. Cambridge, MA: MIT Press, pp. 303–332.

Deeb, S. S. 2004. Molecular genetics of colour vision deficiencies. *Clinical and Experimental Optometry*, **87**(4-5), 224–229.

Deeb, S. S. 2005. The molecular basis of variation in human color vision. *Clinical Genetics*, **67**(5), 369–377.

Deeb, S. S. 2006. Genetics of variation in human color vision and the retinal cone mosaic. *Current Opinions in Genetics & Development*, **16**(3), 301–307.

Deeb, S. S., and Kohl, Susanne. 2003. Genetics of color vision deficiencies. *Developments in Ophthalmology*, **37**, 170–187.

Deriziotis, P., and Fisher, S. E. 2013. Neurogenomics of speech and language disorders: the road ahead. *Genome Biology*, **14**(4), 204.

Devlin, B, and Roeder, K. 1999. Genomic control for association studies. *Biometrics*, **55**(4), 997–1004.

De Vos, C. 2011. A signers' village in Bali, Indonesia. *Minpaku Anthropology Newsletter*, **33**, 4–5.

De Vos, C. 2012. Kata Kolok: An updated sociolinguistic profile. In: Zeshan, U., and De Vos, C. (eds), *Village Sign Languages: Anthropological and Linguistic Insights*. Berlin: Mouton de Gruyter, pp. 381–386.

De Vos, C. forthcoming. *Sign-Spatiality in Kata Kolok: How a Village Sign Language of Bali Inscribes its Signing Space*. Ph.D. thesis, Max Planck Institute for Psycholinguistics, Nijmegen, The Netherlands.

Diamond, J. 1998. *Guns, Germs and Steel: A Short History of Everybody for the Last 13,000 Years*. London: Vintage.

Diamond, J. 2012. *The World Until Yesterday: What Can We Learn from Traditional Societies?* Viking.

Diamond, J., and Bellwood, P. 2003. Farmers and their languages: the first expansions. *Science*, **300**, 597–603.

Dobson, J. M. 2013. Breed-predispositions to cancer in pedigree dogs. *ISRN Veterinary Science*, 941275.

Dobzhansky, T. 1973. Nothing in biology makes sense except in the light of evolution. *American Biology Teacher*, **35**, 125–129.

Doolittle, W. F., and Bapteste, E. 2007. Pattern pluralism and the Tree of Life hypothesis. *Proceedings of the National Academy of the Sciences of the USA*, **104**(7), 2043–2049.

Dorr, J., Doring, A., and Paul, F. 2013. Can we prevent or treat multiple sclerosis by individualised vitamin D supply? *The EPMA Journal*, **4**(1), 4.

Douglass, S., Ali, S., Meeson, A. P., Browell, D., and Kirby, J. A. 2012. The role of FOXP3 in the development and metastatic spread of breast cancer. *Cancer Metastasis Reviews*, **31**(3-4), 843–854.

Douzery, E. J. P., Delsuc, F., Stanhope, M. J., and Huchon, D. 2003. Local molecular clocks in three nuclear genes: divergence times for rodents and other mammals and incompatibility among fossil calibrations. *Journal of Molecular Evolution*, **57 Suppl 1**, S201–213.

Dudoit, S., Shaffer, J. P., and Boldrick, J. C. 2003. Multiple hypothesis testing in microarray experiments. *Statistical Science*, 71–103.

Dugard, P., File, P., and Todman, J. 2011. *Single-case and Small-n Experimental Designs: A Practical Guide To Randomization Tests.* 2 edn. Oxford: Routledge Academic.

Dunn, M., Levinson, S. C., Lindström, E., Reesink, G., and Terrill, A. 2008. Structural phylogeny in historical linguistics: methodological explorations applied in island Melanesia. *Language*, **84**, 710–759.

Dunn, M., Greenhill, S. J., Levinson, S. C., and Gray, R. D. 2011. Evolved structure of language shows lineage-specific trends in word-order universals. *Nature*, **473**, 79–82.

Durand, C., and Rappold, G. A. 2013. Height matters-from monogenic disorders to normal variation. *Nature Reviews Endocrinology*, **9**(3), 171–177.

Edgington, E., and Onghena, P. 2007. *Randomization Tests*, 4th edn. Chapman and Hall/CRC.

Enard, Wolfgang. 2011. FOXP2 and the role of cortico-basal ganglia circuits in speech and language evolution. *Current Opinion in Neurobiology*, **21**(3), 415–424.

Enard, W., Przeworski, M., Fisher, S. E., Lai, C. S. L., Wiebe, V., Kitano, T., Monaco, A. P., et al. 2002. Molecular evolution of FOXP2, a gene involved in speech and language. *Nature*, **418**(6900), 869–872.

Enard, W., Gehre, S., Hammerschmidt, K., Hlter, S. M., Blass, T., Somel, M., Brckner, M. K., et al. 2009. A humanized version of Foxp2 affects cortico-basal ganglia circuits in mice. *Cell*, **137**(5), 961–971.

Evans, N., and Levinson, S. C. 2009. The myth of language universals: Language diversity and its importance for cognitive science. *Behavioral and Brain Sciences*, **32**, 429–492.

Evans, T. A., and Bashaw, G. J. 2010. Axon guidance at the midline: of mice and flies. *Current Opinion in Neurobiology*, **20**(1), 79–85.

Ewens, W. J., and Spielman, R. S. 1995. The transmission/disequilibrium test: history, subdivision, and admixture. *American Journal of Human Genetics*, **57**(2), 455–464.

Falconer, D. S, and Mackay, T. F. C. 1996. *Introduction to Quantitative Genetics*. Harlow: Longman.

Falush, D., Stephens, M., and Pritchard, J. K. 2003. Inference of population structure using multilocus genotype data: linked loci and correlated allele frequencies. *Genetics*, **164**(4), 1567–1587.

Faul, F., Erdfelder, E., Lang, A.-G., and Buchner, A. 2007. G*Power 3: A flexible statistical power analysis for the social, behavioral, and biomedical science. *Behavior Research Methods*, **39**, 175–191.

Feldman, M. W., and Aoki, K. 1992. Assortative mating and grandparental transmission facilitate the persistence of a sign language. *Theoretical Population Biology*, **42**(2), 107–116.

Feldman, M. W., and Laland, K. N. 1996. Gene-culture coevolutionary theory. *Trends in Ecology & Evolution*, **11**(11), 453–457.

Felsenfeld, S. 2002. Finding susceptibility genes for developmental disorders of speech: the long and winding road. *Journal of Communication Disorders*, **35**(4), 329–345.

Felsenstein, J. 2004. *Inferring Phylogenies*. Sunderland, MA: Sinauer.

Feuk, L., Kalervo, A., Lipsanen-Nyman, M., Skaug, J., Nakabayashi, K., Finucane, B., Hartung, D., et al. 2006. Absence of a paternally inherited FOXP2 gene in developmental verbal dyspraxia. *American Journal of Human Genetics*, **79**(5), 965–972.

Field, A. 2009. *Discovering Statistics Using SPSS*. 3rd edn. London: Sage.

Field, A. 2013. *Discovering Statistics using IBM SPSS Statistics*. London: Sage.

Field, A., Miles, J., and Field, Z. 2012. *Discovering Statistics Using R*. London: Sage.

Fisch, L. 1969. Causes of congenital deafness. *Public Health*, **83**(2), 68–74.

Fisher, R. A. 1918. The correlation between relatives on the supposition of Mendelian inheritance. *Philosophical Transactions of the Royal Society of Edinburgh*, **52**, 399–433.

Fisher, S. E. 2006. Tangled webs: tracing the connections between genes and cognition. *Cognition*, **101**(2), 270–297.

Fisher, S. E. 2010. Genetic susceptibility to stuttering. *New England Journal of Medicine*, **362**(8), 750–752.

Fisher, S. E., and Scharff, C. 2009. FOXP2 as a molecular window into speech and language. *Trends in Genetics*, **25**(4), 166–177.

Fisher, S. E., Vargha-Khadem, F., Watkins, K. E., Monaco, A. P., and Pembrey, M. E. 1998. Localisation of a gene implicated in a severe speech and language disorder. *Nature Genetics*, **18**(2), 168–170.

Fleming, L. 2010. From patrilects to performatives: linguistic exogamy and language shift in the Northwest Amazon. *Dissertations available from ProQuest*, Jan., 1–327.

Fortunato, L., and Jordan, F. 2010. Your place or mine? A phylogenetic comparative analysis of marital residence in Indo-European and Austronesian societies. *Philosophical Transactions of the Royal Society B*, **365**(1559), 3913–3922.

Fowler, K. B., and Boppana, S. B. 2006. Congenital cytomegalovirus (CMV) infection and hearing deficit. *Journal of Clinical Virology*, **35**(2), 226–231.

Fraga, M. F., Ballestar, E., Paz, M. F., Ropero, S., Setien, F., Ballestar, M. L., Heine-Suer, D., et al. 2005. Epigenetic differences arise during the lifetime of monozygotic twins. *Proceedings of the National Academy of the Sciences of the USA*, **102**(30), 10604–10609.

Francks, C., Paracchini, S., Smith, S. D., Richardson, A. J., Scerri, T. S., Cardon, L. R., Marlow, A. J., et al. 2004. A 77-kilobase region of chromosome 6p22.2 is associated with dyslexia in families from the United Kingdom and from the United States. *American Journal of Human Genetics*, **75**(6), 1046–1058.

Freedman, M. L., Reich, D., Penney, K. L., McDonald, G. J., Mignault, A. A., Patterson, N., Gabriel, S. B., et al. 2004. Assessing the impact of population stratification on genetic association studies. *Nature Genetics*, **36**(4), 388–393.

French, C. A., Jin, X., Campbell, T. G., Gerfen, E., Groszer, M., Fisher, S. E., and Costa, R. M. 2012. An aetiological Foxp2 mutation causes aberrant striatal activity and alters plasticity during skill learning. *Molecular Psychiatry*, **17**(11), 1077–1085.

Friedman, T. B., Liang, Y., Weber, J. L., Hinnant, J. T., Barber, T. D., Winata, S., Arhya, I. N., and Asher, J. H. 1995. A gene for congenital, recessive deafness DFNB3 maps to the pericentromeric region of chromosome 17. *Nature Genetics*, **9**(1), 86–91.

Fu, Y.-X., and Li, W.-H. 1993. Statistical tests of neutrality of mutations. *Genetics*, **133**(3), 693–709.

Fumagalli, M., Sironi, M., Pozzoli, U., Ferrer-Admettla, A., Pattini, L., and Nielsen, R. 2011. Signatures of environmental genetic adaptation pinpoint pathogens as the main selective pressure through human evolution. *PLoS Genetics*, **7**(11), e1002355.

Furr, R. Michael, and Bacharach, V. R. 2007. *Psychometrics: An Introduction*. 1st edn. Thousand Oaks, CA: Sage.

Futuyma, D. J. 2013. *Evolution*. Sunderland, MA: Sinauer.

Gabel, L. A, Gibson, C. J., Gruen, J. R., and LoTurco, J. J. 2010. Progress towards a cellular neurobiology of reading disability. *Neurobiology of Disease*, **38**(2), 173–180.

Gagnon, A., Beise, J., and Vaupel, J. W. 2005. Genome-wide identity-by-descent sharing among CEPH siblings. *Genetic Epidemiology*, **29**(3), 215–224.

Gathercole, S. E. 2006. Nonword repetition and word learning: the nature of the relationship. *Applied Psycholinguistics*, **27**(4), 513.

Gathercole, S. E., Willis, C. S., Baddeley, A. D., and Emslie, H. 1994. The Children's Test of Nonword Repetition: a test of phonological working memory. *Memory (Hove, England)*, **2**(2), 103–127.

Gauderman, W. J., Murcray, C., Gilliland, F., and Conti, D. V. 2007. Testing association between disease and multiple SNPs in a candidate gene. *Genetic Epidemiology*, **31**(5), 383–395.

Gavara, N., Manoussaki, D., and Chadwick, R. S. 2011. Auditory mechanics of the tectorial membrane and the cochlear spiral. *Current Opinion in Otolaryngology & Head and Neck Surgery*, **19**(5), 382.

Gavrilets, S. 1997. Evolution and speciation on holey adaptive landscapes. *Trends in Ecology & Evolution*, **12**, 307–312.

Gavrilets, S. 2003. Evolution and speciation in a hyperspace: the roles of neutrality, selection, mutation and random drift. In Crutchfield, J., and Schuster, P. (eds), *Towards a Comprehensive Dynamics of Evolution – Exploring the Interplay of Selection, Neutrality, Accident, and Function*. Oxford: Oxford University Press, pp. 135–162.

Gehring, W. J., and Ikeo, K. 1999. *Pax 6*: mastering eye morphogenesis and eye evolution. *Trends in Genetics*, **15**(9), 371–377.

Gelman, A., and Hill, J. 2006. *Data Analysis Using Regression and Multilevel/Hierarchical Models*. 1st edn. Cambridge: Cambridge University Press.

Gelman, A., and Shalizi, C. R. 2013. Philosophy and the practice of Bayesian statistics. *British Journal of Mathematical and Statistical Psychology*, **66**(1), 8–38.

Gerbault, P., Liebert, A., Itan, Y., Powell, A., Currat, M., Burger, J., Swallow, D. M., et al. 2011. Evolution of lactase persistence: an example of human niche construction. *Philosophical Transactions of the Royal Society B*, **366**(1566), 863–877.

Gialluisi, A., Dediu, D., Francks, C., and Fisher, S. E. 2013. Persistence and transmission of recessive deafness and sign language: new insights from village sign languages. *European Journal of Human Genetics*, **21**, 894–896.

Gibson, G. 2012. Rare and common variants: twenty arguments. *Nature Reviews Genetics*, **13**(2), 135–145.

Glansdorff, N., Xu, Y., and Labedan, B. 2008. The last universal common ancestor: emergence, constitution and genetic legacy of an elusive forerunner. *Biology Direct*, **3**, 29.

Gopnik, M. 1990. Feature-blind grammar and dysphasia. *Nature*, **344**, 715.

Gottlieb, G. 1991. Experiential canalization of behavioral development: results. *Developmental Psychology*, **27**(1), 35.

Gottlieb, G. 1997. *Synthesizing Nature-nurture: Prenatal Roots of Instinctive Behavior*. Hove: Psychology Press.

Götz, M., and Huttner, W. B. 2005. The cell biology of neurogenesis. *Nature Reviews. Molecular Cell Biology*, **6**(10), 777–788.

Gratten, J., Wray, N. R., Keller, M. C., and Visscher, P. M. 2014. Large-scale genomics unveils the genetic architecture of psychiatric disorders. *Nature Neuroscience*, **17**(6), 782–790.

Gray, R. D., Bryant, D., and Greenhill, S. J. 2010. On the shape and fabric of human history. *Philosophical Transactions of the Royal Society B*, **365**(1559), 3923–3933.

Green, R. E., Krause, J., Briggs, A. W., Maricic, T., Stenzel, U., Kircher, M., Patterson, N., et al. 2010. A draft sequence of the Neandertal genome. *Science*, **328**, 710–722.

Greer, E. L, and Shi, Y. 2012. Histone methylation: a dynamic mark in health, disease and inheritance. *Nature Reviews Genetics*, **13**(5), 343–357.

Groszer, M., Keays, D. A., Deacon, R. M. J., de Bono, J. P., Prasad-Mulcare, S., Gaub, S., Baum, M. G., et al. 2008. Impaired synaptic plasticity and motor learning in mice with a point mutation implicated in human speech deficits. *Current Biology: CB*, **18**(5), 354–362.

Haesler, S., Rochefort, C., Georgi, B., Licznerski, P., Osten, P., and Scharff, C. 2007. Incomplete and inaccurate vocal imitation after knockdown of FoxP2 in songbird basal ganglia nucleus Area X. *PLoS Biology*, **5**(12), e321.

Haflidadóttir, B. S., Larne, O., Martin, M., Persson, M., Edsjö, A., Bjartell, A., and Ceder, Y. 2013. Upregulation of miR-96 enhances cellular proliferation of prostate cancer cells through FOXO1. *PLoS ONE*, **8**(8), e72400.

Halliburton, R. 2004. *Introduction to Population Genetics*. (International Edition): Upper Saddle River, NJ: Pearson.

Hamdan, F. F., Daoud, H., Rochefort, D., Piton, A., Gauthier, J., Langlois, M., Foomani, G., et al. 2010. De novo mutations in FOXP1 in cases with intellectual disability, autism, and language impairment. *American Journal of Human Genetics*, **87**(5), 671–678.

Hamer, D., and Sirota, L. 2000. Beware the chopsticks gene. *Molecular Psychiatry*, **5**(1), 11–13.

Hanaoka, M., Droma, Y., Basnyat, B., Ito, M., Kobayashi, N., Katsuyama, Y., Kubo, K., et al. 2012. Genetic variants in EPAS1 contribute to adaptation to high-altitude hypoxia in Sherpas. *PLoS ONE*, **7**(12), e50566.

Hancock, A. M., Witonsky, D. B., Gordon, A. S., Eshel, G., Pritchard, J. K., Coop, G., and Di Rienzo, A. 2008. Adaptations to climate in candidate genes for common metabolic disorders. *PLoS Genetics*, **4**(2), e32.

Hancock, A. M., Witonsky, D. B., Alkorta-Aranburu, G., Beall, C. M., Gebremedhin, A., Sukernik, R., Utermann, G., et al. 2011. Adaptations to climate-mediated selective pressures in humans. *PLoS Genetics*, **7**(4), e1001375.

Hannula-Jouppi, K., Kaminen-Ahola, N., Taipale, M., Eklund, R., Nopola-Hemmi, J., Kriinen, H., and Kere, J. 2005. The axon guidance receptor gene ROBO1 is a candidate gene for developmental dyslexia. *PLoS Genetics*, **1**(4), e50.

Hartman, M. A., and Spudich, J. A. 2012. The myosin superfamily at a glance. *Journal of Cell Science*, **125**(7), 1627–1632.

Haworth, C. M. A., and Plomin, R. 2010. Quantitative genetics in the era of molecular genetics: learning abilities and disabilities as an example. *Journal of the American Academy of Child and Adolescent Psychiatry*, **49**(8), 783–793.

Haworth, C. M. A., Wright, M. J., Luciano, M., Martin, N. G., Geus, E. J. C. de, Beijsterveldt, C. E. M. van, Bartels, M., et al. 2009. The heritability of general cognitive ability increases linearly from childhood to young adulthood. *Molecular Psychiatry*, **15**(11), 1112–1120.

Hayiou-Thomas, M. E. 2008. Genetic and environmental influences on early speech, language and literacy development. *Journal of Communication Disorders*, **41**(5), 397–408.

Hensler, B. S., Schatschneider, C., Taylor, J., and Wagner, R. K. 2010. Behavioral genetic approach to the study of dyslexia. *Journal of Developmental & Behavioral Pediatrics*, **31**(7), 525–532.

Herculano-Houzel, S. 2012. The remarkable, yet not extraordinary, human brain as a scaled-up primate brain and its associated cost. *Proceedings of the National Academy of Sciences of the USA*, **109**(Suppl 1), 10661–10668.

Hernán, M. A., and Robins, J. M. manuscript. *Causal Inference*. Available at www.hsph.harvard.edu.

Hershberg, R., and Petrov, D. A. 2008. Selection on codon bias. *Annual Review of Genetics*, **42**(1), 287–299.

Hey, J. 2004. What's so hot about recombination hotspots? *PLoS Biology*, **2**(6), e190.

Higgins, J., Midgley, C., Bergh, A.-M., Bell, S. M., Askham, J. M., Roberts, E., Binns, R. K., et al. 2010. Human ASPM participates in spindle organisation, spindle orientation and cytokinesis. *BMC Cell Biology*, **11**, 85.

Hildebrand, M. S., Witmer, P. D., Xu, S., Newton, S. S., Kahrizi, K., Najmabadi, H., Valle, D., et al. 2010. miRNA mutations are not a common cause of deafness. *American Journal of Medical Genetics. Part A*, **152A**(3), 646–652.

Hindorff, L. A., Sethupathy, P., Junkins, H. A., Ramos, E. M., Mehta, J. P., Collins, F. S., and Manolio, T. A. 2009. Potential etiologic and functional implications of genome-wide association loci for human diseases and traits. *Proceedings of the National Academy of Sciences of the USA*, **106**(23), 9362–9367.

Hirschhorn, J. N., and Daly, M. J. 2005. Genome-wide association studies for common diseases and complex traits. *Nature Reviews Genetics*, **6**(2), 95–108.

Ho, M.-R., Tsai, K.-W., and Lin, W.-c. 2012. A unified framework of overlapping genes: towards the origination and endogenic regulation. *Genomics*, **100**(4), 231–239.

Hoffman, J. I. E, and Kaplan, S. 2002. The incidence of congenital heart disease. *Journal of the American College of Cardiology*, **39**(12), 1890–1900.

Holden, M., Deng, S., Wojnowski, L., and Kulle, B. 2008. GSEA-SNP: applying gene set enrichment analysis to SNP data from genome-wide association studies. *Bioinformatics*, **24**(23), 2784–2785.

Holm, S. 1979. A simple sequentially rejective multiple test procedure. *Scandinavian Journal of Statistics*, **6**, 65–70.

Holsinger, K. E, and Weir, B. S. 2009. Genetics in geographically structured populations: defining, estimating and interpreting F(ST). *Nature Reviews Genetics*, **10**(9), 639–650.

Horn, D., Kapeller, J., Rivera-Brugus, N., Moog, U., Lorenz-Depiereux, B., Eck, S., Hempel, M., et al. 2010. Identification of FOXP1 deletions in three unrelated patients with mental retardation and significant speech and language deficits. *Human Mutation*, **31**(11), E1851–1860.

Huerta-Sánchez, E., DeGiorgio, M., Pagani, L., Tarekegn, A., Ekong, R., Antao, T., Cardona, A., et al. 2013. Genetic signatures reveal high-altitude adaptation in a set of Ethiopian populations. *Molecular Biology and Evolution*, **30**(8), 1877–1888.

Hurford, J. R. 1990. Nativist and functional explanations in language acquisition. In Roca, I. M. (ed), *Logical Issues in Language Acquisition*. Dordrecht: Foris, pp. 85–136.

Hurst, L. D. 2009. Fundamental concepts in genetics: genetics and the understanding of selection. *Nature Reviews Genetics*, **10**(2), 83–93.

Huttner, W. B, and Kosodo, Y. 2005. Symmetric versus asymmetric cell division during neurogenesis in the developing vertebrate central nervous system. *Current Opinion in Cell Biology*, **17**(6), 648–657.

Ioannidis, J. P. A. 2005. Why most published research findings are false. *PLoS Medicine*, **2**(8), e124.

Ioannidis, J. P., Ntzani, E. E., Trikalinos, T. A., and Contopoulos-Ioannidis, D. G. 2001. Replication validity of genetic association studies. *Nature Genetics*, **29**(3), 306–309.

Ishikawa, T., Toyoda, Y., Yoshiura, K.-i., and Niikawa, N. 2013. Pharmacogenetics of human ABC transporter ABCC11: new insights into apocrine gland growth and metabolite secretion. *Frontiers in Pharmacogenetics and Pharmacogenomics*, **3**, 306.

Itan, Y., Powell, A., Beaumont, M. A., Burger, J., and Thomas, M. G. 2009. The origins of lactase persistence in Europe. *PLoS Computational Biology*, **5**(8), e1000491.

Jackson, J. E. 1983. *The Fish People: Linguistic Exogamy and Tukanoan Identity in Northwest Amazonia*. Cambridge: Cambridge University Press.

Jaeger, T. F., Graff, P., Croft, W., and Pontillo, D. 2011. Mixed effect models for genetic and areal dependencies in linguistic typology. *Linguistic Typology*, **15**, 281–319.

Jaffee, S. R., and Price, T. S. 2008. Genotype-environment correlations: implications for determining the relationship between environmental exposures and psychiatric illness. *Psychiatry*, **7**(12), 496–499.

Jakobsson, M., Scholz, S. W., Scheet, P., Gibbs, J. R., VanLiere, J. M., Fung, H.-C., Szpiech, Z. A., et al. 2008. Genotype, haplotype and copy-number variation in worldwide human populations. *Nature*, **451**(7181), 998–1003.

Jameson, K. A., and Komarova, N. L. 2009a. Evolutionary models of color categorization. I. Population categorization systems based on normal and dichromat observers. *Journal of the Optical Society of America A Optics, Image Science, and Vision*, **26**(6), 1414–1423.

Jameson, K. A., and Komarova, N. L. 2009b. Evolutionary models of color categorization. II. Realistic observer models and population heterogeneity. *Journal of the Optical Society of America A Optics, Image Science, and Vision*, **26**(6), 1424–1436.

Jenkins, K. J., Correa, A., Feinstein, J. A., Botto, L., Britt, A. E., Daniels, S. R., Elixson, M., et al. 2007. Noninherited risk factors and congenital cardiovascular defects: current knowledge: a scientific statement from the American Heart Association Council on Cardiovascular Disease in the Young: endorsed by the American Academy of Pediatrics. *Circulation*, **115**(23), 2995–3014.

Jensen-Seaman, M. I, and Hooper-Boyd, K. A. 2001. Molecular clocks: determining the age of the human-chimpanzee divergence. In *eLS*. Chichester: John Wiley & Sons.

Jiang, Y., Bolnick, D., and Kirkpatrick, M. 2013. Assortative mating in animals. *The American Naturalist*, **181**(6), E125–E138.

Jobling, M. A., Hollox, E., Hurles, M., Kivisild, T., and Tyler-Smith, C. 2013. *Human Evolutionary Genetics*. New York: Garland Science.

Jolliffe, I. T. 2002. *Principal Component Analysis*. 2nd edn. Springer Series in Statistics. New York: Springer.

Jordan, F. M., Gray, R. D., Greenhill, S. J., and Mace, R. 2009. Matrilocal residence is ancestral in Austronesian societies. *Proceedings of the Royal Society B*, **276**(1664), 1957–1964.

Jordan, G., Deeb, S. S., Bosten, J. M, and Mollon, J. D. 2010. The dimensionality of color vision in carriers of anomalous trichromacy. *Journal of Vision*, **10**(8), 12.

Jørgensen, A. L., Philip, J., Raskind, W. H., Matsushita, M., Christensen, B., Dreyer, V., and Motulsky, A. G. 1992. Different patterns of X inactivation in MZ twins discordant for red-green color-vision deficiency. *American Journal of Human Genetics*, **51**(2), 291–298.

Kaindl, A. M., Passemard, S., Kumar, P., Kraemer, N., Issa, L., Zwirner, A., Gerard, B., et al. 2009. Many roads lead to primary autosomal recessive microcephaly. *Progress in Neurobiology*, **90**(3), 363–383.

Kang, C., and Drayna, D. 2012. A role for inherited metabolic deficits in persistent developmental stuttering. *Molecular Genetics and Metabolism*, **107**(3), 276–280.

Kang, C., Riazuddin, S., Mundorff, J., Krasnewich, D., Friedman, P., Mullikin, J. C., and Drayna, D. 2010. Mutations in the lysosomal enzyme-targeting pathway and persistent stuttering. *The New England Journal of Medicine*, **362**(8), 677–685.

Kaplan, J. 2008. The end of the adaptive landscape metaphor? *Biology & Philosophy*, **23**(5), 625–638.

Karlsson, F. 2007. Constraints on multiple center-embedding of clauses. *Journal of Linguistics*, **43**(02), 365–392.

Katoh, M., Igarashi, M., Fukuda, H., Nakagama, H., and Katoh, M. 2013. Cancer genetics and genomics of human FOX family genes. *Cancer Letters*, **328**(2), 198–206.

Kelemen, O., Convertini, P., Zhang, Z., Wen, Y., Shen, M., Falaleeva, M., and Stamm, S. 2013. Function of alternative splicing. *Gene*, **514**(1), 1–30.

Kelley, J. L., and Swanson, W. J. 2008. Positive selection in the human genome: from genome scans to biological significance. *Annual Review of Genomics and Human Genetics*, **9**, 143–160.

Kendal, J., Tehrani, J. J., and Odling-Smee, J. 2011. Human niche construction in interdisciplinary focus. *Philosophical Transactions of the Royal Society B*, **366**(1566), 785–792.

Kidd, T, Brose, K, Mitchell, K J, Fetter, R D, Tessier-Lavigne, M, Goodman, C S, and Tear, G. 1998. Roundabout controls axon crossing of the CNS midline and defines a novel subfamily of evolutionarily conserved guidance receptors. *Cell*, **92**(2), 205–215.

Kiikeri, M., and Kokkonen, T. 2007. Biological notion of innateness and explanation of language acquisition. In Persson, J., and Ylikoski, P. (eds), *Rethinking Explanation*. Boston Studies in the Philosophy of Science, no. 252. Dordrecht: Springer, pp. 177–192.

Kimura, M. 1968. Evolutionary rate at the molecular level. *Nature*, **217**(5129), 624–626.

Kimura, R., Fujimoto, A., Tokunaga, K., and Ohashi, J. 2007. A practical genome scan for population-specific strong selective sweeps that have reached fixation. *PLoS ONE*, **2**(3), e286.

King, J. L., and Jukes, T. H. 1969. Non-Darwinian evolution. *Science*, **164**(3881), 788–798.

Kirby, S., Dowman, M., and Griffiths, T. L. 2007. Innateness and culture in the evolution of language. *Proceedings of the National Academy of the Sciences of the USA*, **104**(12), 5241–5245.

Kirby, S., Cornish, H., and Smith, K. 2008. Cumulative cultural evolution in the laboratory: an experimental approach to the origins of structure in human language. *Proceedings of the National Academy of Sciences of the USA*, **105**(31), 10681–10686.

Klein, R. G. 2009. *The Human Career: Human Biological and Cultural Origins*. 3rd edn. Chicago, IL: University of Chicago Press.

Knafo, A., and Jaffee, S. R. 2013. Gene-environment correlation in developmental psychopathology. *Development and Psychopathology*, **25**(Special Issue 01), 1–6.

Knowler, W C, Williams, R C, Pettitt, D J, and Steinberg, A G. 1988. Gm3;5,13,14 and type 2 diabetes mellitus: an association in American Indians with genetic admixture. *American Journal of Human Genetics*, **43**(4), 520–526.

Kokotas, H, Petersen, M B, and Willems, P J. 2007. Mitochondrial deafness. *Clinical Genetics*, **71**(5), 379–391.

Komlos, J., and Lauderdale, B. E. 2007. The mysterious trend in American heights in the 20th century. *Annals of Human Biology*, **34**(2), 206–215.

Koonin, E. V. 2009. On the origin of cells and viruses: primordial virus world scenario. *Annals of the New York Academy of Sciences*, **1178**, 47–64.

Koonin, E. V. 2012. *The Logic of Chance: The Nature and Origin of Biological Evolution*. Upper Saddle River, NJ: Pearson.

Kovas, Y., and Plomin, R. 2007. Learning abilities and disabilities: generalist genes, specialist environments. *Current Directions in Psychological Science*, **16**(5), 284–288.

Kozak, M. 1999. Initiation of translation in prokaryotes and eukaryotes. *Gene*, **234**(2), 187–208.

Kraft, P., Zeggini, E., and Ioannidis, J. P. A. 2009. Replication in genome-wide association studies. *Statistical Science*, **24**(4), 561–573.

Krause, J., Lalueza-Fox, C., Orlando, L., Enard, W., Green, R. E., Burbano, H. A., Hublin, J.-J., et al. 2007. The derived FOXP2 variant of modern humans was shared with Neandertals. *Current Biology*, **17**(21), 1908–1912.

Kruglyak, L., Daly, M. J., Reeve-Daly, M. P., and Lander, E. S. 1996. Parametric and nonparametric linkage analysis: a unified multipoint approach. *American Journal of Human Genetics*, **58**(6), 1347–1363.

Ku, C.-S., Naidoo, N., and Pawitan, Y. 2011. Revisiting Mendelian disorders through exome sequencing. *Human Genetics*, **129**(4), 351–370.

Kumar, A. 2009. An overview of nested genes in eukaryotic genomes. *Eukaryotic Cell*, **8**(9), 1321–1329.

Kurt, S., Fisher, S. E., and Ehret, G. 2012. Foxp2 mutations impair auditory-motor association learning. *PLoS ONE*, **7**(3), e33130.

Kutschera, U., and Niklas, K. J. 2005. Endosymbiosis, cell evolution, and speciation. *Theory in Biosciences*, **124**(1), 1–24.

Ladd, D. R., Dediu, D., and Kinsella, A. R. 2008. Languages and genes: reflections on biolinguistics and the nature-nurture question. *Biolinguistics*, **2**(1), 114–126.

Lai, C. S. L., Fisher, S. E., Hurst, J. A., Levy, E. R., Hodgson, S., Fox, M., Jeremiah, S., et al. 2000. The SPCH1 region on human 7q31: genomic characterization of the critical interval and localization of translocations associated with speech and language disorder. *American Journal of Human Genetics*, **67**(2), 357–368.

Lai, C. S., Fisher, S. E., Hurst, J. A., Vargha-Khadem, F., and Monaco, A. P. 2001. A forkhead-domain gene is mutated in a severe speech and language disorder. *Nature*, **413**(6855), 519–523.

Laland, K. N., Odling-Smee, J., and Myles, S. 2010. How culture shaped the human genome: bringing genetics and the human sciences together. *Nature Reviews Genetics*, **11**(2), 137–148.

Lalueza-Fox, C., Römpler, H., Caramelli, D., Stubert, C., Catalano, G., Hughes, D., Rohland, N., et al. 2007. A melanocortin 1 receptor allele suggests varying pigmentation among Neanderthals. *Science*, **318**(5855), 1453–1455.

Lamminmäki, S., Massinen, S., Nopola-Hemmi, J., Kere, J., and Hari, R. 2012. Human ROBO1 regulates interaural interaction in auditory pathways. *Journal of Neuroscience*, **32**(3), 966–971.

Langergraber, K. E., Siedel, H., Mitani, J. C., Wrangham, R. W., Reynolds, V., Hunt, K., and Vigilant, L. 2007. The genetic signature of sex-biased migration in patrilocal chimpanzees and humans. *PLoS ONE*, **2**(10), e973.

Lao, O, de Gruijter, J M, van Duijn, K, Navarro, A, and Kayser, M. 2007. Signatures of positive selection in genes associated with human skin pigmentation as revealed from analyses of single nucleotide polymorphisms. *Annals of Human Genetics*, **71**(Pt 3), 354–369.

Larnkaer, A., Schrøder, S. A., Schmidt, I. M., Jørgensen, M. H., and Michaelsen, K. F. 2006. Secular change in adult stature has come to a halt in northern Europe and Italy. *Acta Paediatrica*, **95**(6), 754–755.

Lee, T. I., and Young, R. A. 2013. Transcriptional regulation and its misregulation in disease. *Cell*, **152**(6), 1237–1251.

Lennon, P A, Cooper, M L, Peiffer, D A, Gunderson, K L, Patel, A, Peters, Sarika, Cheung, S W, and Bacino, C A. 2007. Deletion of 7q31.1 supports involvement of FOXP2 in language impairment: clinical report and review. *American Journal of Medical Genetics. Part A*, **143A**(8), 791–798.

Lettre, G. 2009. Genetic regulation of adult stature. *Current Opinion in Pediatrics*, **21**(4), 515–522.

Leung, S. S., Lau, J. T., Xu, Y. Y., Tse, L. Y., Huen, K. F., Wong, G. W., Law, W. Y., et al. 1996. Secular changes in standing height, sitting height and sexual maturation of Chinese–the Hong Kong Growth Study, 1993. *Annals of Human Biology*, **23**(4), 297–306.

Levinson, S. C, and Dediu, D. 2013. The interplay of genetic and cultural factors in ongoing language evolution. In Richerson, P. J., and Christiansen, M. H. (eds), *Cultural Evolution: Society, Technology, Language, and Religion*. Strüngmann Forum Reports, vol. 12. Cambridge, MA: MIT Press, pp. 219–232.

Levinson, S. C., and Evans, N. 2010. Time for a sea-change in linguistics: Response to comments on 'The myth of language universals'. *Lingua*, **120**, 2733–2758.

Lewin, B., and Krebs, J. 2011. *Lewin's GENES X*. Sudbury, MA: Jones & Bartlett Learning.

Lewis, C. M., and Knight, J. 2012. Introduction to genetic association studies. *Cold Spring Harbor Protocols*, **2012**(3), 297–306.

Lewis, M. A., Quint, E., Glazier, A. M., Fuchs, H., De Angelis, M. H., Langford, C., van Dongen, S., et al. 2009. An ENU-induced mutation of miR-96 associated with progressive hearing loss in mice. *Nature Genetics*, **41**(5), 614–618.

Li, G., Wang, J., Rossiter, S. J., Jones, G., and Zhang, S. 2007. Accelerated FoxP2 evolution in echolocating bats. *PLoS ONE*, **2**(9), e900.

Li, H., Gu, S., Cai, X., Speed, W. C., Pakstis, A. J., Golub, E. I., Kidd, J. R., et al. 2008. Ethnic related selection for an ADH Class I variant within East Asia. *PLoS ONE*, **3**(4), e1881.

Li, M., Reilly, M. P., Rader, D. J., and Wang, L.-S. 2010. Correcting population stratification in genetic association studies using a phylogenetic approach. *Bioinformatics*, **26**(6), 798–806.

Li, Q., and Yu, K. 2008. Improved correction for population stratification in genome-wide association studies by identifying hidden population structures. *Genetic Epidemiology*, **32**(3), 215–226.

Li, R., Montpetit, A., Rousseau, M., Wu, S. Y. M., Greenwood, C. M. T., Spector, T. D., Pollak, M., et al. 2013. Somatic point mutations occurring early in development: a monozygotic twin study. *Journal of Medical Genetics*, **51**(1), 28–34.

Li, S., Zhou, D., Lu, M. M., and Morrisey, E. E. 2004. Advanced cardiac morphogenesis does not require heart tube fusion. *Science (New York, N.Y.)*, **305**(5690), 1619–1622.

Li, S., Wang, Y., Zhang, Y., Lu, M. M., DeMayo, F. J., Dekker, J. D., Tucker, P. W., et al. 2012. Foxp1/4 control epithelial cell fate during lung development and regeneration through regulation of anterior gradient 2. *Development (Cambridge, England)*, **139**(14), 2500–2509.

Liang, Y., Wang, A., Probst, F. J., Arhya, I. N., Barber, T. D., Chen, K. S., Deshmukh, D., et al. 1998. Genetic mapping refines DFNB3 to 17p11.2, suggests multiple alleles of DFNB3, and supports homology to the mouse model shaker-2. *American Journal of Human Genetics*, **62**(4), 904–915.

Ligeois, Fr., Baldeweg, T., Connelly, A., Gadian, D. G., Mishkin, M., and Vargha-Khadem, F. 2003. Language fMRI abnormalities associated with FOXP2 gene mutation. *Nature Neuroscience*, **6**(11), 1230–1237.

Liu, C. C., and Schultz, P. G. 2010. Adding new chemistries to the genetic code. *Annual Review of Biochemistry*, **79**(1), 413–444.

Liu, Y.-J., Papasian, C. J., Liu, J.-F., Hamilton, J., and Deng, H.-W. 2008. Is replication the gold standard for validating genome-wide association findings? *PLoS ONE*, **3**(12), e4037.

Loat, C. S., Asbury, K., Galsworthy, M. J., Plomin, R., and Craig, I. W. 2004. X inactivation as a source of behavioural differences in monozygotic female twins. *Twin Research*, **7**(1), 54–61.

Lobo, I., and Shaw, K. 2008. Discovery and types of genetic linkage. *Nature Education*, **1**(1), 139.

Lodish, H., Berk, A., Kaiser, C. A., Krieger, M., Bretscher, A., Ploegh, H., Amon, A., et al. 2012. *Molecular Cell Biology*. San Francisco, CA: W. H. Freeman.

Loebrich, S., and Nedivi, E. 2009. The function of activity-regulated genes in the nervous system. *Physiological Reviews*, **89**(4), 1079–1103.

Lordkipanidze, D., Len, M. S. P. de, Margvelashvili, A., Rak, Y., Rightmire, G. P., Vekua, A., and Zollikofer, C. P. E. 2013. A complete skull from Dmanisi, Georgia, and the evolutionary biology of early Homo. *Science*, **342**(6156), 326–331.

Lynch, M. 2007. *The Origins of Genome Architecture*. 1st edn. Sunderland, MA: Sinauer.

Lynch, M., and Walsh, B. 1998. *Genetics and Analysis of Quantitative Traits*. Sunderland, MA: Sinauer.

MacDermot, K. D., Bonora, E., Sykes, N., Coupe, A.-M., Lai, C. S. L., Vernes, S. C., Vargha-Khadem, F., et al. 2005. Identification of FOXP2 truncation as a novel cause of developmental speech and language deficits. *American Journal of Human Genetics*, **76**(6), 1074–1080.

Mahmood, S., Ahmad, W., and Hassan, M. J. 2011. Autosomal recessive primary microcephaly (MCPH): clinical manifestations, genetic heterogeneity and mutation continuum. *Orphanet Journal of Rare Diseases*, **6**(1), 39.

Majumder, P. P. 2010. The human genetic history of South Asia. *Current Biology*, **20**(4), R184–R187.

Makalowska, I. 2004. Overlapping genes in the human genome. In: *Encyclopedia of Genetics, Genomics, Proteomics and Bioinformatics*. New York: John Wiley & Sons.

Mameli, M. 2008. On innateness: the clutter hypothesis and the cluster hypothesis. *Journal of Philosophy*, **105**(12), 719.

Mameli, M., and Bateson, P. 2006. Innateness and the sciences. *Biology and Philosophy*, **21**, 155–188.

Mameli, M., and Bateson, P. 2011. An evaluation of the concept of innateness. *Philosophical Transactions of the Royal Society B*, **366**(1563), 436–443.

Mangs, A. H., and Morris, B. J. 2007. The human pseudoautosomal region (PAR): origin, function and future. *Current Genomics*, **8**(2), 129–136.

Manica, A., Amos, W., Balloux, F., and Hanihara, T. 2007. The effect of ancient population bottlenecks on human phenotypic variation. *Nature*, **448**(7151), 346–348.

Manolio, T. A., Collins, F. S., Cox, N. J., Goldstein, D. B., Hindorff, L. A., Hunter, D. J., McCarthy, M. I., et al. 2009. Finding the missing heritability of complex diseases. *Nature*, **461**(7265), 747–753.

Manor, U., Disanza, A., Grati, M., Andrade, L., Lin, H., Di Fiore, P. P., Scita, G., et al. 2011. Regulation of stereocilia length by myosin XVa and whirlin depends on the actin-regulatory protein Eps8. *Current Biology : CB*, **21**(2), 167–172.

Margulis, L., and Sagan, D. 1997. *Microcosmos: Four Billion Years of Microbial Evolution*. Oakland, CA: University of California Press.

Maricic, T., Günther, V., Georgiev, O., Gehre, S., Curlin, M., Schreiweis, C., Naumann, R., et al. 2013. A recent evolutionary change affects a regulatory element in the human FOXP2 gene. *Molecular Biology and Evolution*, **30**(4), 844–852.

Matuli, M., Sopta, M., and Rubelj, I. 2007. Telomere dynamics: the means to an end. *Cell Proliferation*, **40**(4), 462–474.

McBrearty, S., and Brooks, A. S. 2000. The revolution that wasn't: a new interpretation of the origins of modern human behavior. *Journal of Human Evolution*, **39**, 453–563.

McDonald, J. H., and Kreitman, M. 1991. Adaptive protein evolution at the Adh locus in Drosophila. *Nature*, **351**(6328), 652–654.

McEvoy, B., Beleza, S., and Shriver, M. D. 2006. The genetic architecture of normal variation in human pigmentation: an evolutionary perspective and model. *Human Molecular Genetics*, **15**(suppl 2), R176–R181.

McGarvey, S. T. 1991. Obesity in Samoans and a perspective on its etiology in Polynesians. *The American Journal of Clinical Nutrition*, **53**(6), 1586S–1594S.

McGue, M. 1997. A behavioral-genetic perspective on children of alcoholics. *Alcohol Health and Research World*, **21**, 210–217.

Meir, I., Sandler, W., Padden, C., and Aronoff, M. 2010. Emerging sign languages. In Marschark, M., and Spencer, P. E. (eds), *Oxford Handbook of Deaf Studies, Language, and Education*, vol. 2. Oxford: Oxford University Press.

Mencía, A., Modamio-Høybjr, S., Redshaw, N., Morín, M., Mayo-Merino, F., Olavarrieta, L., Aguirre, L. A., et al. 2009. Mutations in the seed region of human miR-96 are responsible for nonsyndromic progressive hearing loss. *Nature Genetics*, **41**(5), 609–613.

Meyer, M., Kircher, M., Gansauge, M.-T., Li, H., Racimo, F., Mallick, S., Schraiber, J. G., et al. 2012. A high-coverage genome sequence from an archaic Denisovan individual. *Science*, **338**(6104), 222–226.

Mihola, O., Trachtulec, Z., Vlcek, C., Schimenti, J. C, and Forejt, J. 2009. A mouse speciation gene encodes a meiotic histone H3 methyltransferase. *Science*, **323**(5912), 373–375.

Mithen, S. 1996. *The Prehistory of the Mind: A Search for the Origins of Art, Science and Religion*. London: Thames & Hudson.

Mithen, S. 2003. *After the Ice: A Global Human History*. London: Orion Books.

Montgomery, D. C., Peck, E. A., and Vining, G. G. 2007. *Introduction to Linear Regression Analysis, 4th edition Student Solutions Manual*. 4th edn. New York: Wiley-Interscience.

Montgomery, D. C., Peck, E. A., and Vining, G. G. 2013. *Introduction to Linear Regression Analysis*. New York: John Wiley & Sons.

Montgomery, S. H., and Mundy, N. I. 2010. Brain evolution: microcephaly genes weigh in. *Current Biology: CB*, **20**(5), R244–246.

Montreuil, B., Bendavid, Y., and Brophy, J. 2005. What is so odd about odds? *Canadian Journal of Surgery*, **48**(5), 400–408.

Moorjani, P., Thangaraj, K., Patterson, N., Lipson, M., Loh, P.-R., Govindaraj, P., et al. 2013. Genetic evidence for recent population mixture in India. *The American Journal of Human Genetics*, **93**(3), 422–438.

Moorman, S., Mello, C. V., and Bolhuis, J. J. 2011. From songs to synapses: molecular mechanisms of birdsong memory. *BioEssays*, **33**(5), 377–385.

Mooseker, M. S., and Cheney, R. E. 1995. Unconventional myosins. *Annual Review of Cell and Developmental Biology*, **11**(1), 633–675.

Morton, N. E. 1955. Sequential tests for the detection of linkage. *American Journal of Human Genetics*, **7**(3), 277–318.

Morton, V., and Torgerson, D. J. 2003. Effect of regression to the mean on decision making in health care. *BMJ : British Medical Journal*, **326**(7398), 1083–1084.

Mustapha, M., Weil, D., Chardenoux, S., Elias, S., El-Zir, E., Beckmann, J. S., Loiselet, J., et al. 1999. An alpha-tectorin gene defect causes a newly identified autosomal recessive form of sensorineural pre-lingual non-syndromic deafness, DFNB21. *Human Molecular Genetics*, **8**(3), 409–412.

Myers, R. H., Montgomery, D. C., Vining, G. G., and Robinson, T. J. 2012. *Generalized Linear Models: with applications in engineering and the sciences*. New York: John Wiley & Sons.

Myles, S., Lea, R. A., Ohashi, J., Chambers, G. K., Weiss, J. G., Hardouin, E., Engelken, J., et al. 2011. Testing the thrifty gene hypothesis: the Gly482Ser variant in PPARGC1A is associated with BMI in Tongans. *BMC Medical Genetics*, **12**(1), 10.

Nance, W. E., and Kearsey, M. J. 2004. Relevance of connexin deafness (DFNB1) to human evolution. *American Journal of Human Genetics*, **74**(6), 1081–1087.

Nance, W E, Liu, X Z, and Pandya, A. 2000. Relation between choice of partner and high frequency of connexin-26 deafness. *Lancet*, **356**(9228), 500–501.

Nassir, R., Kosoy, R., Tian, C., White, P. A., Butler, L. M., Silva, G., Kittles, R., et al. 2009. An ancestry informative marker set for determining continental origin: validation and extension using human genome diversity panels. *BMC Genetics*, **10**, 39.

Neel, J V. 1962. Diabetes mellitus: a "thrifty" genotype rendered detrimental by "progress"? *American Journal of Human Genetics*, **14**, 353–362.

Nelson, C. S., Fuller, C. K., Fordyce, P. M., Greninger, A. L., Li, H., and DeRisi, J. L. 2013. Microfluidic affinity and ChIP-seq analyses converge on a conserved FOXP2-binding motif in chimp and human, which enables the detection of evolutionarily novel targets. *Nucleic Acids Research*, **41**(12), 5991–6004.

Neumann, H., Wang, K., Davis, L., Garcia-Alai, M., and Chin, J. W. 2010. Encoding multiple unnatural amino acids via evolution of a quadruplet-decoding ribosome. *Nature*, **464**(7287), 441–444.

Newbury, D. F., Paracchini, S., Scerri, T. S., Winchester, L., Addis, L., Richardson, A. J., Walter, J., et al. 2011. Investigation of dyslexia and SLI risk variants in reading- and language-impaired subjects. *Behavior Genetics*, **41**(1), 90–104.

Newmeyer, F. J. 2005. *Possible and Probable Languages: A Generative Perspective on Linguistic Typology*. Oxford: Oxford University Press.

Ng, S. B., Turner, E. H., Robertson, P. D., Flygare, S. D., Bigham, A. W., Lee, C., Shaffer, T., et al. 2009. Targeted capture and massively parallel sequencing of 12 human exomes. *Nature*, **461**(7261), 272–276.

Nilsen, T. W., and Graveley, B. R. 2010. Expansion of the eukaryotic proteome by alternative splicing. *Nature*, **463**(7280), 457–463.

Nopola-Hemmi, J., Myllyluoma, B., Haltia, T., Taipale, M., Ollikainen, V., Ahonen, T., Voutilainen, A., et al. 2001. A dominant gene for developmental dyslexia on chromosome 3. *Journal of Medical Genetics*, **38**(10), 658–664.

Norio, R. 2003a. Finnish disease heritage I: characteristics, causes, background. *Human Genetics*, **112**(5-6), 441–456.

Norio, Reijo. 2003b. Finnish disease heritage II: population prehistory and genetic roots of Finns. *Human Genetics*, **112**(5-6), 457–469.

Norio, Reijo. 2003c. The Finnish disease heritage III: the individual diseases. *Human Genetics*, **112**(5-6), 470–526.

Norton, H. L., Kittles, R. A., Parra, E., McKeigue, P., Mao, X., Cheng, K., Canfield, V. A., et al. 2007. Genetic evidence for the convergent evolution of light skin in Europeans and East Asians. *Molecular Biology and Evolution*, **24**(3), 710–722.

Novembre, J., and Han, E. 2012. Human population structure and the adaptive response to pathogen-induced selection pressures. *Philosophical Transactions of the Royal Society B*, **367**(1590), 878–886.

Novembre, J., Johnson, T., Bryc, K., Kutalik, Z., Boyko, A. R., Auton, A., Indap, A., et al. 2008. Genes mirror geography within Europe. *Nature*, **456**, 98–101.

Nunnally, J., and Bernstein, I. 1994. *Psychometric Theory*. 3rd edn. New York: McGraw-Hill.

O'Connor, C. 2008. Human chromosome translocations and cancer. *Nature Education*, **1**(1), 56.

Odling-Smee, F. J., Laland, K. N., and Feldman, M. W. 2003. *Niche Construction: The Neglected Process in Evolution*. Princeton, NJ: Princeton University Press.

Ohashi, J., Naka, I., and Tsuchiya, N. 2011. The impact of natural selection on an ABCC11 SNP determining earwax type. *Molecular Biology and Evolution*, **28**(1), 849–857.

Ohta, T. 1973. Slightly deleterious mutant substitutions in evolution. *Nature*, **246**(5428), 96–98.

Ohta, T. 1992. The nearly neutral theory of molecular evolution. *Annual Review of Ecology and Systematics*, **23**(1), 263–286.

Okada, H, Kuhn, C, Feillet, H, and Bach, J-F. 2010. The 'hygiene hypothesis' for autoimmune and allergic diseases: an update. *Clinical and Experimental Immunology*, **160**(1), 1–9.

Okasha, S. 2008. *Evolution and the Levels of Selection*. New York: Oxford University Press.

O'Rawe, J., Jiang, T., Sun, G., Wu, Y., Wang, W., Hu, J., Bodily, P., et al. 2013. Low concordance of multiple variant-calling pipelines: practical implications for exome and genome sequencing. *Genome Medicine*, **5**(3), 28.

O'Roak, B. J, Deriziotis, P., Lee, C., Vives, L., Schwartz, J. J., Girirajan, S., Karakoc, E., et al. 2011. Exome sequencing in sporadic autism spectrum disorders identifies severe de novo mutations. *Nature Genetics*, **43**(6), 585–589.

Oyama, S., Griffiths, P. E, and Gray, R. D. 2003. *Cycles of Contingency: Developmental Systems and Evolution*. Cambridge, MA: MIT Press.

Palka, C., Alfonsi, M., Mohn, A., Cerbo, R., Guanciali Franchi, P., Fantasia, D., Morizio, E., et al. 2012. Mosaic 7q31 deletion involving FOXP2 gene associated with language impairment. *Pediatrics*, **129**(1), e183–188.

Paracchini, S., Scerri, T., and Monaco, A. P. 2007. The genetic lexicon of dyslexia. *Annual Review of Genomics and Human Genetics*, **8**, 57–79.

Parvanov, E. D., Petkov, P. M., and Paigen, K. 2010. Prdm9 controls activation of mammalian recombination hotspots. *Science*, **327**(5967), 835.

Paschou, P., Lewis, J., Javed, A., and Drineas, P. 2010. Ancestry informative markers for fine-scale individual assignment to worldwide populations. *Journal of Medical Genetics*, **47**(12), 835–847.

Passemard, S., Kaindl, A. M., Titomanlio, L., Gerard, B., Gressens, P., and Verloes, A. 1993. Primary autosomal recessive microcephaly. In Pagon, R. A., Bird, T. D., Dolan, C. R., Stephens, K., and Adam, M. P. (eds), *GeneReviews*. Seattle, WA: University of Washington.

Patel, M., and Hu, B. H. 2012. MicroRNAs in inner ear biology and pathogenesis. *Hearing Research*, **287**(1–2), 6–14.

Patel, S. R., Frame, J. M., Larkin, E. K., and Redline, S. 2008. Heritability of upper airway dimensions derived using acoustic pharyngometry. *European Respiratory Journal*, **32**(5), 1304–1308.

Pearl, J. 2000. *Causality: Models, Reasoning, and Inference*. Cambridge: Cambridge University Press.

Peng, Y., Shi, H., Qi, X.-b., Xiao, C.-j., Zhong, H., Ma, R. Z., and Su, B. 2010. The ADH1B Arg47His polymorphism in East Asian populations and expansion of rice domestication in history. *BMC Evolutionary Biology*, **10**, 15.

Peng, Y., Yang, Z., Zhang, H., Cui, C., Qi, X., Luo, X., Tao, X., et al. 2011. Genetic variations in Tibetan populations and high-altitude adaptation at the Himalayas. *Molecular Biology and Evolution*, **28**(2), 1075–1081.

Penny, D., and Poole, A. 1999. The nature of the last universal common ancestor. *Current Opinion in Genetics & Development*, **9**(6), 672–677.

Pérez-Cadahía, B., Drobic, B., and Davie, J. R. 2011. Activation and function of immediate-early genes in the nervous system. *Biochemistry and Cell Biology*, **89**(1), 61–73.

Perkel, J. 2008. SNP genotyping: six technologies that keyed a revolution. *Nature Methods*, **5**(5), 447–453.

Perry, G. H., Dominy, N. J., Claw, K. G., Lee, A. S., Fiegler, H., Redon, R., Werner, J., et al. 2007. Diet and the evolution of human amylase gene copy number variation. *Nature Genetics*, **39**(10), 1256–1260.

Pinel, P., Fauchereau, F., Moreno, A., Barbot, A., Lathrop, M., Zelenika, D., Le Bihan, D., et al. 2012. Genetic variants of FOXP2 and KIAA0319/TTRAP/THEM2 locus are associated with altered brain activation in distinct language-related regions. *Journal of Neuroscience*, **32**(3), 817–825.

Plomin, R., and Kovas, Y. 2005. Generalist genes and learning disabilities. *Psychological Bulletin*, **131**(4), 592–617.

Plomin, R., and Simpson, M. A. 2013. The future of genomics for developmentalists. *Development and Psychopathology*, **25**(4 0 2), 1263–1278.

Plomin, R., DeFries, J. C., and Loehlin, J. C. 1977. Genotype-environment interaction and correlation in the analysis of human behavior. *Psychological Bulletin*, **84**(2), 309–322.

Plomin, R., DeFries, J. C., McClearn, G. E., and McGuffin, P. 2001. *Behavioral Genetics*. 4th edn. New York: Worth.

Plomin, R., DeFries, J. C., McClearn, G. E., and McGuffin, P. 2008. *Behavioral Genetics*. 5th edn. New York: Worth.

Plomin, R., Haworth, C. M. A., Meaburn, E. L., Price, T. S., and Davis, O. S. P. 2013. Common DNA markers can account for more than half of the genetic influence on cognitive abilities. *Psychological Science*, **24**(4), 562–568.

Posthuma, D. 2009. Multivariate genetic analysis. In: Kim, Y.-K. (ed), *Handbook of Behaviour Genetics*. Berlin: Springer, pp. 47–60.

Press, S. James. 2003. *Subjective and Objective Bayesian Statistics*. 2nd edn. Wiley Series in Probability and Statistics. New York: John Wiley & Sons.

Price, A. L., Patterson, N. J., Plenge, R. M., Weinblatt, M. E., Shadick, N. A., and Reich, D. 2006. Principal components analysis corrects for stratification in genome-wide association studies. *Nature Genetics*, **38**(8), 904–909.

Pritchard, J. K., Pickrell, J. K., and Coop, G. 2010. The genetics of human adaptation: hard sweeps, soft sweeps, and polygenic adaptation. *Current Biology : CB*, **20**(4), R208–R215.

Ptak, S. E., Enard, W., Wiebe, V., Hellmann, I., Krause, J., Lachmann, M., and Pbo, S. 2009. Linkage disequilibrium extends across putative selected sites in FOXP2. *Molecular Biology and Evolution*, **26**(10), 2181–2184.

Puigbò, P., Wolf, Y. I., and Koonin, E. V. 2009. Search for a 'Tree of Life' in the thicket of the phylogenetic forest. *Journal of Biology*, **8**(6), 59.

Pullum, G. K., and Scholz, B. C. 2002. Empirical assessment of stimulus poverty arguments. *The Linguistic Review*, **18**(1-2), 9–50.

Purcell, S., Neale, B., Todd-Brown, K., Thomas, L., Ferreira, M. A. R., Bender, D., Maller, J., et al. 2007. PLINK: a tool set for whole-genome association and population-based linkage analyses. *American Journal of Human Genetics*, **81**(3), 559–575.

Quintana-Murci, L., and Clark, A. G. 2013. Population genetic tools for dissecting innate immunity in humans. *Nature Reviews Immunology*, **13**(4), 280–293.

R Development Core Team. 2010. *R: A Language and Environment for Statistical Computing*. R Foundation for Statistical Computing, Vienna, Austria.

Rabbani, B., Mahdieh, N., Hosomichi, K., Nakaoka, H., and Inoue, I. 2012. Next-generation sequencing: impact of exome sequencing in characterizing Mendelian disorders. *Journal of Human Genetics*, **57**(10), 621–632.

Raj, T., Kuchroo, M., Replogle, J. M., Raychaudhuri, S., Stranger, B. E., and De Jager, P. L. 2013. Common risk alleles for inflammatory diseases are targets of recent positive selection. *American Journal of Human Genetics*, **92**(4), 517–529.

Rasmussen, M., Li, Y., Lindgreen, S., Pedersen, J. S., Albrechtsen, A., Moltke, I., Metspalu, M., et al. 2010. Ancient human genome sequence of an extinct Palaeo-Eskimo. *Nature*, 757–762.

Reed, P. W., Davies, J. L., Copeman, J. B., Bennett, S. T., Palmer, S. M., Pritchard, L. E., Gough, S. C., et al. 1994. Chromosome-specific microsatellite sets for fluorescence-based, semi-automated genome mapping. *Nature Genetics*, **7**(3), 390–395.

Rees, D. C, Williams, T. N., and Gladwin, M. T. 2010. Sickle-cell disease. *Lancet*, **376**(9757), 2018–2031.

Rees, J. L. 2003. Genetics of hair and skin color. *Annual Review of Genetics*, **37**, 67–90.

Regier, T., and Kay, P. 2009. Language, thought, and color: Whorf was half right. *Trends in Cognitive Sciences*, **13**(10), 439–446.

Regier, T., Kay, P., and Khetarpal, N. 2007. Color naming reflects optimal partitions of color space. *Proceedings of the National Academy of the Sciences of the USA*, **104**(4), 1436–1441.

Riaz, N., Steinberg, S., Ahmad, J., Pluzhnikov, A., Riazuddin, S., Cox, N. J., and Drayna, D. 2005. Genomewide significant linkage to stuttering on chromosome 12. *American Journal of Human Genetics*, **76**(4), 647–651.

Rice, G. M., Raca, G., Jakielski, K. J., Laffin, J. J., Iyama-Kurtycz, C. M., Hartley, S. L., Sprague, R. E., et al. 2012. Phenotype of FOXP2 haploinsufficiency in a mother and son. *American Journal of Medical Genetics. Part A*, **158A**(1), 174–181.

Richardson, G. P., Lukashkin, A. N., and Russell, I. J. 2008. The tectorial membrane: one slice of a complex cochlear sandwich. *Current Opinion in Otolaryngology & Head and Neck Surgery*, **16**(5), 458.

Richerson, P. J., and Boyd, R. 2008. *Not By Genes Alone: How culture transformed human evolution*. Chicago, IL: University of Chicago Press.

Richerson, P. J., and Christiansen, M. (eds). 2013. *Cultural Evolution: Society, Technology, Language, and Religion*. Strungmann Forum Reports, vol. 12. Cambridge, MA: MIT Press.

Richerson, P. J., Boyd, R., and Henrich, J. 2010. Gene-culture coevolution in the age of genomics. *Proceedings of the National Academy of the Sciences of the USA*, **107 Suppl 2**, 8985–8992.

Riethoven, J.-J. M. 2010. Regulatory regions in DNA: promoters, enhancers, silencers, and insulators. In Ladunga, I. (ed), *Computational Biology of Transcription Factor Binding*. Methods in Molecular Biology, no. 674. New York: Humana Press, pp. 33–42.

Rimol, L. M., Agartz, I., Djurovic, S., Brown, A. A., Roddey, J. C., Khler, A. K., Mattingsdal, M., et al. 2010. Sex-dependent association of common variants of microcephaly genes with brain structure. *Proceedings of the National Academy of the Sciences of the USA*, **107**(1), 384–388.

Roberts, S., and Winters, J. 2013. Linguistic diversity and traffic accidents: lessons from statistical studies of cultural traits. *PLoS ONE*, **8**(8), e70902.

Rodenas-Cuadrado, P., Ho, J., and Vernes, S. C. 2013. Shining a light on CNTNAP2: complex functions to complex disorders. *European Journal of Human Genetics*, **22**, 171–178.

Rohrer, J. D., and Schott, J. M. 2011. Primary progressive aphasia: defining genetic and pathological subtypes. *Current Alzheimer Research*, **8**(3), 266–272.

Rohrer, J. D., Ridgway, G. R., Crutch, S. J., Hailstone, J., Goll, J. C., Clarkson, M. J., Mead, S., et al. 2010. Progressive logopenic/phonological aphasia: erosion of the language network. *Neuroimage*, **49**(1), 984–993.

Rohs, R., Jin, X., West, S. M., Joshi, R., Honig, B., and Mann, R. S. 2010. Origins of specificity in protein-DNA recognition. *Annual Review of Biochemistry*, **79**, 233–269.

Rosenthal, A. Z., and Elowitz, M. B. 2011. Following evolution of bacterial antibiotic resistance in real time. *Nature Genetics*, **44**(1), 11–13.

Roth, C., Rastogi, S., Arvestad, L., Dittmar, K., Light, S., Ekman, D., and Liberles, D. A. 2007. Evolution after gene duplication: models, mechanisms, sequences, systems, and organisms. *Journal of Experimental Zoology. Part B, Molecular and Developmental Evolution*, **308**(1), 58–73.

Rousso, D. L., Pearson, C. A., Gaber, Z. B., Miquelajauregui, A., Li, S., Portera-Cailliau, C., Morrisey, E. E., et al. 2012. Foxp-mediated suppression of N-cadherin regulates neuroepithelial character and progenitor maintenance in the CNS. *Neuron*, **74**(2), 314–330.

Ruxton, G. D., Sherratt, T. N., and Speed, M. P. 2004. *Avoiding Attack: The Evolutionary Ecology of Crypsis, Warning Signals and Mimicry*. Oxford: Oxford University Press.

Sabeti, P. C., Reich, D. E., Higgins, J. M., Levine, H. Z. P., Richter, D. J., Schaffner, S. F., Gabriel, S. B., et al. 2002. Detecting recent positive selection in the human genome from haplotype structure. *Nature*, **419**(6909), 832–837.

Sabeti, P. C., Schaffner, S. F., Fry, B., Lohmueller, J., Varilly, P., Shamovsky, O., Palma, A., et al. 2006. Positive natural selection in the human lineage. *Science (New York, N.Y.)*, **312**(5780), 1614–1620.

Sagan, L. 1967. On the origin of mitosing cells. *Journal of Theoretical Biology*, **14**(3), 255–274.

Sagoo, G. S., Little, J., and Higgins, J. P. T. 2009. Systematic reviews of genetic association studies. *PLoS Medicine*, **6**(3), e1000028.

Sánchez-Meca, J., Marín-Martnez, F., and Chacón-Moscoso, S. 2003. Effect-size indices for dichotomized outcomes in meta-analysis. *Psychological Methods*, **8**(4), 448–467.

Sandler, W., Meir, I., Padden, C., and Aronoff, M. 2005. The emergence of grammar: Systematic structure in a new language. *Proceedings of the National Academy of Sciences of the USA*, **102**(7), 2661–2665.

Sandler, W., Aronoff, M., Meir, I., and Padden, C. 2011. The gradual emergence of phonological form in a new language. *Natural Language & Linguistic Theory*, **29**(2), 503–543.

Sankararaman, S., Mallick, S., Dannemann, M., Prfer, K., Kelso, J., Pbo, S., Patterson, N. et al. 2014. The genomic landscape of Neanderthal ancestry in present-day humans. *Nature*, **507**, 354–357.

Sanna, C. R., Li, W.-H., and Zhang, L. 2008. Overlapping genes in the human and mouse genomes. *BMC Genomics*, **9**(1), 169.

Santos, M. E., Athanasiadis, A., Leitão, A. B., DuPasquier, L., and Sucena, É. 2011. Alternative splicing and gene duplication in the evolution of the FoxP gene subfamily. *Molecular Biology and Evolution*, **28**(1), 237–247.

Sato, T., Amano, T., Ono, H., Ishida, H., Kodera, H., Matsumura, H., Yoneda, M., et al. 2009. Allele frequencies of the ABCC11 gene for earwax phenotypes among ancient populations of Hokkaido, Japan. *Journal of Human Genetics*, **54**(7), 409–413.

Schmucker, D., Clemens, J. C., Shu, H., Worby, C. A., Xiao, J., Muda, M., Dixon, J. E., et al. 2000. Drosophila Dscam is an axon guidance receptor exhibiting extraordinary molecular diversity. *Cell*, **101**(6), 671–684.

Schwab, R. J, Pasirstein, M., Kaplan, L., Pierson, R., Mackley, A., Hachadoorian, R., Arens, R., et al. 2006. Family aggregation of upper airway soft tissue structures in normal subjects and patients with sleep apnea. *American Journal of Respiratory and Critical Care Medicine*, **173**(4), 453–463.

Scott, D. A., Carmi, R., Elbedour, K., Duyk, G. M., Stone, E. M., and Sheffield, V. C. 1995. Nonsyndromic autosomal recessive deafness is linked to the DFNB1 locus in a large inbred Bedouin family from Israel. *American Journal of Human Genetics*, **57**(4), 965–968.

Sharp, P. M., Averof, M., Lloyd, A. T., Matassi, G., and Peden, J. F. 1995. DNA sequence evolution: the sounds of silence. *Philosophical Transactions of the Royal Society B*, **349**(1329), 241–247.

Shimomura, Y., and Christiano, A. M. 2010. Biology and genetics of hair. *Annual Review of Genomics and Human Genetics*, **11**(Sep), 109–132.

Shriberg, L. D., Ballard, K. J., Tomblin, J. B., Duffy, J. R., Odell, K. H., and Williams, C. A. 2006. Speech, prosody, and voice characteristics of a mother and daughter with a 7;13 translocation affecting FOXP2. *Journal of Speech, Language, and Hearing Research: JSLHR*, **49**(3), 500–525.

Shu, W., Lu, M. M., Zhang, Y., Tucker, P. W., Zhou, D., and Morrisey, E. E. 2007. Foxp2 and Foxp1 cooperatively regulate lung and esophagus development. *Development*, **134**(10), 1991–2000.

Shultz, S., Nelson, E., and Dunbar, R. I. M. 2012. Hominin cognitive evolution: identifying patterns and processes in the fossil and archaeological record. *Philosophical Transactions of the Royal Society B*, **367**(1599), 2130–2140.

Silventoinen, K., Magnusson, P. K. E., Tynelius, P., Kaprio, J., and Rasmussen, F. 2008. Heritability of body size and muscle strength in young adulthood: a study of one million Swedish men. *Genetic Epidemiology*, **32**(4), 341–349.

Silvestre, D. C., and Londoo-Vallejo, A. 2012. Telomere dynamics in mammals. *Genome Dynamics*, **7**, 29–45.

Simon, S D. 2001. Understanding the odds ratio and the relative risk. *Journal of Andrology*, **22**(4), 533–536.

Smith, J. M., and Szathmáry, E. 1995. *The Major Transitions in Evolution*. Oxford: Oxford University Press.

Smith, K., and Binder, P.-M. 2013. *The Language Phenomenon: Human Communication from Milliseconds to Millennia*. Berlin: Springer.

Smith, K., and Wonnacott, E. 2010. Eliminating unpredictable variation through iterated learning. *Cognition*, **116**(3), 444–449.

Snustad, D. P., and Simmons, M. J. 2010. *Principles of Genetics*. 5th edn. New York: John Wiley & Sons.

Speliotes, E. K., Willer, C. J., Berndt, S. I., Monda, K. L., Thorleifsson, G., Jackson, A. U., Allen, H. L., et al. 2010. Association analyses of 249,796 individuals reveal eighteen new loci associated with body mass index. *Nature Genetics*, **42**(11), 937–948.

Spielman, R. S., McGinnis, R. E., and Ewens, W. J. 1993. Transmission test for linkage disequilibrium: the insulin gene region and insulin-dependent diabetes mellitus (IDDM). *American Journal of Human Genetics*, **52**(3), 506–516.

Spirtes, P., Glymour, C. N., and Scheines, R. 2000. *Causation, Prediction, and Search*. Cambridge, MA: MIT Press.

Spiteri, E., Konopka, G., Coppola, G., Bomar, J., Oldham, M., Ou, J., Vernes, S. C., et al. 2007. Identification of the transcriptional targets of FOXP2, a gene linked to speech and language, in developing human brain. *American Journal of Human Genetics*, **81**(6), 1144–1157.

Sproat, R. 2011. Phonemic diversity and the out-of-Africa theory. *Linguistic Typology*, **15**(2), 199–206.

Stover, T., and Diensthuber, M. 2012. Molecular biology of hearing. *GMS Current Topics in Otorhinolaryngology, Head and Neck Surgery*, **10**.

Strachan, D. P. 1989. Hay fever, hygiene, and household size. *BMJ: British Medical Journal*, **299**(6710), 1259–1260.

Strachan, T., and Read, A. P. 1999. *Genetic Mapping of Mendelian Characters*. 2nd edn. New York: Wiley-Liss.

Strauss, K. A., Puffenberger, E. G., Huentelman, M. J., Gottlieb, S., Dobrin, S. E., Parod, J. M., Stephan, D. A., et al. 2006. Recessive symptomatic focal epilepsy and mutant contactin-associated protein-like 2. *New England Journal of Medicine*, **354**(13), 1370–1377.

Stromswold, K. 2001. The heritability of language: A review and metaanalysis of twin, adoption, and linkage studies. *Language*, **77**, 647–723.

Stromswold, K. 2006. Why aren't identical twins linguistically identical? Genetic, prenatal and postnatal factors. *Cognition*, **101**(2), 333–384.

Stroud, J. C., Wu, Y., Bates, D. L., Han, A., Nowick, K., Paabo, S., Tong, H., et al. 2006. Structure of the forkhead domain of FOXP2 bound to DNA. *Structure*, **14**(1), 159–166.

Sturm, R. A. 2009. Molecular genetics of human pigmentation diversity. *Human Molecular Genetics*, **18**(R1), R9–R17.

Super Science High School Consortium. 2009. Japanese map of the earwax gene frequency: a nationwide collaborative study by Super Science High School Consortium. *Journal of Human Genetics*, **54**(9), 499–503.

Tabachnick, B. G., and Fidell, L. S. 2001. *Using Multivariate Statistics*. 4th edn. Needham Heights, MA: Allyn & Bacon.

Tajima, F. 1989. Statistical method for testing the neutral mutation hypothesis by DNA polymorphism. *Genetics*, **123**(3), 585–595.

Tam, W. Y., Leung, C. K. Y., Tong, K. K., and Kwan, K. M. 2011. Foxp4 is essential in maintenance of Purkinje cell dendritic arborization in the mouse cerebellum. *Neuroscience*, **172**, 562–571.

Teer, J. K., and Mullikin, J. C. 2010. Exome sequencing: the sweet spot before whole genomes. *Human Molecular Genetics*, **19**(R2), R145–R151.

The ENCODE Project Consortium. 2012. An integrated encyclopedia of DNA elements in the human genome. *Nature*, **489**(7414), 57–74.

The SLI Consortium. 2002. A genomewide scan identifies two novel loci involved in specific language impairment. *American Journal of Human Genetics*, **70**(2), 384–398.

Tishkoff, S. A., Reed, F. A., Ranciaro, A., Voight, B. F., Babbitt, C. C., Silverman, J. S., Powell, K., et al. 2006. Convergent adaptation of human lactase persistence in Africa and Europe. *Nature Genetics*, **39**, 31–40.

Törnqvist, G., Sandberg, A., Hgglund, A.-C., and Carlsson, L. 2010. Cyclic expression of lhx2 regulates hair formation. *PLoS Genetics*, **6**(4), e1000904.

Townsend, G. C., Richards, L. C., Sekikawa, M., Brown, T., and Ozaki, T. 1990. Variability of palatal dimensions in South Australian twins. *Journal of Forensic Odontostomatology*, **8**(2), 3–14.

Toyoda, Y., and Ishikawa, T. 2010. Pharmacogenomics of human ABC transporter ABCC11 (MRP8): potential risk of breast cancer and chemotherapy failure. *Anti-Cancer Agents in Medicinal Chemistry*, **10**(8), 617–624.

Toyoda, Y., Sakurai, A., Mitani, Y., Nakashima, M., Yoshiura, K.-i., Nakagawa, H., Sakai, Y., et al. 2009. Earwax, osmidrosis, and breast cancer: why does one SNP (538G>A) in the human ABC transporter ABCC11 gene determine earwax type? *FASEB Journal*, **23**(6), 2001–2013.

Tuteja, G., and Kaestner, K. H. 2007a. SnapShot: forkhead transcription factors II. *Cell*, **131**(1), 192–192.e1.

Tuteja, G., and Kaestner, K. H. 2007b. SnapShot: forkhead transcription factors I. *Cell*, **130**(6), 1160.e1–1160.e2.

Vafai, S. B., and Mootha, V. K. 2012. Mitochondrial disorders as windows into an ancient organelle. *Nature*, **491**(7424), 374–383.

van der Bom, T., Zomer, A. C., Zwinderman, A. H., Meijboom, F. J., Bouma, B. J., and Mulder, B. J. M. 2010. The changing epidemiology of congenital heart disease. *Nature Reviews Cardiology*, **8**(1), 50–60.

van der Mei, I. a. F., Ponsonby, A.-L., Dwyer, T., Blizzard, L., Simmons, R., Taylor, B. V., Butzkueven, H., and Kilpatrick, T. 2003. Past exposure to sun, skin phenotype, and risk of multiple sclerosis: case-control study. *BMJ*, **327** (7410), 316.

Vaquerizas, J. M., Kummerfeld, S. K., Teichmann, S. A., and Luscombe, N. M. 2009. A census of human transcription factors: function, expression and evolution. *Nature Reviews Genetics*, **10**(4), 252–263.

Veltman, J. A., and Brunner, H. G. 2012. De novo mutations in human genetic disease. *Nature Reviews Genetics*, **13**(8), 565–575.

Verhoeven, K., Laer, L. V., Kirschhofer, K., Legan, P. K., Hughes, D. C., Schatteman, I., Verstreken, M., et al. 1998. Mutations in the human alpha-tectorin gene cause autosomal dominant non-syndromic hearing impairment. *Nature Genetics*, **19**(1), 60–62.

Vernes, S. C., Nicod, J., Elahi, F. M., Coventry, J. A., Kenny, N., Coupe, A.-M., Bird, L. E., et al. 2006. Functional genetic analysis of mutations implicated in a human speech and language disorder. *Human Molecular Genetics*, **15**(21), 3154–3167.

Vernes, S. C., Spiteri, E., Nicod, J., Groszer, M., Taylor, J. M., Davies, K. E., Geschwind, D. H., et al. 2007. High-throughput analysis of promoter occupancy reveals direct neural targets of FOXP2, a gene mutated in speech and language disorders. *American Journal of Human Genetics*, **81**, 1232–1250.

Vernes, S. C., Newbury, D. F., Abrahams, B. S., Winchester, L., Nicod, J., Groszer, M., Alarcn, M., et al. 2008. A functional genetic link between distinct developmental language disorders. *New England Journal of Medicine*, **359**(22), 2337–2345.

Vernes, S. C., Oliver, P. L., Spiteri, E., Lockstone, H. E., Puliyadi, R., Taylor, J. M., Ho, J., et al. 2011. Foxp2 regulates gene networks implicated in neurite outgrowth in the developing brain. *PLoS Genetics*, **7**(7), e1002145.

Vernot, B., and Akey, J. M. 2014. Resurrecting surviving Neandertal lineages from modern human genomes. *Science*, **343**(6174), 1017–1021.

Vetsigian, K., Woese, C., and Goldenfeld, N. 2006. Collective evolution and the genetic code. *Proceedings of the National Academy of the Sciences of the USA*, **103**(28), 10696–10701.

Villarreal, L. P., and Witzany, G. 2010. Viruses are essential agents within the roots and stem of the tree of life. *Jornal of Theoretical Biology*, **262**(4), 698–710.

Visscher, P. M., Medland, S. E., Ferreira, M. A. R., Morley, K. I., Zhu, G., Cornes, B. K., Montgomery, G. W., et al. 2006. Assumption-free estimation of heritability from genome-wide identity-by-descent sharing between full siblings. *PLoS Genetics*, **2**(3), e41.

Visscher, P. M., Hill, W. G., and Wray, N. R. 2008. Heritability in the genomics era–concepts and misconceptions. *Nature Reviews Genetics*, **9**(4), 255–266.

Visscher, P. M., Yang, J., and Goddard, M. E. 2010a. A commentary on common SNPs explain a large proportion of the heritability for human height by Yang et al. (2010). *Twin Research and Human Genetics*, **13**(06), 517–524.

Visscher, P. M., McEvoy, B., and Yang, J. 2010b. From Galton to GWAS: quantitative genetics of human height. *Genetics Research*, **92**(5-6), 371–379.

Vissers, L. E. L. M., de Ligt, J., Gilissen, C., Janssen, I., Steehouwer, M., de Vries, P., van Lier, B., et al. 2010. A de novo paradigm for mental retardation. *Nature Genetics*, **42**(12), 1109–1112.

Voet, D., and Voet, J. G. 2011. *Biochemistry*. New York: John Wiley & Sons.

Voight, B. F., Kudaravalli, S., Wen, X., and Pritchard, J. K. 2006. A map of recent positive selection in the human genome. *PLoS Biology*, **4**(3), e72.

Wain, H. M., Bruford, E. A., Lovering, R. C., Lush, M. J., Wright, M. W., and Povey, S. 2002. Guidelines for human gene nomenclature. *Genomics*, **79**(4), 464–470.

Wang, A., Liang, Y., Fridell, R. A., Probst, F. J., Wilcox, E. R., Touchman, J. W., Morton, C. C., et al. 1998. Association of unconventional myosin MYO15 mutations with human nonsyndromic deafness DFNB3. *Science*, **280**(5368), 1447–1451.

Wang, J.-k., Li, Y., and Su, B. 2008. A common SNP of MCPH1 is associated with cranial volume variation in Chinese population. *Human Molecular Genetics*, **17**(9), 1329–1335.

Wang, K., Schmied, W. H., and Chin, J. W. 2012. Reprogramming the genetic code: from triplet to quadruplet codes. *Angewandte Chemie International Edition*, **51**(10), 2288–2297.

Wang, S., Lewis, C. M., Jr., Jakobsson, M., Ramachandran, S., Ray, N., Bedoya, G., Rojas, W., et al. 2007. Genetic variation and population structure in Native Americans. *PLoS Genet*, **3**(11), e185.

Wardhaugh, R. 2010. *An Introduction to Sociolinguistics*. New York: John Wiley & Sons.

Watkins, K. E., Dronkers, N. F., and Vargha Khadem, F. 2002. Behavioural analysis of an inherited speech and language disorder: comparison with acquired aphasia. *Brain*, **125**(3), 452–464.

West-Eberhard, M. J. 2003. *Developmental Plasticity and Evolution*. 1st edn. New York: Oxford University Press.

Westfall, P. H., and Troendle, J. F. 2008. Multiple testing with minimal assumptions. *Biometrische Zeitschrift*, **50**(5), 745–755.

White, S. A., Fisher, S. E., Geschwind, D. H., Scharff, C., and Holy, T. E. 2006. Singing mice, songbirds, and more: models for FOXP2 function and dysfunction in human speech and language. *Journal of Neuroscience*, **26**(41), 10376–10379.

Whitehouse, A. J. O., Bishop, D. V. M., Ang, Q. W., Pennell, C. E., and Fisher, S E. 2011. CNTNAP2 variants affect early language development in the general population. *Genes, Brain, and Behavior*, **10**(4), 451–456.

Wichterle, H., Gifford, D., and Mazzoni, E. 2013. Mapping neuronal diversity one cell at a time. *Science*, **341**(6147), 726–727.

Wilkins, J. F., and Marlowe, F. W. 2006. Sex-biased migration in humans: what should we expect from genetic data? *BioEssays*, **28**(3), 290–300.

Williamson, S. J., Rusch, D. B., Yooseph, S., Halpern, A. L., Heidelberg, K. B., Glass, J. I., Andrews-Pfannkoch, C., et al. 2008. The Sorcerer II Global Ocean Sampling Expedition: metagenomic characterization of viruses within aquatic microbial samples. *PLoS One*, **3**(1), e1456.

Wilson, K. M., and Walker, J. M. 2010. *Principles and Techniques of Biochemistry and Molecular Biology*. Cambridge: Cambridge University Press.

Wilson, R. C., and Doudna, J. A. 2013. Molecular mechanisms of RNA interference. *Annual Review of Biophysics*, **42**(1), 217–239.

Winata, S., Arhya, I. N., Moeljopawiro, S., Hinnant, J. T., Liang, Y., Friedman, T. B., and Asher, J. H. 1995. Congenital non-syndromal autosomal recessive deafness in Bengkala, an isolated Balinese village. *Journal of Medical Genetics*, **32**(5), 336–343.

Woese, C. 1998. The universal ancestor. *Proceedings of the National Academy of the Sciences of the USA*, **95**(12), 6854–6859.

Woese, C. R. 2000. Interpreting the universal phylogenetic tree. *Proceedings of the National Academy of the Sciences of the USA*, **97**(15), 8392–8396.

Woese, C. R., Kandler, O., and Wheelis, M. L. 1990. Towards a natural system of organisms: proposal for the domains Archaea, Bacteria, and Eucarya. *Proceedings of the National Academy of the Sciences of the USA*, **87**(12), 4576–4579.

Wong, C. C. Y., Caspi, A., Williams, B., Houts, R., Craig, I. W., and Mill, J. 2011. A longitudinal twin study of skewed X chromosome-inactivation. *PLoS ONE*, **6**, e17873.

Wong, P. C. M., Chandrasekaran, B., and Zheng, J. 2012. The derived allele of ASPM is associated with lexical tone perception. *PLoS One*, **7**(4), e34243.

Worthey, E. A., Raca, G., Laffin, J. J., Wilk, B. M., Harris, J. M., Jakielski, K. J, Dimmock, D. P., et al. 2013. Whole-exome sequencing supports genetic heterogeneity in childhood apraxia of speech. *Journal of Neurodevelopmental Disorders*, **5**(1), 29.

Wrangham, R. 2009. *Catching Fire: How Cooking Made Us Human*. New York: Basic Books.

Wright, S. 1922. Coefficients of inbreeding and relationship. *American Naturalist*, **56**(645), 330–338.

Wu, M. C., Kraft, P., Epstein, M. P., Taylor, D. M., Chanock, S. J., Hunter, D. J., and Lin, X. 2010. Powerful SNP-set analysis for case-control genome-wide association studies. *American Journal of Human Genetics*, **86**(6), 929–942.

Xing, G., Chen, Z., and Cao, X. 2007. Mitochondrial rRNA and tRNA and hearing function. *Cell Research*, **17**(3), 227–239.

Yang, J., Benyamin, B., McEvoy, B. P., Gordon, S., Henders, A. K., Nyholt, D. R., Madden, P. A., et al. 2010. Common SNPs explain a large proportion of the heritability for human height. *Nature Genetics*, **42**(7), 565–569.

Yang, J., Lee, S. H., Goddard, M. E., and Visscher, P. M. 2011. GCTA: A tool for genome-wide complex trait analysis. *American Journal of Human Genetics*, **88**(1), 76–82.

Yang, J., Wang, L., Song, H., and Sokolov, M. 2012. Current understanding of Usher syndrome Type II. *Frontiers in Bioscience*, **17**, 1165–1183.

Yates, L. A., Norbury, C. J., and Gilbert, R. J. C. 2013. The long and short of microRNA. *Cell*, **153**(3), 516–519.

Yi, X., Liang, Y., Huerta-Sanchez, E., Jin, X., Cuo, Z. X. P., Pool, J. E., Xu, X., et al. 2010. Sequencing of fifty human exomes reveals Adaptation to high altitude. *Science*, **329**(5987), 75–78.

Yip, M. 2002. *Tone*. Cambridge Textbooks in Linguistics. Cambridge: Cambridge University Press.

Yoshiura, K.-i., Kinoshita, A., Ishida, T., Ninokata, A., Ishikawa, T., Kaname, T., Bannai, M., et al. 2006. A SNP in the ABCC11 gene is the determinant of human earwax type. *Nature Genetics*, **38**(3), 324–330.

Zeshan, U., and de Vos, C. 2012. *Sign Languages in Village Communities: Anthropological and Linguistic Insights*. Berlin: Walter de Gruyter.

Zhang, J., Webb, D. M., and Podlaha, O. 2002. Accelerated protein evolution and origins of human-specific features: Foxp2 as an example. *Genetics*, **162**(4), 1825–1835.

Zong, X.-N., Li, H., and Zhu, Z.-H. 2010. Secular trends in height and weight for healthy Han children aged 0-7 years in China, 1975–2005. *American Journal of Human Biology*, **23**(2), 209–215.

Index

Glossary

action potential see **neural impulse**. 272

active genotype–environment correlation a type of **genotype–environment correlation** which arises because individuals actively change their environment to suit their genotype; for example, people with high verbal abilities might actively look for experiences involving language. 41, 265

admixture occurs when two populations that have been separated exchange migrants, resulting in an admixed population. 189, 272

adoption study a design whereby adopted children are compared to their biological family members and to their adoptive family members, allowing one to distinguish the influence of genetic (biological family) and environmental (adoptive family) factors. 35

AIM see **Ancestry Informative Marker**. 114, 257

Al-Sayyid Bedouin Sign Language or ABSL, a sign language that emerged spontaneously in a village in the Negev desert, Israel, due to a high frequency of **recessive congenital** hearing loss due to a mutation at the *DFNB1* locus. 204

allele possible variants of a given genetic **locus** (such as a **gene** or **SNP**) that might or might not affect the **phenotype**, and its phenotypic effects (if any) might or might not be "visible" to **selection**. 16, 32, 53, 161, 257, 258, 260–270, 272, 274–278

alternative hypothesis H_1 represents the "interesting" case when there is a relationship between the variables of interest (e.g., they are correlated); **statistical tests** oppose it to the **null hypothesis**. 273, 279

amino acid component of **proteins**, playing essential roles throughout the living world; there are 20 amino acids encoded in the **genetic code**. 49, 260, 264, 273, 276, 280, 281

ancestral allele the original **allele** at a **locus**; as opposed to the new or **derived allele** resulting from a **mutation** at the locus. 169, 184, 260

Ancestry Informative Marker abbreviated **AIM**, a **locus** that has different **allele** frequencies in different populations, thus distinguishing these populations from a genetic point of view; please note that in general many such **AIMs** are required for a reliable distinction between populations given the high genetic uniformity of our species and the gradual and multivariate nature of the remaining genetic diversity. 114, 192, 257

animal model refers to the general use of non-human animals for studies that cannot be conducted using human participants; there are several issues concerning this methodology including ethical concerns and, very important here, the transferability of the findings to humans; for example, mice and songbirds have been extensively used to understand the phenotypic effects of ***FOXP2*** but obviously these animals do not have language, requiring indirect and complex inferences based on evolutionary conservatism. 121, 259, 263

archaeon a group of **prokaryotes** distinct from **bacteria**, probably more closely related to the **eukaryotes**; some **archaea** are extremophiles, living in very inhospitable environments such as hot vents. 46, 258, 261, 281

artificial selection a type of selection where differential survival and reproduction are determined by breeders aiming to produce various **phenotypes** such as increased milk production or body shape in dogs. 170, 278

ASPM a gene involved in **Primary Autosomal Recessive Microcephaly**; a normal **allele** of ***ASPM*** was proposed to result in a **bias** affecting **linguistic tone**. 140, 258

association study a method for finding **loci** involved in a **phenotype** using non-related participants, the idea being that the causative **alleles** should occur in the participants that show the phenotype ("affected") but not in the others ("non-affected"); just like **linkage studies**, **association studies** can be used not only for binary (presence/absence) **phenotypes** but also for **quantitative phenotypes**. 94, 258, 265, 268, 275

assortative mating occurs when mating partners are more similar with respect to a **phenotype** than expected by chance; in humans, intelligence and socio-economic status are examples; see also **disassortative mating**. 188, 261, 275

atom composed of a positively charged nucleus (different from the cell **nucleus**!) made up of protons and neutrons, orbited by negatively charged electrons; atoms are electrically neutral (there are as many electrons as protons) but they can become electrically charged **ions**; atoms are of different types with different chemical and physical properties (e.g, oxygen, carbon, iron or uranium). 48, 259, 268, 270

autism a disorder characterized by difficulties in communication and social interaction and repetitive behaviours; there is in fact a continuum encompassed by the autism spectrum. 121, 260

axon a long, tube-like projection from the **neuron**'s body essential for transmitting the **neural impulse** to other **neurons**. 260, 272, 280

bacterium very widespread and diverse group of **prokaryotes** distinct from **archaea**. 46, 258, 261, 262, 267, 281

balancing selection a type of **selection** whereby several **alleles** are maintained in the population, resulting in increased **genetic diversity**; examples include **heterozygote advantage** and negative **frequency-dependent selection**. 261, 264, 266, 278

basal ganglia subcortical structures playing important roles in many aspects of information processing and behaviour control. 259

bias see **genetically biased cultural evolution**. 258, 269

Bonferroni correction a popular but very conservative method of **multiple testing correction** where the adjusted **significance level** $\alpha_{adjusted} = \alpha/N$, where α is the standard 1% (or 5%) level and N the number of tests. 99, 263, 271

brain the main **organ** of the **nervous system** where information is processed and stored and essential for cognition and language; composed of **grey matter** and **white matter** and structured into **cerebral cortex** and various subcortical structures such as the **hippocampus** and the **basal ganglia**. 259, 269, 272, 273, 275, 279

carbohydrate involved in energy storage but also in the immune system and genetic information (components of **RNA** and **DNA**) inter alia. 48

case an affected, or abnormal, individual showing the pathology or **phenotype** of interest, as opposed to a **control**. 95, 260, 269, 275

catalysis the process of increasing the rate of a chemical reaction mediated by a catalyst which does not take part in the reaction itself; without catalysis (and the **enzymes**) life as we know it would be impossible. 262, 276

cell the fundamental unit composing most living organisms (it is debatable whether **viruses** and certainly **prions** are to be considered alive). 259–261, 266, 268–270, 272–274, 280, 282

cell line an alternative to **animal models** using instead **cells** derived from human tissues; it presents several advantages such as reduced ethical issues, controllability and speed, but has several disadvantages as well including the difficulty of extrapolation to the whole organism and the fact that cell lines are usually different from healthy, normal functional cells; for example, HeLa cells are one of the most widely used cell lines and are derived from cervical cancer cells. 121, 150

cell type in complex multicellular **eukaryotes** such as ourselves, **cells** are differentiated and optimized for various roles such as **neurons** and skin cells; in humans there are hundreds of such types. 48

centimorgan 1 **cM** is the genetic distance between two **loci** that have a 1% chance of **recombination** in one generation; independent **loci** are separated by more than 50 cM. 92, 259, 277

cerebral cortex in mammals (including humans) this is a layer of **grey matter** at the surface of the **brain** and plays essential roles in cognition and language. 259

chemical bond a physical attractive force due to the electrical attraction between positive and negative charges that binds **atoms** together to form **molecules**; there are several types differing in strength. 48, 50, 270

chi-square test or χ^2 is a **statistical test** used to check (a) if the distribution of a categorical variable differs from an expected distribution (e.g., if the actual count of heads and tails in a coin-tossing experiment differs from 50:50) and (b) if two categorical variables are independent (e.g., if there is relationship between flower colour and pollen grain shape in peas). 76

chloroplast an organelle present in plant **cells** that performs **photosynthesis**. 46, 274

chromatin composed of a **DNA** molecule bound to **proteins** such as **histones**; forms the **chromosomes** and is very important for **gene regulation** and **epigenetics**. 53, 266

chromosome a single tightly packed molecule of **DNA** with bound **proteins**; there are 23 pairs of chromosomes in **diploid** human cells. 46, 53, 259, 260, 266, 268, 269, 277, 278, 281

CI see **confidence interval**. 260

cM see **centimorgan**. 259

CNTNAP2 this gene encodes a **neurexin** that has several functions in the **nervous system**; it is down-regulated (its expression is reduced) by ***FOXP2*** and has been recently implicated in **autism**, **SLI** and even normal language variation. 121, 263

CNV see **Copy Number Variant**. 169, 260

codon the unit composed of three **nucleotides** that uniquely identifies a single **amino acid**. 63, 264, 273, 276, 281

cone a type of **photoreceptor** specialized for colour perception; usually of three types: S (sensitive to short wavelength or "blue" light), M (medium, "green") and L (long, "red"). 85, 274

confidence interval denoted **CI**, measures the degree of confidence in an estimated value, given as a range of values and the probability that the real one is contained therein; it is closely related to ***p*-values** and **significance level**: for example, a 95%CI of (−0.5,0.5) for a **Pearson correlation** coefficient means that with a 95% probability the true correlation is between −0.5 and +0.5 and, because this interval contains 0, we cannot reject the **null hypothesis** of no correlation at an α-level of 0.05. 108, 259, 279

congenital present at birth. 80, 257, 268, 271

control a non-affected, or normal, individual, as opposed to a **case**. 95, 259

Copy Number Variant abbreviated **CNV**, is a type of variation whereby the number of copies of the same section of **DNA** differs between alleles. 260

covariance a quantitative measure of the relationship between two variables, defined as $cov(x, y) = mean[(x - mean(x)) \times (y - mean(y))]$, where x and y are two variables of same length. 22

critical period a developmental stage during which the organism is particularly sensitive to external stimuli critical for the emergence of a particular feature, in this case language; the absence of appropriate input during this period leads to the abnormal acquisition of language. 7

crossover see **recombination**. 58

cultural evolution an approach to understanding culture that proposes that evolutionary processes similar to those affecting biological evolution shape the temporal dynamics and distribution of cultural phenomena. 199, 264, 265

cytoskeleton plays essential roles in the structural integrity of **cells** as well as their mobility and the transport of molecules within them. 46

deleterious as opposed to an advantageous **allele**, results in decreased **fitness** for the carriers. 76, 168, 267, 271, 272

deletion represents the removal of one or more **nucleotides** from a **DNA** molecule; if not a multiple of three it can determine a **frameshift**; see also **indel**. 67, 267, 271

dendrite projections from a **neuron**'s body essential for transmitting the **neural impulse** from other **neurons** for processing and further conduction through the **axon**. 272, 280

derived allele the new **allele** at a **locus** produced by **mutation**; as opposed to the original or **ancestral allele**. 169, 184, 257

diploid an organism (or cell) that has two sets of homologous (or paired) **chromosomes**, as opposed to **haploid** ones; the vast majority of human cells are diploid with 46 chromosomes (23 pairs), while sex cells (**sperm** and **ova**) are haploid with 23 chromosomes. 16, 32, 55, 163, 259, 266

disassortative mating occurs when mating partners are more dissimilar with respect to a **phenotype** than expected by chance; see also **assortative mating**. 188, 258

disruptive selection see **balancing selection**. 172

division **cell** division is the process whereby a "mother" cell produces two (or more) "daughter" cells; there are two main types: **mitosis** and **meiosis**. 269, 270

dizygotic twins that derive from two different fertilized **ova**, thus sharing on average only 50% their **genomes**, just as normal brothers and sisters do. 34, 281

DNA deoxyribonucleic acid, a molecule that encodes genetic information through the linear sequence (possibly extremely long) of the four **nucleotides** A, T, C and G; it plays the role of stable repository of genetic information in most biological organisms. 50, 259–262, 264–267, 270, 273, 276–278, 280–282

DNA polymerase the **enzyme** that ensures the **replication** of the genetic information from the original **DNA** molecule to its two "daughter" DNA molecules. 51, 277

domain of life the highest grouping of organisms based on descent from a common ancestor; currently three such domains are recognized: **Archaea**, **Bacteria** and **Eukaryotes**, and while these are ultimately related, the exact relationships are currently unclear; see also **Tree of Life** and **endosymbiotic theory**. 46, 281

dominant a dominant **allele**, usually denoted using upper-case letters such as *A*, expresses its effect even when paired with a **recessive** allele at the same **locus**; thus **homozygous** *AA* and **heterozygous** *Aa* individuals have the same **phenotype** and different from **homozygous** *aa*; **DVD** caused by certain mutations in ***FOXP2*** is a dominant speech and language pathology. 33, 73, 276, 280

dosage compensation a type of **gene regulation** that ensures that the **genes** on the X chromosome are expressed in the same amount in females (having two copies of X) and males (only one copy). 84

down-regulation **gene regulation** that results in less of the target **gene** being expressed; also known as negative regulation or repression. 149

DVD Developmental Verbal Dyspraxia or Childhood Apraxia of Speech, a rare autosomal dominant pathology with a complex manifestation also affecting language and speech, caused by mutations in the ***FOXP2*** gene. For more info see OMIM 602081 (`http://omim.org/entry/602081`). 10, 145, 261, 263, 266, 275, 277

dyslexia a relatively frequent difficulty in reading and spelling despite normal intelligence and educational opportunities. 123, 137, 275, 278

effect size for a **statistical test** this gives a measure of the magnitude of the relationship tested (i.e., the strength of the linear relationship as given by the **Pearson correlation**); while the **statistical significance** of a test will depend on sample size, the effect size does not. 105, 279

endogamy the mating within a group (such as a village, caste or ethnic group); as opposed to **exogamy**. 187, 262

endoplasmic reticulum or ER, an organelle of **eukaryotic cells** that is involved in **protein** synthesis (**transcription**) and other functions. 46, 136, 196, 265, 271

endosymbiosis see **endosymbiotic theory**. 281

endosymbiotic theory proposes that the **Eukaryotes** descend from two (or more) ancestors through **symbiosis**; for example, the present-day **mitochondria** are descendants of a **bacterium** that instead of being digested and destroyed was co-opted to produce energy within the cell. 48, 133, 261, 280

enhancer a region on the **DNA** that controls the **transcription** of a particular **gene** or **genes**; this might not necessarily be close to these gene(s). 70, 277, 281

enzyme highly specialized biological molecules involved in the **catalysis** of reactions essential to life. 49, 259, 261, 265, 270, 271, 278

epigenetic see **epigenetics**. 35, 274

epigenetics in general concerns the transmission of information across generations (of organisms or cells within organisms) in the absence of changes to the **DNA** message; most commonly chemical marks on the DNA itself or on the **histones** modify **gene expression**. 259, 262, 266

epistasis non-linear interaction between different loci, such as involving **transcription factors** and their target **genes**. 33, 117

eukaryote uni- or multicellular organisms (including humans) where the cell(s) contain various substructures bounded by **membranes** such as the nucleus, which contains the **DNA** molecules. 258, 259, 261, 262, 267, 270, 273, 276, 281

eukaryotic see **eukaryote**. 46, 261, 269, 276

evocative genotype–environment correlation a type of **genotype–environment correlation** which arises because individuals with different genotypes generate different reactions from other people; for example, people with high verbal abilities might somehow make their language teachers give them more attention or feedback. 41, 265

evolution in biology, refers to changes across generations in characters that can be inherited; in a genetic context it is usually taken to mean changes in **allele** frequencies across generations; note that it includes both adaptive (e.g. **selection**) and non-adaptive (e.g., **genetic drift**) processes and in general also considers inheritance with a non-genetic component such as culture and **niche construction**. 160, 264, 266

exaptation or pre-adaptation represents the process whereby a trait that evolved for one function (or had none) becomes involved in another function through a change in selective pressures; one well-known example is represented by feathers, initially involved in thermoregulation and possibly sexual display, and only later becoming selected for flight; likewise, the vocal tract was initially selected for its functions in breathing and food processing and the selective pressures related to speech emerged only later. 262

exapted see **exaptation**. 176

exogamy the mating across group boundaries; as opposed to **endogamy**. 188, 261

exome the set of all the **exons** in a **genome**. 69, 153

exon **genes** in **eukaryotes** do not usually contain their message in a continuous string of **nucleotides** but in discrete regions, the **exons**, separated by non-coding regions called **introns**; alternative **splicing** can produce multiple **proteins** from the **gene** by retaining different sets of **exons**. 68, 180, 262, 267, 279

expression represents the transformation of the genetic message in the **DNA** into a **phenotype** such as a **protein** product or a more abstract feature. 264, 281

extended phenotype an extension of the view of **phenotypes** as properties of individual organisms to include phenomena that go beyond the individual such as the beaver's dam and even human culture. 177

False Discovery Rate or **FDR** is a way of **multiple testing correction** that controls the proportion of **false positives** out of the significant tests only (as opposed to all tests performed); it is less conservative than **Bonferroni correction**. 101, 263, 271

false negative or type II error is the probability of not rejecting the **null hypothesis** when this is in fact false and is usually denoted β; **statistical power** is given by $1 - \beta$. 98, 279

false positive or type I error is the probability of falsely rejecting the **null hypothesis** when this is in fact true, denoted usually by α, the **significance level**. 97–99, 263, 271, 275, 278

FDR see **False Discovery Rate**. 101, 263

fertilization the process that initiates the development of a new organism through the fusion of a **sperm** and an **ovum**. 56

fitness represents the relative (dis)advantage of an organism or **allele** relative to other organisms or **alleles** in a given environment; can be understood as the expected contribution to the next generation relative to the competitors, and can be visualized using the **fitness landscape** metaphor; **alleles** with equal **fitness** are called **neutral** and are affected mainly by **genetic drift**; see also **neutral theory** and **nearly neutral theory**. 170, 260, 263, 272, 275, 281

fitness landscape a useful but limited (and sometimes even misleading) metaphor which visualizes the different relative **fitness** of different genotypes as hills with different heights; on such a **fitness landscape** selection "pushes" a population of individuals higher and higher every generation by increasing the frequency of fitter genotypes. 174, 263

fixation the process whereby one **allele** at a **locus** reaches 100% frequency in a population (every individual has it, being thus **fixed**), replacing all other alleles and leading to loss of **genetic diversity**; can be due to **genetic drift** or **positive selection**. 166, 263, 265, 269, 275, 278

fixed an allele is fixed in a population if everybody has it (100% frequency); resulting from **fixation**. 263, 271

founder effect an extreme case of **genetic drift** where a very small new "daughter" population splits off a "mother" population, resulting in a dramatic loss of genetic diversity in this new population; usually happens during the (intentional or not) colonization of new habitats, such as the expansion of modern humans into the Americas. 166, 191, 265

FOXP2 Forkhead box protein P2, a **transcription factor** seen as a "molecular window" into the genetic foundations of speech and language discovered due to its involvement in a rare speech and language pathology (**DVD**) affecting the members of the British KE family; **animal models** suggest that it is involved in motor control, communicative behaviour (such as birdsong) and neural development; it regulates several genes such as ***CNTNAP2***. 10, 75, 185, 258, 260, 261, 266, 268, 275, 277, 281

frameshift a type of change affecting the protein-coding sequence of a **gene** whereby a number of **nucleotides** not a multiple of three are added or removed (**indel**); given that the unit of genetic message is the **codon** composed of three **nucleotides**, such changes completely alter its meaning. 168, 260, 267

frequency-dependent selection a type of **selection** whereby the fitness of an **allele** is dependent on its own (and possibly other **alleles**) frequency in the population; in negative **frequency-dependent selection** the fitness decreases with increasing frequency, resulting in **balancing selection** and the maintenance of genetic diversity (some components of the immune system might be such examples), while in positive **frequency-dependent selection** the fitness increases with increasing frequency. 172, 258, 264

F_{ST} the fixation index represents a much used measure of **genetic diversity** and genetic differences between populations, ranging between 0 (no **population structure** or **panmixia**) to 1 (the populations are completely separate). 181, 264, 275

gamete or sex cell is a cell specialized for reproduction; in humans they can be male (**sperm**) or female (**ovum**) and are **haploid**. 55, 163, 168, 265, 269, 274, 279

ganglion component of the peripheral **nervous system** composed of neural bodies. 272

GCTA see **Genome-wide Complex Trait Analysis**. 37, 265

gene a complex concept with multiple and fluid meanings, essentially referring to the unit of genetic transmission and expression. 257, 261–265, 267, 271, 272, 274, 276–278, 280, 281

gene duplication the process whereby **genes** (or other sections of the **genome**) are present in more copies than in the ancestor organisms; can lead to **non-functionalization**, **neo-functionalization** or **sub-functionalization** and plays a major role in the evolution of biological complexity; gene duplication leads to the emergence of **gene families**. 264, 272, 273, 276, 280

gene expression represents the transformation of the genetic message encoded as the **gene**'s **DNA** into a **phenotype** such as a **protein** product or a more abstract feature. 262, 277

gene family a set of genes that descend from a single ancestral gene through **gene duplication** and that usually have relatively similar functions. 135, 264

gene flow represents the flow of **alleles** between two or more populations. 189, 267

gene regulation the multiple manners in which the **expression** of a gene is altered (turned on or off, increased or decreased); can act during or after **transcription** and during **translation** and involves **transcription factors** and **miRNAs** inter alia, and creates **regulatory networks**. 33, 53, 259, 261, 266, 270, 275, 277, 281

gene–culture co-evolution the proposal that genetic (biological) **evolution** and **cultural evolution** interact, mutually shaping each other; potent examples are represented by language and agriculture. 268

genetic code represents the mapping between **codons** and **amino acids**; it is degenerate in the sense that several **codons** might code for the same **amino acid**. 63, 257, 270, 271, 280, 281

genetic diversity the amount of genetic variants present in a population, quantified using various measures such as the **F_{ST}**; in general **genetic drift**, **positive selection** and **negative selection** reduce diversity, while **mutation** and **balancing selection** increase it. 258, 263, 264, 269, 275

genetic drift the stochastic process affecting **allele** frequencies in a population across generations, invariably leading to **fixation** or **loss** in finite populations; it is more powerful in small populations and plays a major role in **population bottlenecks** and **founder effects**. 262–264, 269, 271, 272, 275

genetic marker see **marker**. 269

genetically biased cultural evolution the proposal that genetic influences might result in biases that affect **cultural evolution**, resulting in universal tendencies or cultural diversity. 258

genome the totality of an organism's genetic material; given that the differences (if any) from **genotype** are not clear and there seems to be no consistent use, we will consider them interchangeable. 153, 161, 261, 262, 264, 265, 269, 270, 273

Genome-Wide Association Study abbreviated **GWAS**, an **association study** where **genotype** data for many (usually hundreds of thousands) **loci** across the whole **genome** is obtained for very large numbers of participants; such studies are very useful for discovering new **loci** involved in a given **phenotype** but suffer from issues of **statistical power** and **multiple testing correction**. 37, 98, 265

Genome-wide Complex Trait Analysis abbreviated **GCTA**, a method for estimating **heritability** from **GWAS** data. 37, 264

genotype the totality of an organism's genetic material; given that the differences (if any) from **genome** are not clear and there seems to be no consistent use, we will consider them interchangeably. 18, 73, 265, 266, 269

genotype–environment correlation the greater than expected co-occurrence of certain genotypes and environments; of three types: **passive genotype–environment correlation**, **active genotype–environment correlation** and **evocative genotype–environment correlation**. 257, 262, 274

genus a biological category larger than the **species** and containing one or more of these; an example is our own genus, ***Homo***. 266

germ line the set of cells that produce the **gametes**, thus passing their **DNA** to the offspring; **mutations** occurring in them will affect the next generations, as opposed to **somatic mutations**. 168

GLM the Generalized Linear Model is a generalization of simple linear **regression** by allowing the DV to be related to the IVs through a link function. See regression. 112

glossogenetic glossogeny refers to the processes of language change across generations. 5

GNPTAB together with ***GNPTG*** and ***NAGPA***, this **gene** is involved in transportation of digestive **enzymes** from the **endoplasmic reticulum** to the **lysosome**; surprisingly, mutations in these genes are involved in **stuttering**. 136, 265, 271, 280

GNPTG together with ***GNPTAB*** and ***NAGPA***, this **gene** is involved in transportation of digestive **enzymes** from the **endoplasmic reticulum** to the **lysosome**; surprisingly, mutations in these genes are involved in **stuttering**. 136, 265, 271, 280

grey matter in the **nervous system**, is composed mainly of the **neurons**' cell bodies; distinguished from **white matter**. 259, 282

GWAS see **Genome-Wide Association Study**. 37, 101, 265

hair cell specialized **cells** in the inner ear that convert the mechanical energy ultimately derived from the sound waves that hit the eardrum into **neural impulses**. 131, 280

haploid an organism (or cell) that has one set of non-paired (i.e., dissimilar or non-homologous) **chromosomes**, as opposed to **diploid** ones; human sex cells (**sperm** and **ova**) are haploid with 23 chromosomes. 18, 55, 163, 260, 264

haploinsufficiency the situation where one copy of a **diploid** gene is inactivated due to **mutation** and the other remaining normal copy is not enough to result in a normal **phenotype**; **DVD** in the KE family is a case where the normal copy of ***FOXP2*** does not produce enough of the normal FOXP2 **protein** to result in normal development. 138, 146

haplotype the combination of **alleles** present at different **loci** on the same **DNA** molecule. 59, 92, 126

Hardy-Weinberg equilibrium abbreviated **HWE**, describes the **evolution** of **allele** and **genotype** frequencies at a **locus** in a **population** across generations when several conditions are met such as infinite population size, absence of **mutation**, **selection** and **migration**, and **panmixia**; when these conditions are met **HWE** shows that the allele and genotype frequencies do not change across generations. 162, 267

hemizygous the situation where there is a single copy of a **chromosome** in an otherwise **diploid** organism, such as the X chromosome in males. 86

heritability the proportion of variation in a phenotype among a set of individuals explained by variation in their genotypes, varying between 0 (all variation is due to variation in the environment) and 1 (all variation is due to variation in the genes). 12, 31, 265, 270, 271

heterozygote advantage a type of **balancing selection** occurring when at a given **locus** the **heterozygous** individuals have higher fitness than both types of **homozygous** individuals; a well-known example is represented by **sickle cell anaemia**. 173, 258, 278

heterozygous an individual that has the different **alleles** at both homologous **loci**; for example, if only two alleles *A* and *a* are possible at a locus, both *Aa* and *aA* are heterozygous at this locus, in the vast majority of cases expressing the same **phenotype**, thus making *aA* and *Aa* equivalent (except for **parent of origin effects**). 72, 161, 261, 266, 277, 278

high-throughput sequencing **DNA sequencing** methods capable of producing large amounts of sequence data at relatively low cost; see **next-generation sequencing** and **third-generation sequencing**. 153, 272, 278, 280

hippocampus subcortical structure involved in memory and spatial navigation. 259

histogram visual representation of a statistical distribution showing the number (or frequency) or values in a given bin. 13

histone **proteins** bound to **DNA** forming the **chromatin**; important in **gene regulation** and **epigenetics**. 53, 61, 259, 262

hitch-hiking the increase in frequency of an **allele** (usually **neutral**) at a locus simply because it is **linked** to an advantageous **allele** at another **locus**. 278

Homo our own **genus**, appearing about 1.8 **mya** in Africa. 191, 265, 266

Homo erectus the earliest **species** of our **genus *Homo*** appearing in Africa about 1.8 **mya** and probably the ancestor of all other members of this genus, including ourselves and the **Neandertals**. 191

homozygous an individual that has the same **allele** at both homologous **loci**; for example, if only two alleles *A* and *a* are possible at a locus, both *aa* and *AA* are homozygous at this locus, usually expressing different **phenotypes**. 72, 161, 261, 266, 267, 277, 278

horizontal genetic transfer or horizontal gene transfer, abbreviated HGT, refers to the complex process whereby genetic material can jump between biological lineages; this is akin to **horizontal transmission** in culture; an important example is represented by the acquisition of antibiotic resistance across different species of **bacteria**. 281

horizontal transmission refers to the transmission of culture among peers, as opposed to the **vertical transmission** from parents to offspring and **oblique transmission** from non-parental adults to children. 267, 273, 282

HWE see **Hardy-Weinberg equilibrium**. 162, 266

hybridization the mating between populations that are considered to belong to different **species**; might result in **introgression**. 189, 267

immediate early gene a **gene** that is activated extremely quickly in response to stimuli such as hormones, ionizing radiation or viruses. 143

inbred line produced through repeated **mating** between closely related individuals across many generations and resulting in a set of individuals that are genetically almost identical to each other as well as **homozygous** for most **loci**. 91

inbreeding the mating between genetically related organisms; can be associated with inbreeding depression due to **recessive deleterious alleles** being more frequently in **homozygous** form. 187

incomplete penetrance **alleles** with less than 100% **penetrance**. 274

indel an **insertion** or a **deletion**; can have very large effects if it determines a **frameshift**. 67, 260, 264, 267

insertion represents the addition of one or more **nucleotides** to a **DNA** molecule; if not a multiple of three it can determine a **frameshift**; see also **indel**. 67, 267, 271

in silico as opposed to ***in vivo*** and ***in vitro***, usually means an experiment or study using computers (simulations, data analysis, etc.). 149, 267, 268

inter-rater reliability measures the **reliability** of a measurement instrument across different raters (e.g., estimates by several clinicians of a language pathology in an individual). 277

introgression **gene flow** between different **species** involving **hybridization** and the backcrossing of these hybrids into one of the source populations. 189, 267

intron **genes** in **eukaryotes** do not usually contain their message in a continuous string of **nucleotides** but in discrete regions called **exons** separated by non-coding regions, the **introns**, that can contain **regulatory elements** or even other **genes**. 68, 262, 267, 275, 279

inversion a type of **mutation** whereby a section of the **DNA** molecule is rotated such that an initial order -1-2-3-4- becomes -1-3-2-4-. 168, 271

in vitro as opposed to ***in vivo*** and ***in silico***, an experiment or study done using biological components or units (cells, molecules) outside their normal biological context. 267, 268

in vivo as opposed to ***in vitro*** and ***in silico***, an experiment or study done using whole living organisms. 267

ion an electrically charged **atom** or **molecule** with more (negative charge) or fewer (positive charge) electrons than protons. 258

Kata Kolok a sign language that emerged spontaneously in the village of Bengkala on the island of Bali, Indonesia, due to a high frequency of **recessive congenital** hearing loss due to a mutation in the ***MYO15A*** gene. 80, 134, 204, 271

lactose tolerant the phenotype characterized by the retention into adulthood of the capacity to digest lactose, a main ingredient of fresh milk; its selection in some agricultural populations is one of the best cases of **gene–culture co-evolution**. 203

Last Universal Common Ancestor the proposed common ancestor of all cellular life, the root of the **Tree of Life**. 46

Law of Independent Assortment the second of **Mendel's laws**, postulating that **loci** for different **phenotypes** are independently transmitted to the new generation. 56, 268, 269

Law of Segregation the first of **Mendel's laws**, postulating that the two parental **alleles** at a given **locus** have equal chances of being transmitted to the new generation. 56, 269, 278

linguistic tone the phenomenon whereby languages make use of variation in voice pitch to convey lexical or grammatical meaning. A classic case is Mandarin Chinese, where, for example, the same syllable /ma/ can mean four different things depending on its tone: high /mā/ = mother, middle rising /má/ = hemp, low dipping /mǎ/ = horse, and high falling /mà/ = scold, but about half the world's languages are tone languages with the complexity of their tone system ranging from some words having two possible accents (for example some dialects of Swedish and Japanese) to five or more tones potentially for each syllable (for example Thai). By opposition, languages such as English or French use voice pitch variation across larger units of speech through intonation. 6, 209, 258, 269

linkage disequilibrium see **linked**; linkage disequilibrium is essential in **linkage studies** and most **association studies**. 59, 90, 93, 101, 268

linkage study a method for finding **loci** involved in a **phenotype** using families showing the phenotype, the idea being that the causative **alleles** at these **loci** should occur in the same family members that show the phenotype ("affected") but not in the others ("non-affected"); an important example of a gene discovered through linkage in a large family is ***FOXP2***. 94, 258, 268

linked **loci** (or **phenotypes**) deviate from the expected **Law of Independent Assortment**; **loci** physically close together on the same **chromosome** might be linked, with the strength of linkage being roughly inversely proportional to the distance due to the fact that **recombination** events that might break the physical connection have a lower probability of occurring between them; see **linkage disequilibrium**. 266, 268, 278

lipid essential components of the **cell membrane** but also important for signalling and energy storage. 48

locus a specific position in the **genome** that can be occupied by one or more **alleles**. 16, 32, 58, 161, 257–261, 263, 265–272, 274, 275, 277, 278

log odds ratios denoted **LOR**, are the **logarithms** of the ratio of the odds of being a **case** in two conditions, such as when carrying **allele** *A* versus the alternative allele *a* at a **locus** or when smoking versus not smoking. 108, 269

logarithm of a number x, $log(x)$ is the power z at which the so-called Euler's number $e \approx 2.7182\ldots$ must be raised to give x; thus $e^{log(x)} = x$; for example $log(1) = 0$, $log(e) = 1$, $log(2) \approx 0.6932\ldots$ and $log(10) \approx 2.3026\ldots$. 269

logistic regression a type of **regression** used when the dependent variable is dichotomous (binary). 112

LOR see **log odds ratios**. 269

loss when referring to an allele, the opposite process to **fixation**, resulting in the elimination of the allele (0% frequency) and a reduction of **genetic diversity**; can be caused by **genetic drift** or **negative selection**. 265, 272

lysosome an organelle of **eukaryotic cells** that is involved in breaking down anything from waste molecules to cell invaders. 46, 136, 265, 271

MAF see **minor allele frequency**. 168, 270

magnetoencephalography abbreviated **MEG**, a **brain** imaging technique which records the magnetic fields generated by the brain's activity; has very good temporal resolution. 138, 269

marker or **genetic marker** is a landmark on the **genotype** with a known position and which shows observable variation among individuals (at the level of the **phenotype** or the **genotype**), allowing the relative localization of other **loci**; **SNPs** are widely used such genetic markers. 93, 265

mating pairing of different-sex organisms (males and females) with the aim of producing offspring. 267, 274–276, 279

MCPH see **Primary Autosomal Recessive Microcephaly**. 139, 275

MCPH1 a gene involved in **Primary Autosomal Recessive Microcephaly**; a normal **allele** of ***MCPH1*** was proposed to result in a **bias** affecting **linguistic tone**. 140, 269

MDS see **Multidimensional Scaling**. 115, 271

mean the average of a set of numbers, formally defined as $mean(x) = \sum_{i=1}^{N} x_i/N$ where x_i are N numbers. 14, 273, 274, 279, 282

MEG see **magnetoencephalography**. 138, 269

meiosis a type of **cell division** where the resulting "daughter" cells carry only half the number of **chromosomes** of the "mother" cell; this type of division is involved in the production of **gametes**. 261

membrane an essential component of **cells** that separates them from their environment as well as various intra-cellular components (such as the **nucleus** and the **mitochondria**) from the rest of the cell; they also regulate the exchange of matter, energy and information between the interior and the exterior. 45, 262, 268, 272

Mendel's laws laws of inheritance empirically discovered by Gregor Mendel, such as the **Law of Independent Assortment** and the **Law of Segregation**. 268

Mendelian genetics a branch of genetics studying simple (usually binary, presence/absence) traits transmitted in simple patterns across generations; see **quantitative genetics**. 30, 276

metabolism chemical reactions driven by **enzymes** and taking place within living **cells** that make life possible, such as energy generation by oxidizing within **mitochondria**. 45

micrometre see **micron**. 60

micron 1 micron (denoted 1 μm) represents 1×10^{-6} m; there are 1 million μm in 1 m. 60, 270

migration the process whereby new individuals arrive in a **population** and, through mating, can introduce new **alleles** or change the frequencies of those already present. 266

minor a minor **allele** is the one with the lowest frequency (**minor allele frequency**) among all the alleles present at the **locus** of interest; a conventional threshold of 1% or 5% marks the difference between a **polymorphism** and a variant. 95, 270

minor allele frequency denoted **MAF**, the frequency of the **minor allele**; MAF larger than a conventional threshold of 1% (or 5%) defines a **polymorphism** versus a variant. 269, 270

miRNA microRNAs are small molecules of **RNAs** that do not code for **proteins** but instead play essential roles in **gene regulation** by binding to **mRNAs** and altering their stability or their **translation**. 151, 264, 275

missing heritability the observation that the known **loci** involved in various complex traits (such as height) do not seem to account for the **heritability** of those traits as estimated by **quantitative genetics** methods such as **twin studies**. 155

mitochondrion an organelle of **eukaryote cells** primarily responsible for energy generation; it has its own tiny **DNA** that uses a slightly different **genetic code** than the **DNA** in the **cell nucleus**; see **mtDNA**. 46, 133, 262, 269–271, 276

mitosis a type of **cell division** where the resulting "daughter" cells are more or less identical to the "mother" cell, including in the amount of genetic information contained; this is the dominant type of cell division during growth, development and maintenance of the organism. 51, 261

mobile element pieces of **DNA** that can spread across the **genome** or change their position therein, potentially causing **mutations**. 271

molecular clock the rate of evolution of **neutral alleles** is constant, allowing the estimation of time knowing the **mutation** rate and the number of accumulated mutations; it has been used to date past events but newer models are better able to deal with rate variability. 178, 272

molecule a compound comprising several **atoms** bound together through **chemical bonds**; can range from the very simple (such as O_2, composed of two oxygen atoms) to extremely complex **proteins** composed of tens of thousands of diverse **atoms**. 48, 259, 268, 273

monozygotic twins that derive from the same fertilized **ovum** sharing thus all (100%) their **genomes** (except for post-separation **mutations**). 34, 84, 281

mRNA the messenger **RNA** transfers the genetic message from the **DNA** to the **ribosome**, where its **translation** takes place. 62, 68, 270, 276–281

mtDNA or mitochondrial DNA, the genetic information of the **mitochondria** using its own slightly different **genetic code**; mtDNA is transmitted maternally (i.e., mothers transmit their own mtDNA to their offspring but fathers' mtDNA is not transmitted) allowing the reconstruction of the history of the female line. 270

MTRNR1 mutations in this mitochondrial gene can cause hearing loss, for example, through interaction with certain antibiotics. 133

multi-level regression also known as mixed-effects or hierarchical, is a type of **regression** that allows for observations to be nested within higher-order units and that can account for non-independence within such units; a classic example is predicting performance of a test on other variables for children within schools given that the children enrolled in the same school potentially share more characteristics than those at different schools such as teaching style and quality, parental socio-economic status, etc, sharing that needs to be taken into account. 112

Multidimensional Scaling abbreviated **MDS**, a technique that projects a matrix of distances among points in a multi-dimensional space to a lower-dimensional space with minimal distortion; much used to visualize (2D or 3D) highly dimensional structures such as distances between populations computed using many **loci**. 115, 269

multiple testing correction when conducting multiple **statistical tests** at the same time, the **significance level** must be lowered to reflect the increased **false positive** rate; a very conservative approach is represented by the **Bonferroni correction**, while the **False Discovery Rate** is a different approach. 258, 263, 265

mutation a novel variant at a locus that can be of several types (e.g., **point mutation**, **insertion**, **deletion**, **inversion**) and that can arise through several process (**replication** or **repair** errors or the **insertion** of **mobile elements**); initially mutations are at a very low frequency in the population but they can spread through **genetic drift** or various forms of **selection** to become **polymorphisms** or even **fixed**, but the vast majority are (slightly) **deleterious** and will remain at low frequency or be completely removed by **negative selection**. 53, 168, 257, 260, 264–267, 270, 273, 275, 277, 279–281

mya shorthand for million years ago. 266, 272

MYO15A mutations in this gene are the cause of **congenital** hearing loss involved in the emergence of **Kata Kolok**. 134, 268

NAGPA together with ***GNPTAB*** and ***GNPTG***, this **gene** is involved in transportation of digestive **enzymes** from the **endoplasmic reticulum** to the **lysosome**; surprisingly, mutations in these genes are involved in **stuttering**. 136, 265, 280

narrow-sense heritability a specific aspect of **heritability** quantifying the amount of phenotypic variance explained by the additive genetic variance, usually denoted h^2. 33, 281

natural selection a type of selection where differential survival and reproduction is due to selective pressures generated by the interaction of the organism with its environment. 170, 278

Neandertal also sometimes spelled Neanderthal, probably our closest evolutionary cousins having evolved in western Eurasia around 0.5 **mya**, being replaced by modern humans about 40,000 years ago with some limited **admixture**; there is controversy concerning their status as a separate **species** and their cognitive capacities, including language. 191, 266

nearly neutral theory an extension of the **neutral theory** that also describes **alleles** with small **deleterious fitness** effects: they evolve neutrally in small populations (i.e., **selection** does not "see" them) but non-neutrally in large populations (i.e., **selection** does "see" them here). 179, 263, 272

negative selection a type of **selection** whereby one (or more) deleterious **alleles** (i.e., with a lower **fitness** than the competing alleles) decrease in frequency across generations because their carriers tend, on average, to leave fewer offspring; negative selection can result in the **loss** of these deleterious **alleles**. 264, 269, 271, 278, 279

neo-functionalization following **gene duplication**, one of the copies sometimes gains a new function and experiences new selective pressures, becoming a new **gene**. 135, 264

nerve component of the peripheral **nervous system** composed of **axons**, essential for information transmission. 272

nerve impulse see **neural impulse**. 272

nervous system is specialized in information processing and is composed mainly of **neurons**; it can be divided into the central nervous system comprising the **brain** and the **spinal cord**, and the peripheral nervous system comprising various **nerves** and **ganglia**. 259, 260, 264, 265, 272, 279, 280, 282

neural impulse **nerve impulse** or **action potential**, represents an electrical signal that is transmitted across the **neuron**'s cell **membrane** carrying information; it is important to note that this is a discrete signal (present or absent, on or off). 257, 258, 260, 266, 272, 280

neurexin a family of **proteins** that play essential roles in the **nervous system** such as the **synapse**; CNTNAP2 is a member of this family. 121, 150, 260

neuron a **cell** type specialized for information processing (coming in several subtypes found in various parts of the **nervous system**), forming complex networks where information is encoded, transmitted, processed and stored; they usually have a cell body, a long **axon** and several **dendrites**, information flowing from the dendrites towards the axon as an electrical **neural impulse** and passing to the next neuron through **synapses** by means of molecules called **neurotransmitters**. 258–260, 265, 272, 280, 282

neurotransmitter a molecule that transmits information across the **synapse**; there are many types with different properties, functions and localizations in the **nervous system**. 49, 272, 280

neutral **allele**, **locus** or **phenotype** that is not affected by **selection** as the possible variants have equal **fitness**; their evolution is described by the **neutral theory** and **nearly neutral theory**. 178, 263, 266, 270, 272

neutral theory describes the evolution of **neutral alleles** under **genetic drift** and represents a null hypothesis in evolutionary theory; such **alleles** evolve at a constant rate, resulting in a **molecular clock**. 179, 263, 272

next-generation sequencing a class of **high-throughput sequencing** methods that uses a massively parallel approach allowing fast and relatively cheap sequencing. 266, 280

Nicaraguan Sign Language or NSL, a sign language that emerged spontaneously when deaf children were brought together in special schools in Nicaragua. 204

niche construction the process whereby an organism inherits not only its **genome** from the previous generation but also components of the environment (niche) that it inhabits, establishing a parallel type of inheritance; a classic example is the beaver's dam, which is built by the beavers but also drastically changes the environment experienced by the animals (across multiple generations) and requires specific phenotypes as well (building, maintaining, etc.). 177, 262

non-functionalization following **gene duplication**, one of the copies sometimes loses function, becoming a **pseudogene**. 134, 264, 276

non-synonymous mutations are **mutations** that result in an **amino acid** other than the original one being incorporated in the resulting **protein**; see also **synonymous mutations**. 180, 280

Non-Word Repetition a standardized task where a participant (usually a child) has to repeat nonsense words (such as "glistow") of varying difficulty, the score being based on the number and type of errors; used in the study of **SLI** and as a measure of phonological working memory, but it is currently unclear what exactly it measures. 123, 138, 273

normal distribution a very important type of statistical distribution, bell-shaped and symmetric, completely determined by its **mean** and **standard deviation**. 16

nucleic acid a very large biological **molecule** essential for genetic information storage and transmission; **DNA** and **RNA** are nucleic acids. 48, 49, 273

nucleotide the basic unit of the **nucleic acids DNA** and **RNA**, allowing the encoding of information in their sequence. DNA contains four types (adenine, A; cytosine, C; guanine, G and thymine, T) while RNA replaces thymine (T) by uracil (U). 49, 260–264, 267, 275, 278, 281

nucleus an organelle of **eukaryote cells** that hosts the genetic material in the form of one or more **DNA** molecules. 46, 258, 269, 270, 276

null hypothesis H_0, represents the "boring" (or default) case when there is no relationship between the variables of interest (e.g., there is no correlation between them); **statistical tests** oppose it to the **alternative hypothesis**, but note that the null hypothesis can never be "proven": it can only be rejected or fail to be rejected. 97, 99, 257, 260, 263, 276, 279

NWR see **Non-Word Repetition**. 123, 278

oblique transmission refers to the transmission of culture from adults to non-related children, as opposed to the **horizontal transmission** among peers and **vertical transmission** from parents to offspring. 267, 282

ontogenetic ontogeny refers to the development of an individual, here a specific human being. 5

open reading frame a **reading frame** in which a sequence of nucleotides does not contain a stop **codon**. 67, 273

ORF see **open reading frame**. 67

organ a structure composed of several **tissues** that accomplish a given function (e.g., the heart or the **brain**). 48, 259, 280

orthologous genes or orthologs are **genes** that derive from the same ancestral gene. 190

ovum the female sex cell (or **gamete**), usually immobile and much larger than the **sperm**. 55, 260, 261, 263, 264, 266, 270

panmictic population in such a population any individual has equal chances of **mating** with any other individual. 274, 275

panmixia see **panmictic population**. 187, 264, 266

parent of origin effects rare cases where **epigenetic** marks allow the same allele at a locus to have different phenotypic effects depending on the parent (the mother or the father) it was inherited from. 266

passive genotype–environment correlation a type of **genotype–environment correlation** which arises because individuals inherit both their genes and the environment; for example, people with high verbal abilities might inherit both the genes for high verbal ability from their parents as well as the environment favouring high verbal ability that the parents have constructed and maintained. 41, 265

pathway in genetics refers to a set of **genes** that are involved in a meaningful set of metabolic paths, developmental or other type of processes. 118

PCA see **Principal Component Analysis**. 115, 275

Pearson correlation a quantitative measure of the strength and direction of the linear relationship between two variables, defined as $cor(x, y) = \frac{cov(x,y)}{sd(x) \times sd(y)}$, where x and y are two variables of the same length. 23, 260, 261

pedigree depiction of genetic relationships between a set of related individuals as well as some extra information concerning the state of one or more **phenotypes** and **alleles** at a set of **loci**. 73

penetrance the proportion of individuals carrying a given **allele** that show the **phenotype** of interest; **incomplete penetrance** refers to less than 100% penetrance. 133, 267

permutation a shuffling of the data; in the context of **statistical tests**, a set of methods that compares a summary (such as the **mean**) of the original data with the distribution of the same summaries computed for many random permutations of the data: if the original value is very extreme compared to the permuted values (more so than 5% or 1% of them) then the test is declared significant; permutation tests have fewer assumptions than standard tests and are more flexible, but require more computer resources and usually programming skills. 101

phenotype any property of interest, usually referring to individual organisms (e.g., height, vocabulary size or having dyslexia or not), but also used to refer to **genes** in the sense of their protein product (if any) and to supra-individual systems such as the beavers' dam. 8, 257–259, 261–269, 272, 274–278, 281

photoreceptor the light-sensitive **cells** of the **retina**, coming in two types: **rods** and **cones**. 85, 260, 277, 278

photosynthesis the conversion of water and carbon dioxide (CO_2) into more complex molecules using energy from light, realized by **chloroplasts**. 46, 259

phylogenetic phylogeny refers to the evolution of groups of related biological organisms such as species, including our own species, *Homo sapiens*. 5

point mutation a type of **mutation** where a single **nucleotide** is changed, such as the change from a G to an A at a certain position in ***FOXP2*** causing **DVD**. 53, 168, 271

polymorphism the presence in a **population** of at least two **alleles** of the same **locus** at frequencies greater than a given threshold, usually 5% or 1%. 153, 270, 271, 278

population in genetics, a group of organisms that can breed and produce offspring. 161, 266, 270, 275

population bottleneck an extreme case of **genetic drift** where a population's size is drastically reduced for one or more generations, with significant impact on its **genetic diversity**; can be caused by many factors such as habitat destruction, climate change or infectious diseases. 166, 265

population stratification results in differences in **allele** frequency between populations; it is a potential source of **false positives** in genetic **association studies** and must be taken into account. 114, 281

population structure as opposed to a **panmictic population**, there is genetic structure due to non-random **mating**; can be generated for example by geographic or cultural barriers, such as mountain ranges or rules governing acceptable marriages, or **assortative mating**; can be measured using for instance $\mathbf{F_{ST}}$. 264

positive selection a type of **selection** whereby one (or more) advantageous **alleles** (i.e., with a higher **fitness** than the competing alleles) increases in frequency across generations because its carriers tend, on average, to leave more offspring; positive selection can result in the **fixation** of these advantageous **alleles** and might produce **selective sweeps**. 171, 263, 264, 278, 279

post-transcriptional regulation a type of **gene regulation** which acts on the transcribed **RNA** through **intron splicing** and alterations of the transcript stability by **miRNAs** among other factors. 280

poverty of the stimulus an influential argument for the existence of an innate language-specific faculty, claiming that there is insufficient data in the linguistic input received by children to allow them to learn all the intricacies of the actually attested languages. 9

Primary Autosomal Recessive Microcephaly abbreviated **MCPH**, **brain** development pathologies characterized by very small head size with mild to severe mental retardation but without other malformations or neurological problems. 139, 258, 269

Principal Component Analysis abbreviated **PCA**, a technique that computes a new coordinate system (the principal components, PCs) such that the first PC explains most of the variance, the second PC most of the remaining variance, and so on. 115, 194, 274

prion a **protein** capable of replication in the sense that it can convert other proteins to its own conformation; sometimes involved in transmissible pathologies such as mad cow disease and Creutzfeldt-Jakob disease (in humans). 45, 259

proband usually the **case** or the individual showing the **phenotype** of interest that comes to the attention of the researchers, later allowing the inclusion of others of his/her family members as well; for example a child might be first diagnosed with **dyslexia** due to difficulties in school (the proband)

and later the parents might be included in a study of the genetics of dyslexia. 119

prokaryote usually unicellular organisms lacking internal structures such as the **eukaryotic nucleus** and **mitochondria**. 258, 276

prokaryotic see **prokaryote**. 46

promoter region on the **DNA** that initiates the **transcription** of a particular **gene** or **genes**. 62, 70, 277, 281

protein an important class of large biological molecules composed of one or more linear chains of **amino acids**, playing multiple roles such as structural (main components of skin, bones, etc.), metabolic (**catalysis** of chemical reactions) and informational (**transcription** and **translation** of genetic information carried by the **DNA**). 48, 257, 259, 261, 262, 264, 266, 270, 272, 273, 275–277, 280–282

proteome the entire set of **proteins** expressed in a given context and time. 68

protist a generic (but biologically problematic) term encompassing various groups of mostly unicellular **eukaryotes**. 46

proximate as opposed to **ultimate**, a mechanism or explanation in terms of immediate causes of a phenomenon; e.g., the proximate causes for running away when encountering a predator are given by the various neural and cognitive processes of perceiving the predator, experiencing fear and activating motor patterns that result in running; answers the question "how?". 160, 281

pseudogene ancient **genes** that are no longer functional; usually resulting from relaxed **selection** or **gene duplication** that leads to **non-functionalization**. 134, 273

psychometrics concerns psychological measurement, its properties and ways of conducting and analysing such measurements; see also **reliability** and **validity**. 39, 277

Punnett square tabular depiction of possible types of **mating** showing the possible types of resulting offspring. 74

***p*-value** for a given **statistical test**, the probability of obtaining such a test score if the **null hypothesis** were true; the ***p*-value** is usually compared against the standard **significance level** and the test is deemed **statistically significant** if $p < \alpha$. 260, 276, 279

quantitative genetics a highly mathematical branch of genetics studying **quantitative traits**; see **Mendelian genetics**. 30, 270, 276

quantitative phenotype a **phenotype** where variation is continuous as opposed to discrete as in the presence or absence of a pathology; examples are height, IQ and vocabulary size. 110, 258

quantitative trait a trait or **phenotype** that shows a relatively large number of variants (in principle infinite or continuous variation); investigated by **quantitative genetics**. 30, 276

reading frame one of three possible ways of reading a strand of **DNA** (or **mRNA**) in three-letter non-overlapping contiguous **codons**; for example, the sequence GCCACCAUG can be read in three frames as GCC:ACC:AUG, GC:CAC:CAU:G and G:CCA:CCA:UG. 65, 273

recessive a recessive **allele**, usually denoted using lower-case letters such as *a*, expresses its effect only when paired with itself but not with a **dominant**

allele at the same **locus**; thus **homozygous** *aa* individuals have a different **phenotype** than the **heterozygous** *Aa* and **homozygous** *AA* individuals; many alleles causing pathologies are recessive. 33, 73, 257, 261, 267, 268, 280

recombination the process whereby two homologous **chromosomes** pair up and exchange genetic material, producing new combinations of **alleles** at different **loci**; given two **loci** on the same **chromosome**, the probability of a recombination occurring between them roughly increases with the physical distance, measured in **centimorgans**. 259, 260, 268

regression a set of statistical techniques for estimating the relationship between an outcome (or dependent variable; DV) and one or more predictors (independent variables or co-variates; IVs); best known is linear regression, which quantifies the linear relationship between the DV and the IVs plus unaccounted-for noise: $DV = \alpha + \beta_1 IV_1 + \beta_2 IV_2 + \cdots + \epsilon$, where α represents the intercept, β's are the coefficients of the predictors and ϵ the residual error not explained by the regression. 24, 110, 265, 269, 271, 277

regression to the mean the statistical phenomenon that a selected sample extreme on one variable is not necessarily as extreme on another, imperfectly correlated variable; for example, repeated measurements that involve a great deal of randomness show that the best/worst performers do not stay as extreme the second time, their averages regressing towards the mean; this is a widespread phenomenon that must be carefully considered in the design and interpretation of experimental results; related to **regression**. 28

regulatory element a region on the **DNA** molecule than can affect **gene expression** by binding other different molecules such as **transcription factors**; includes **promoters**, **enhancers** and **silencers**. 70, 267

regulatory network a set of **genes** and other regions on the **DNA** that interact through **gene regulation**; for example, **gene** 1 might inhibit gene 2, which might enhance gene 3, which in turn inhibits gene 2. 264

reliability a fundamental concept in **psychometrics**, referring to the consistency of a measurement instrument, such as **test-retest reliability** and **inter-rater reliability**; it is different from but related to **validity**. 39, 267, 276, 280, 281

repair mechanisms that ensure that the error rate of **DNA replication** is extremely small by identifying and correcting differences between the original "mother" and the "daughter" **DNA** molecules. 271, 277

replication is the fundamental process whereby the genetic information is transmitted across generations, creating two (quasi) identical **DNA** "daughter" molecules from one "mother" **DNA** molecule through the activity of molecular machinery including the **DNA polymerase**; it has a very high accuracy ensured by error-correction mechanisms (**repair**) but still errors happen, resulting in **mutations**. 51, 261, 271, 277

retina the light-sensitive layer of cells (the **photoreceptors** at the back of the eye responsible for perceiving light). 84, 274

ribosome the protein-producing machinery of the cell where the message in the **mRNA** is used to assemble the corresponding **protein** through **translation**; it is composed of both **proteins** and **RNA**. 65, 133, 270

risk allele an **allele** at a **locus** that causes (or is associated with) a **phenotype** of interest (usually a pathology); for example, the specific ***FOXP2*** mutation in the KE family is a risk allele for **DVD**. 122

RNA ribonucleic acid is a molecule that encodes genetic information through the linear sequence of the four **nucleotides** A, U, C and G; it also has enzymatic capacities, plays important roles in expressing the genetic information encoded by the DNA (**transcription** and **translation**), and is the main repository of genetic information in only a few viruses. 50, 259, 270, 273, 275, 277, 281, 282

RNA polymerase the **enzyme** that copies the genetic message from the **DNA** onto the **mRNA**. 62, 280

ROBO1 a gene involved in **dyslexia** but also apparently in the normal variation of phonological working memory as measured by **NWR**. 123

rod a type of **photoreceptor** specialized for low-light monochromatic vision. 85, 274

scatterplot a graphical representation of the relationship between two (or rarely three) variables, where each variable is represented on one of the axes of the plot and each point has coordinates corresponding to its values on the variables. 22

segregation the process whereby the two homologous parental **chromosomes** separate; see also **Law of Segregation**. 56

selection the differential survival and reproduction usually of individuals, under various selective pressures; one can distinguish **natural selection**, **artificial selection** and **sexual selection**, as well as **positive selection**, **negative selection** and **balancing selection**. 257, 258, 262, 264, 266, 271, 272, 275, 276

selective sweep occurs when an advantageous **allele** is driven to **fixation** by **positive selection**, carrying with it **alleles** at other **linked loci** (a phenomenon called **hitch-hiking**). 182, 275, 279

sequencing the process of finding out the sequence (order) of **nucleotides** in a **DNA** molecule; there are several methods available that differ in their capacity, quality and cost, currently the most used being various types of **high-throughput sequencing**. 266

sexual selection a type of selection where differential reproduction is due to attracting and securing mates; sometimes generates extravagant **phenotypes** that can be deleterious from the point of view of natural selection such as the peacock's tail. 170, 278

sickle cell anaemia a severe pathology caused by **homozygous** *HbS*/*HbS* genotypes at a specific locus; interestingly, it has high prevalence in areas affected by malaria, maintained by **balancing selection** due to **heterozygote advantage** (**heterozygous** *HbS*/*HbA* better resist malaria, while the **homozygous** *HbS*/*HbS* is affected by **sickle cell anaemia** and the normal **homozygous** *HbA*/*HbA* is affected by malaria). 173, 266, 278

significance level α is the probability of a **false positive**, conventionally taken to be 0.01 (1%) or 0.05 (5%). 258, 260, 263, 271, 276, 279

silencer region on the **DNA** that can inhibit the **transcription** of a particular **gene** or **genes** when certain **transcription factors** are bound. 277, 281

SLI see **Specific Language Impairment**. 39, 260, 273, 279

SNP Single Nucleotide Polymorphism, pronounced "snip", one of the simplest types of **polymorphisms** where variation is restricted to a single **nucleotide**. 53, 101, 168, 257, 269

somatic mutation a **mutation** that occurs in a somatic cell (i.e., a cell which will not result in a **gamete**); it is thus limited to the individual in which it occurred and will not be transmitted to any of the individual's offspring (if any). 35, 143, 168, 265

speciation the process whereby new **species** arise, either through splitting of an old species into two or more "daughter" species (cladogenesis), or the transformation of an old species into a new one through time (anagenesis). 93, 189

species a fundamental unit of biological diversity (and of its classification) generally understood as a group of organisms capable of **mating** and producing viable offspring, and that have a unitary evolutionary history; however, there are at least 20 definitions of species. 265–267, 272, 279

Specific Language Impairment SLI; abnormal development of language in the absence of other developmental delays or hearing loss and despite normal intelligence and educational opportunities. 39, 121, 278

sperm the male sex cell (or **gamete**), usually highly mobile. 55, 260, 263, 264, 266, 274

spinal cord an important component of the central **nervous system** with major roles in information transmission to and from the **brain** as well as processing and reflex control. 272

splicing the process happening after **transcription** of removing **introns** and joining up the **exons** to produce the mature **mRNA**; a single precursor **mRNA** can produce multiple mature **mRNAs** through alternative **splicing** whereby different sets of **exons** are retained. 68, 262, 275, 279

stabilizing selection see **negative selection**. 172

standard deviation a more useful measure of the spread around the **mean** for a set of numbers than their **variance**, formally defined as $sd(x) = \sqrt{var(x_i)}$ where x_i are N numbers. 15, 273

standing variation in the context of soft **selective sweeps**, genetic variation already present in the population that becomes advantageous (under **positive selection**) due to a change in environment, etc. 184

statistical power of a **statistical test** is the probability of correctly rejecting the **null hypothesis** when it is false, denoted $1 - \beta$, where β is the probability of a **false negative**; bigger sample sizes increase the power to detect small **effect sizes**. 98, 103, 263, 265

statistical significance see **statistically significant**; this is usually contrasted to the "real-world" significance highlighting the fact that statistically significant tests might sometimes have too small **effect sizes** to make any practical difference. 261

statistical test usually a test checks whether the **null hypothesis** of no effect can be rejected in favour of the **alternative hypothesis** with a given **significance level**; most tests produce a ***p*-value** or a **confidence interval**. 257, 259, 261, 271, 273, 274, 276, 279

statistically significant a statistical test is statistically significant if its ***p*-value** is smaller than a conventional **significance level**. 276, 279

`STRUCTURE` a method (and associated software package) for inferring population structure from individual-level genetic data. 114

stuttering a speech disfluency characterized by involuntary repetitions or prolongations of segments of speech or by interruptions; the **genes *GNPTAB*, *GNPTG*** and ***NAGPA*** seem to be involved at least in some cases. 136, 265, 271

sub-functionalization following **gene duplication**, the two copies sometimes become specialized in different facets of the original function. 135, 264

symbiosis refers to the durable and close interaction between different species; a well-known example is represented by the lichens, which are composed of a fungus and an alga living in close mutualism; see also **endosymbiotic theory**. 262

synapse a structure where a **neuron**'s **axon** transmits information to another **neuron**'s **dendrite** by means of chemical **neurotransmitters**; can be inhibitory or excitatory and play an essential role in memory by changing the ability to transmit information; synapses convert the electrical **neural impulse** coming down the **axon** into a chemical signal transmitted by **neurotransmitters** and again into an electrical **neural impulse** in the **dendrite** (if a certain threshold is achieved). 272

synonymous mutations are **mutations** that do not result in a change of **amino acid** in the resulting **protein**; this is due to the degeneracy of the **genetic code**; see also **non-synonymous mutations**. 64, 180, 273

system a set of **organs** with a shared function (e.g., the **nervous system** or the circulatory system). 48

TDT see **Transmission Disequilibrium Test**. 118, 281

TECTA a gene involved in the structural properties of the **tectorial membrane**; different mutations in this gene can result in either **recessive** or **dominant** hearing loss. 132

tectorial membrane a membrane of the inner ear that transmits the mechanical energy ultimately derived from the sound waves that hit the eardrum to the **hair cells**, thus playing an essential role in hearing. 131, 280

test-retest reliability measures the **reliability** of a measurement instrument across applications; for example, measuring the height of the same individuals twice in short succession should result in very similar estimates; it is a fundamental requirement for a good phenotypic measure for genetic studies. 277

third-generation sequencing a class of **high-throughput sequencing** methods more advanced than **next-generation sequencing** that can deliver faster results at lower costs and require an even smaller quantity of **DNA** to sequence. 266

tissue a structure composed of similar **cells** that realize a given function (e.g., muscle or neural tissue). 48, 273

trade-off results from the non-independence of traits and the fact that organisms are simultaneously under many selective pressures, resulting in the apparent sub-optimality of some traits. 176

transcription the process of copying the genetic message encoded in the **DNA** to the **mRNA**, mediated by **RNA polymerase**, which allows **transcriptional regulation** and **post-transcriptional regulation** to take place. 62, 261, 262, 264, 276, 278, 279, 281

transcription factor proteins that bind to specific sequences of **DNA nucleotides** (such as **enhancers** and **promoters**) and can affect the **expression** of other **genes** by modifying their **transcription**; a very important **transcription factor** for language and speech is ***FOXP2***. 116, 144, 262–264, 277, 278, 281

transcriptional regulation a type of **gene regulation** which modifies the **transcription** rates through **transcription factors, enhancers** and **silencers** among other mechanisms. 280

transfer RNA denoted **tRNA**, an **RNA** molecule that associates the correct **codon** with the **amino acid** and is essential for **translation**; the set of **tRNAs** embodies the **genetic code**. 67, 281

transgenic an organism that contains **genes** from another biological lineage or genes that have been artificially modified; a genetically modified organism. 91

translation the process whereby the message carried by the **mRNA** is used to assemble the corresponding **protein** using the **genetic code**. 62, 264, 270, 276–278, 281

translocation a type of **mutation** caused by the exchange of pieces of **DNA** between non-corresponding **chromosomes**. 137, 168

Transmission Disequilibrium Test abbreviated **TDT**, simultaneously tests for linkage and association and overcomes issues related to **population stratification**; this is a family-based test using for example trios composed of one affected child and his/her parents. 114, 118, 280

Tree of Life denoted TOL, a proposal connecting the three **domains of life** suggesting that **Archaea** and **Eukaryotes** are more closely related than both are to **Bacteria**; however, **horizontal genetic transfer** and **endosymbiosis** represent major issues for this simple metaphor. 46, 261, 268

tRNA see **transfer RNA**. 67, 281

twin studies a method for estimating **narrow-sense heritability** comparing the similarity between pairs of **monozygotic** and **dizygotic** twins: decreased similarity between the latter compared to the former is taken to point to genetic (as opposed to environmental) influences on the **phenotype** of interest. 34, 270

ultimate as opposed to **proximate**, an explanation that looks for the evolutionary forces that have shaped the phenomenon; e.g. the ultimate cause for running away when encountering a predator is given by the selective advantage (i.e., higher **fitness**) of those past organisms that did versus those that didn't, translated into the former leaving more descendants than the latter, so that present-day organisms inherit those genes that make them run away from predators; observe that the **proximate** mechanisms implementing this are not specified; answers the question "why?". 276

universals (of language) are proposed properties shared by all (or most) languages and which constrain the amount of cross-linguistic variation. 9

up-regulation gene regulation that results in more of the target **gene** being expressed; also known as positive regulation or activation. 149

validity represents the capacity of a measurement instrument to actually measure what it is supposed to; it is different from but related to **reliability**. 39, 276, 277

variance is the spread around the **mean** for a set of numbers, formally defined as $var(x) = \sum_{i=1}^{N} (x_i - mean(x))^2/(N-1)$ where x_i are N numbers. 15, 279

vertical transmission refers to the transmission of culture from parents to offspring, as opposed to the **horizontal transmission** among peers and **oblique transmission** from non-parental adults to children. 267, 273

virus replicating molecules (usually **DNA** but also **RNA**) encapsulated within a **protein** coat that need to hijack the informational and metabolic machinery of a living **cell** in order to reproduce; sometimes cause pathologies such as AIDS (the HIV virus) or the common flu. 45, 259

white matter in the **nervous system**, composed mainly of the **neurons**' axons, which ensure neural connectivity; distinguished from **grey matter**. 259, 265